AF698228

The Centrosome

CELL BIOLOGY: A Series of Monographs

EDITORS

D. E. BUETOW
Department of Physiology
and Biophysics
University of Illinois
Urbana, Illinois

I. L. CAMERON
Department of Cellular and
Structural Biology
The University of Texas
Health Science Center at San Antonio
San Antonio, Texas

G. M. PADILLA
Department of Physiology
Duke University Medical Center
Durham, North Carolina

A. M. ZIMMERMAN
Department of Zoology
University of Toronto
Toronto, Ontario, Canada

Volumes published since 1983

John Morrow. EUKARYOTIC CELL GENETICS, 1983
John F. Hartmann (editor). MECHANISM AND CONTROL OF ANIMAL FERTILIZATION, 1983
Gary S. Stein and **Janet L. Stein** (editors). RECOMBINANT DNA AND CELL PROLIFERATION, 1984
Prasad S. Sunkara (editor). NOVEL APPROACHES TO CANCER CHEMOTHERAPY, 1984
B. G. Atkinson and **D. B. Walden** (editors). CHANGES IN EUKARYOTIC GENE EXPRESSION IN RESPONSE TO ENVIRONMENTAL STRESS, 1985
Reginald M. Gorczynski (editor). RECEPTORS IN CELLULAR RECOGNITION AND DEVELOPMENTAL PROCESSES, 1986
Govindjee, Jan Amesz, and **David Charles Fork** (editors). LIGHT EMISSION BY PLANTS AND BACTERIA, 1986
Peter B. Moens (editor). MEIOSIS, 1986
Robert A. Schlegel, Margaret S. Halleck, and **Potu N. Rao** (editors). MOLECULAR REGULATION OF NUCLEAR EVENTS IN MITOSIS AND MEIOSIS, 1987
Monique C. Braude and **Arthur M. Zimmerman** (editors). GENETIC AND PERINATAL EFFECTS OF ABUSED SUBSTANCES, 1987
E. J. Rauckman and **George M. Padilla** (editors). THE ISOLATED HEPATOCYTE: USE IN TOXICOLOGY AND XENOBIOTIC BIOTRANSFORMATIONS, 1987
Heide Schatten and **Gerald Schatten** (editors). THE MOLECULAR BIOLOGY OF FERTILIZATION, 1989
Heide Schatten and **Gerald Schatten** (editors). THE CELL BIOLOGY OF FERTILIZATION, 1989
Anwar Nasim, Paul Young, and **Byron F. Johnson** (editors). MOLECULAR BIOLOGY OF THE FISSION YEAST, 1989
Mary P. Moyer and **George Poste** (editors). COLON CANCER CELLS, 1990
Gary S. Stein and **Jane B. Lian** (editors). MOLECULAR AND CELLULAR APPROACHES TO THE CONTROL OF PROLIFERATION AND DIFFERENTIATION, 1991
Vitauts I. Kalnins (editor). THE CENTROSOME, 1992
Carl M. Feldherr (editor). NUCLEAR TRAFFICKING, 1992, *In preparation*

The Centrosome

Edited by

Vitauts I. Kalnins

Department of Anatomy and Cell Biology
University of Toronto
Toronto, Ontario, Canada

ACADEMIC PRESS, INC.
Harcourt Brace Jovanovich, Publishers
San Diego New York Boston London Sydney Tokyo Toronto

This book is printed on acid-free paper. ♾

Copyright © 1992 by ACADEMIC PRESS, INC.
All Rights Reserved.
No part of this publication may be reproduced or transmitted in any form or by any means, electronic or mechanical, including photocopy, recording, or any information storage and retrieval system, without permission in writing from the publisher.

Academic Press, Inc.
1250 Sixth Avenue, San Diego, California 92101

United Kingdom Edition published by
Academic Press Limited
24–28 Oval Road, London NW1 7DX

Library of Congress Cataloging-in-Publication Data

The Centrosome / edited by Vitauts I. Kalnins.
p. cm. -- (Cell biology)
Includes bibliographical references and index.
ISBN 0-12-394770-7
1. Centrosomes. I. Kalnins, Vitauts I. II. Series.
QH597.C46 1992
574.87'34--dc20 91-35190
CIP

PRINTED IN THE UNITED STATES OF AMERICA
92 93 94 95 96 97 BB 9 8 7 6 5 4 3 2 1

Contents

4 Microtubule and Intermediate Filament Patterns around the Centrosome in Interphase Cells

I. B. Alieva, E. S. Nadezhdina, E. A. Vaisberg, and I. A. Vorobjev

5 Monoclonal Antibodies to Microtubule-Organizing Center Antigens

Ryoko Kuriyama

6 Role of Centrin in Spindle Pole Dynamics

Andre T. Baron and Jeffrey L. Salisbury

7 Spindle Pole Body of *Saccharomyces cerevisiae*: A Model for Genetic Analysis of the Centrosome Cycle

Mark Winey and Breck Byers

Contributors

Numbers in parentheses indicate the pages on which the authors' contributions begin.

Guenter Albrecht-Buehler (69), Department of Cellular, Molecular and Structural Biology, Northwestern University Medical School, Chicago, Illinois 60611

I. B. Alieva (103), A. N. Belozersky Institute of Physico-Chemical Biology, Moscow State University, Moscow, USSR 119899

Andre T. Baron (167), Laboratory for Cell Biology, Department of Biochemistry and Molecular Biology, Mayo Clinic Foundation, Rochester, Minnesota 55905

Michel Bornens (1), Centre de Génétique Moléculaire du CNRS, 91198 Gif-sur-Yvette, France

David L. Brown (261), Department of Biology, University of Ottawa, Ottawa K1N 6N5, Canada

Breck Byers (197), Department of Genetics, University of Washington, Seattle, Washington 98195

David M. Glover (219), Cell Cycle Genetics Group, Cancer Research Campaign Laboratories, Department of Biochemistry, University of Dundee, Dundee DD1 4HN, Scotland

V. I. Kalnins (287), Department of Anatomy and Cell Biology, University of Toronto, Toronto, Ontario M5S 1A8, Canada

J. David Knox (261), Department of Biology, University of Ottawa, Ottawa K1N 6N5, Canada

Ryoko Kuriyama (131), Department of Cell Biology and Neuroanatomy, University of Minnesota, Minneapolis, Minnesota 55455

Mark A. McNiven (313), Department of Biochemistry and Molecular Biology, Center for Basic Research in Digestive Diseases, Mayo Clinic Foundation, Rochester, Minnesota 55905

E. S. Nadezhdina (103), A. N. Belozersky Institute of Physico-Chemical Biology, Moscow State University, Moscow, USSR 119899

Micheline Paulin-Levasseur (261), Department of Biology, University of Ottawa, Ottawa K1N 6N5, Canada

Keith R. Porter (313), Department of Biology, University of Pennsylvania, Philadelphia, Pennsylvania 19104

J. B. Rattner (45), Departments of Anatomy and Medical Biochemistry, The University of Calgary, Calgary, Alberta T2N 4N1, Canada

K. A. Rogers (287), Department of Anatomy, University of Western Ontario, London, Ontario N6A 5C1, Canada

Jeffrey L. Salisbury (167), Laboratory for Cell Biology, Department of Biochemistry and Molecular Biology, Mayo Clinic Foundation, Rochester, Minnesota 55905

Manfred Schliwa[1] (331), Department of Molecular and Cell Biology, University of California, Berkeley, Berkeley, California 94720

[1] Present address: Institute for Cell Biology, University of Munich, 8000 Munich 2, Germany.

Greenfield Sluder (235), Worcester Foundation for Experimental Biology, Shrewsbury, Massachusetts 01545

E. A. Vaisberg (103), Institute of Protein Research, Academy of Science, USSR, Pushchino, USSR 142292

I. A. Vorobjev (103), A. N. Belozersky Institute of Physico-Chemical Biology, Moscow State University, Moscow, USSR 119899

Mark Winey[2] (197), Department of Genetics, University of Washington, Seattle, Washington 98195

[2]Present address: Department of Molecular, Cellular and, Developmental Biology, Porter Biosciences, Campus Box 347, University of Colorado, Boulder, Boulder, Colorado 80309-0347.

Preface

The centrosome, discovered over a century ago, has lost little of its fascination over time and could very well become the organelle of the nineties. Located in the central part of the cell, the centrosome is the region from which most microtubules assemble. Because the general organization of the other cytoplasmic organelles depends to a large extent on the distribution of microtubules, the centrosome from which they radiate is in a strategic position to control various cell functions. The answers to some of the most central and challenging problems in cell biology may well depend on a better understanding of the centrosome. It clearly deserves to occupy a central position in our view of the cell as a unified, well-integrated entity. Hopefully, as more information accumulates, some of the mystery surrounding this organelle will give way to rational explanations of its function and role in cell organization. Because of the growing interest in the centrosome, a volume summarizing some of the previous findings and suggesting new directions for research seemed timely and appropriate.

This volume collates in one source the work of scientists actively engaged in studying various aspects of the centrosome, using a wide assortment of experimental approaches, techniques, and model systems. After presenting an overview of a particular area, the articles summarize work from the authors' own laboratories and include new, unpublished material. Emphasis is on the more dynamic aspects of the subject rather than on detailed descriptions. The contributions range from descriptions of the organization of the centrosome at the molecular level to speculations on how the centrosome may affect the behavior of entire cells. Experimental studies are complemented by theoretical considerations to provide added insight into the structure and function of this organelle and by speculations on directions which appear most profitable for future studies. I have encouraged the inclusion of controversial ideas and conflicting hypotheses, which often provide the driving force for new advances.

As will be evident from reading this volume, even defining the centrosome presents difficulties because it has no clear-cut boundaries. The isolation of the centrosomes in relatively pure form has been a major achievement of the past decade and some of the results obtained are reported. The central component of the centrosome, the centriole, seems just as enigmatic as ever and speculations on its formation, organization, and function have been included in this volume. The surrounding region which contains the microtubule initiation sites is described and its structure, chemical composition, and organization in relation to microtubule initiation sites are discussed. In many of the studies a combination of new and sophisticated techniques from many fields such as biochemistry, immunology, genetics, and morphology have been used to generate a wealth of knowledge about the centrosome. With the use of labeled antibody techniques, we have been able to identify and localize some of the molecules in the centrosomal region of the cell with great precision. The availability of antibodies to many centrosomal proteins suggests that with the application of techniques of molecular biology it will soon be feasible to elucidate the functional role of these proteins in the centrosome. Genetic analyses of the centrosome, its duplication, and its inheritance and role in early development in different model systems are represented. Changes in the centrosome during the cell cycle and how these changes are related to other cellular events are examined in a number of articles. Finally, the role of the centrosome in defining the shape and behavior of various types of cells is examined and the participation of the centrosome in establishing and defining cell polarity is discussed.

The book should provide useful background information on the present state of knowledge about the centrosome to researchers and advanced students interested in the organization and behavior of cells. If it provides a deeper understanding and appreciation of the centrosome and stimulates further research on this mysterious and so far rather neglected cell organelle, its goal will have been achieved.

I am grateful to the authors for the time and effort spent in preparing their thoughtful and thought-provoking articles. I am also much indebted to the Cell Biology Series editors, Drs. Dennis Buetow, Ivan Cameron, George Padilla, and especially Arthur Zimmerman for suggesting the idea and for his encouragement and helpful advice throughout this endeavor. Thanks are also due the staff of Academic Press for their assistance in seeing the book through to publication.

Vitauts I. Kalnins

1

Structure and Functions of Isolated Centrosomes

MICHEL BORNENS

Centre de Génétique Moléculaire du CNRS
91198 Gif sur Yvette
France

Recent advances in the isolation of centrosomes have opened the way to experimental studies on this potentially important organelle. In this chapter we will discuss recent data about isolated centrosomes. We will begin with an overview of the subject to show the reasons for our approach to the study of the centrosome and end by reflecting on future directions and prospects.

Copyright © 1992 Academic Press, Inc.
All rights of reproduction in any form reserved.

I. OVERVIEW

A. What Is a Centrosome?

1. Centrosome Is a Central Body

The definition of the centrosome has been largely morphological until now, a criterion that is dependent on the cell system under study. [Everybody "likes to think of his favorite cell as the archetype of the cell" (Mazia, 1987).] Furthermore, eggs or early embryos in which the elementary cell cycle is short have been compared often with somatic cells, in which the cell cycle is more complex, a comparison that requires caution because egg cells might be undergoing a developmental process. The definition of the centrosome as a "polar corpuscle" containing centrioles (Boveri, 1901) has been replaced occasionally by the more restricted definition of a microtubule organizing center (MTOC). In many cells, the pericentriolar material (PCM) is apparently the MTOC, at least morphologically. In acentriolar cells, the arrangement of microtubules is the main criterion used to indicate the position of MTOCs. These "centers," "aggregates," or "osmiophilic regions" can be localized by immunocytochemical means, using spontaneously reacting sera. Many authors consider the staining observed with such sera to be sufficient to identify genuine centrosomes (see Mazia, 1987). We do not share this view.

Before continuing, we should note that this discussion is not a new one: the nature of the centrosome, or even its existence, has been the subject of passionate debates since the beginnings of cell biology; some authors would cite, for example, the "infamous" centrosome as the "dead point in the dynamic movement system of the cell protoplast," whereas others would interpret the centrosome as "just the opposite and in agreement with Bütschli's theory as a kinetic center" (quoted by Pickett-Heaps *et al.*, 1984; see also Wheatley, 1982; Mazia, 1987). Current discussions on the centrosome are less violent, but the debate still goes on more than a century after the centrosome had been discovered and named. This shows that the centrosome is still largely *terra incognita* in cell biology. One of the essential properties of the centrosome implied by the early definitions was its particular dynamics with respect to cell movement, making it appear as a "center." This characteristic led researchers "to attribute a special individuality to this granule" (Lauterborn, 1896; quoted by Pickett-Heaps *et al.*, 1984) which, together with its duplication, credited the centrosome with "autonomy," a rare property indeed.

A frequent element in the debate on the definition of centrosomes concerns the status of the centrioles, since they are not present in the spindle poles of all cells. For those authors who largely equate centrosomes with MTOCs, centrioles or basal bodies are dispensable structures or "passengers" (for a full discussion, see

Bornens and Karsenti, 1984) or even the "Mona Lisa's smile" which should not be confused "with the Lady herself" (Mazia, 1987), an interesting metaphor indeed. For others, centrioles are integral components of the centrosome. Recent advances regarding the structure and functions of centrosomes isolated from somatic animal cells, as well as recent data on the reproductive capacity of centrosomes in marine eggs, do not leave many doubts that the centrioles are functionally important in the centrosomes of those cells that contain centrioles, although the functional significance of these structures has yet to be fully understood.

The opposition of centriole and centrosome might not be the most appropriate or meaningful one at first. If we consider etymology, one of the best examples of "centro-some" should be found in the centroplast of *Heliozoa* (Bardele, 1975; Fèvre-Chevalier and Fèvre, 1984), in which there is a definite "granule" that does not contain centrioles, or in fish melanophores, or erythrophores, in which there is quite an elaborate structure associated with a pair of centrioles (Porter and McNiven, 1982; see Chapter 12). In both cases, the centrosome, which nucleates many microtubules, is a discrete body, corpuscle, or granule, at a single copy number, precisely located at the geometric center of the cell (the nucleus is pushed out of the way to the periphery of the cell). The presence of the microtubule nucleating centers as single copies and the central position in the cell might be good criteria for identifying centrosomes. The typical position of the centrosome in the cell is juxtanuclear, often in the hilus of the kidney-shaped nucleus. This is true for centrosomes with an associated pair of centrioles, as seen in leukocytes (Policard and Bessis, 1953), as well as without an associated pair of centrioles, as seen in diatoms (Pickett-Heaps *et al.*, 1984). When observed in living cells, the nucleus appears to deform around the centrosome, which maintains itself at the cell center. The centrosome does not appear to sit in a pre-established depression of the nucleus (Policard and Bessis, 1953; Rondanelli *et al.*, 1968). In cells that are artificially enucleated by cytochalasins, the centrosome sits precisely at the geometric center of the cytoplast (Wise and Prescott, 1973). Therefore, the juxtacentrosomal location of the nucleus is as accurate a concept as the juxtanuclear location of the centrosome. The mechanism by which the centrosome maintains itself at the cell center is suspected to be linked to its microtubule nucleating activity.

2. *Centrosome Is Tightly Associated with the Nucleus*

The proximity of the nucleus and the centrosome has a structural basis: they are tightly bound to each other. This fact was demonstrated in several animal cells (Bornens, 1977; Nadezhdina *et al.*, 1978; Maro and Bornens, 1980; Fais *et al.*, 1984), but had been suspected (Bessis and Breton-Gorius, 1967) and was even explicitly predicted by Lauterborn in 1896 for diatoms (quoted by Pickett-Heaps *et al.*, 1984): "The clear evidence for a very tight relationship between centrosome and nucleus, perhaps involving a structural connection between the two, . . . is very clear when the nucleus is isolated. . . . The centrosome always stays associated with the nucleus, although the protoplasm and everything else has been torn away." A centrosome–nucleus association is, therefore, the rule, regardless of whether or not the centrosome possesses centrioles. Many unicellular flagellates demonstrate a strong association between basal bodies and the nucleus; a nucleus–basal body connector has been described in some cases (Wright *et al.*, 1985; Salisbury *et al.*, 1987; see Chapter 6). The myxomycete *Physarum polycephalum* is interesting because the ameba form displays a composite centrosome in which the two centrioles are tightly associated with the nucleus through a "granule" that is the real MTOC and resembles the centrosome of a diatom (Wright *et al.*, 1980; Roobol *et al.*, 1982). In many lower eukaryotic cells, the centrosome contains a definite structure (Heath, 1981). The nucleus-associated body (NAB) of *Dictyostelium discoideum* and related organisms has a multilayered paracrystalline core that has a matchbox shape (Roos, 1975; Omura and Fukui, 1985). In many cases, local differentiations of the nuclear envelope, the so-called "plaques" or "spindle pole bodies," correspond to the centrosomes (Heath, 1981; Zickler, 1970, see Chapter 7).

The structural association between the centrosome and the nucleus has obvious functional implications for cell polarity and for mitosis (see, for example, Kitanishi-Yamura *et al.*, 1985; Omura and Fukui, 1985; Wright *et al.*, 1985). It may also manifest an evolutionary process linked to the acquisition of cellular properties associated with cell movement.

3. Centrosome Duplicates Once during the Cell Cycle

The centrosome or single-copy nucleus-associated "granule" or "corpuscle" has a controlled cycle of duplication, the two centrosomes produced functioning as mitotic poles during subsequent cell division. To maintain the polarized organization in the daughter cells, a cell has to pass through a transient bipolar mitotic stage. The centrosome is instrumental in this process.

This centrosome cycle, which has been described in great detail at the ultrastructural level in somatic cells (Robbins and Gonatas, 1964; Rieder and Borisy, 1982; Vorobjev and Chentsov, 1982) and in marine eggs (Sluder *et al.*, 1986; see Chapter 9), is not yet understood at the molecular level. Even the general mechanisms that govern the mode of inheritance of the centrosome are unknown. The idea that a basal body/centriolar DNA exists has been revived recently by the study of a unique chromosome in *Chlamydomonas reinhardtii* (Hall *et al.*, 1989; see, however, Johnson and Rosenbaum, 1990). Other approaches, using isolated centrosomes and the parthenogenetic assay in *Xenopus* eggs, have attempted to unravel the mode of assembly of new centrosomes (Klotz *et al.*, 1990) and will be discussed in Section II.

Another important issue is the mutual interaction between the nuclear and the centrosome cycles. Our understanding of the cell cycle has advanced considerably in recent years (Nurse, 1990). Some progress also has been made regarding the links between the centrosome cycle and the mitotic clock. This aspect will be discussed in Section II (see also Chapter 7).

B. Are Centrosomes and Microtubule Organizing Centers Equivalent?

A distinction must be maintained between centrosomes and MTOCs. Simply stated, *a centrosome is necessarily an MTOC but the reverse is not true.* The difference appears more clearly in the following examples.

1. In some postmitotic differentiated animal cells, the disappearance of centrioles is a specific feature of the differentiation program. Acentriolar MTOCs with a specific cellular distribution can then be observed. One example has been documented during myogenesis *in vitro* in which, after the loss of centrioles, the peripheral regions of the numerous nuclei in the myotube become sites for microtubule nucleation (Tassin *et al.*, 1985a). An ultrastructural study of epidermal cells of *Drosophila* wings provides another example and points to the plasma membrane as a possible site for microtubule nucleation (Tucker *et al.*, 1986; Mogensen and Tucker, 1987). These cells possess numerous MTOCs. Do they, then, possess one or several centrosomes? Lauterborn's answer (see Pickett-Heaps *et al.*, 1984) probably would be neither, since he would find it difficult "to attribute a special individuality" to these MTOCs. They apparently cannot redistribute in an independent manner and the microtubular network that is formed from these MTOCs cannot be reoriented. Even in the case of the myotubes in which the MTOCs fulfill the criteria of a central position and a tight association with the nucleus, they are associated with the nuclear periphery in a manner that does not allow any preferential distribution of microtubules toward any particular sector, as would a spindle pole body, a NAB, or a centrosome. Subsequently, implications for cell polarity and cell movement will be demonstrated.
2. Another model of terminal postmitotic differentiation in culture involves, as does myogenesis, cell fusion—this time of monocytes into a syncytium resembling osteoclasts. In this case, centrioles are retained and the individual centrosomes from precursor cells are distributed in a regular manner in the giant cytoplasm (Moudjou *et al.*, 1989a). The centrosomes are capable of changing their location from the cell periphery to the vicinity of the nuclei, which are grouped in a central area of the cell and with which they probably maintain a structural link.

Are these structures MTOCs or centrosomes? They seem to meet Lauterborn's criterion of "individuality" for centrosomes. Although numerous, they retain a one-to-one relationship with the nuclei.

In all these cases of terminal differentiation, the criteria of duplication or the presence of mitotic poles are indeed irrelevant to identifying centrosomes since the cells do not cycle.

3. Most often the aster of the egg is contributed by the sperm. This is apparently not the case in some mammals, for example, the mouse (Szollozi *et al.*, 1972; Calarco-Gillam *et al.*, 1983; Maro *et al.*, 1985; Schatten and Schatten, 1986) in which there is a structural discontinuity (absence of centrioles), that is, numerous discrete MTOCs scattered within the oocyte participate in the formation of the acentriolar meiotic spindle poles and of several successive acentriolar mitotic spindle poles. Use of a spontaneous human "anticentrosome" serum has demonstrated that these MTOCs contain material immunologically related to that in centrosomes. This observation has led some investigators to consider these structures as genuine centrosomes (Schatten and Schatten, 1986; Mazia, 1987). The criterion of single copy per nucleus is apparently not met, but the multifocal pattern seen in interphase could be argued to correspond to some form of unfolding of the centrosome, although there is no evidence that these foci are linked together. However, one basic criterion for the centrosome in a cycling cell—the capacity to duplicate at some point during the cell cycle—has not been established. Instead, there is evidence against the duplication of the microtubule-nucleating material during the early cycles of the developing mouse embryo. The material forming the maternal MTOC is apparently progressively diluted by successive divisions before centrosomes similar to those of somatic cells are formed (Calarco-Gillam *et al.*, 1983). Why, then, is the material found at meiotic or mitotic poles? The structures may function to stabilize a spindle that is formed by a largely independent mechanism. A case has been made for a specific role of the nucleus in the shaping of the spindle (Karsenti *et al.*, 1984; see also reviews by Bornens and Karsenti, 1984; Karsenti and Maro, 1986). Admittedly, this mechanism is not completely satisfactory, since bipolar division of blastomeres has been observed in marine eggs enucleated chemically (Nagano *et al.*, 1981) or by hand (Picard *et al.*, 1988). However, if it exists, such a mechanism could dominate in the early development of those organisms in which cell polarity is not imposed at fertilization by the sperm aster. Under these circumstances, we would conclude that the early mouse embryo has no functional centrosome.

4. It has been demonstrated that PCM proteins can redistribute to one end of independently assembled microtubules when cells are treated with taxol (Gosti-Testu *et al.*, 1986; see Fig. 1). This treatment is known to decrease dramatically the critical concentration required for tubulin polymerization, leading to the formation of a modified network of microtubules that is no longer anchored to the centrosome (De Brabander *et al.*, 1981). Instead, the microtubules form long

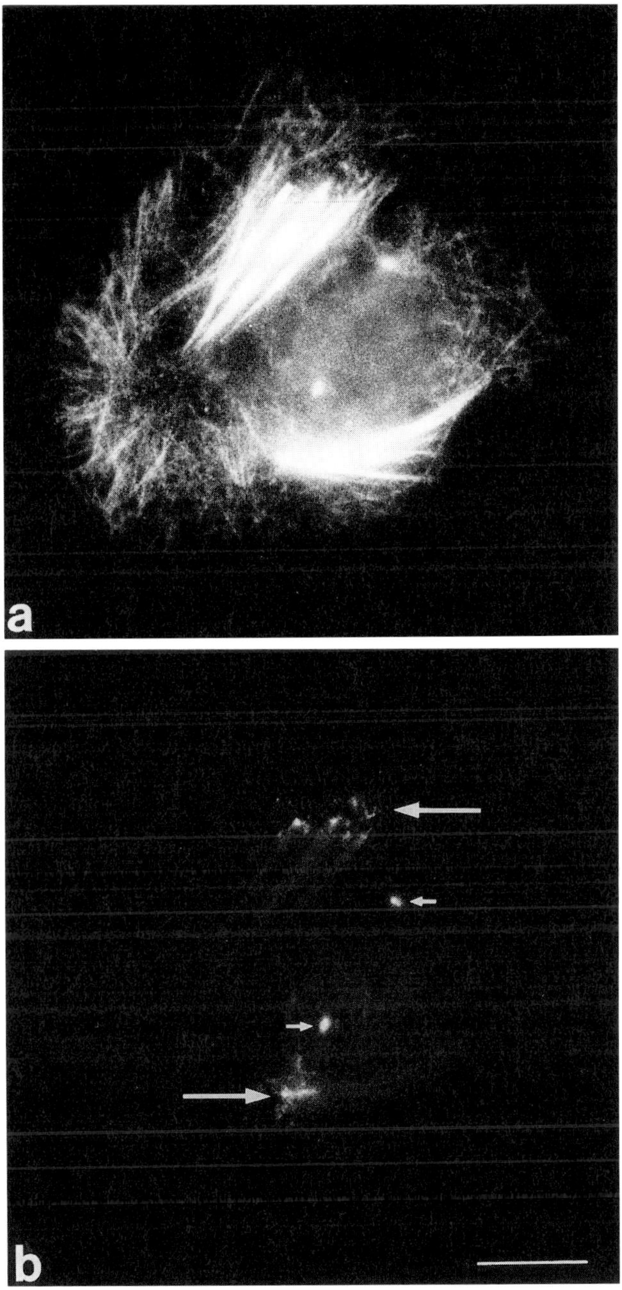

Fig. 1. Double immunofluorescence of a taxol-treated G_2 HeLa cell (4 hr with 5×10^{-6} *M* taxol) with antitubulin antibodies (a) and anticentrosome rabbit 0013 serum (b). The two centrosomes (small arrows) do not nucleate microtubules. Most of the microtubules are organized in thick perinuclear bundles in which individual microtubules apparently have the same polarity. The anticentrosome serum decorates one end of the bundles, probably the minus end (long arrows), where most, but not all, of the pericentriolar material, or "MTOC," has redistributed. The centrosome integrity is lost. Bar: 10 μm.

juxtanuclear bundles in interphase cells and numerous small asters in mitotic cells. In both cases, the PCM relocates to one end of the microtubule arrays and the centrioles, deprived of most of their PCM, may or may not be associated with microtubules.

This artificial situation has two important implications: First, it shows how relative the concept of MTOC is. An "anticentrosome" serum decorating some structure does not necessarily mean that the structure is actually "organizing" microtubules. The microtubules may be organizing the structure, depending on the cellular environment. Microtubules stabilized by any mechanism (there are many examples in eggs, in which microtubule arrays are apparently stabilized by the cortex) will influence the cellular distribution of the "MTOC." In other words, the nucleating activity of a MTOC may depend on other cellular compartments, such as the plasma membrane, the reticulum, or any other organelles that influence the stability of microtubules by interacting with them.

The second implication is that the integrity of the centrosome itself depends on the dynamics of microtubules, at least when the latter are dramatically modified by drugs. The PCM can be dissociated from the centrioles by treating cells with taxol, but also by treating cells with drugs that decrease the assembly of microtubules (Watt and Harris, 1980; Wright and Moisand, 1982; Sellito and Kuriyama, 1988). Such drugs also apparently modify the centroplast of *Heliozoa* (Hauser, 1986). Is there a feedback loop between the centrosome and the microtubules under physiological conditions? This is a pertinent but unanswered question.

5. The possibility that plant cells possess a centrosome has been assumed recently (and considered a "turning point" making it "unnecessary to make any distinction between animals and plants" in terms of centrosomes; Mazia, 1987), because the use of a spontaneous human "anticentrosome" serum showed the staining of some microtubule-associated material (Clayton *et al.*, 1985; Wick, 1985). This result was not observed in all plant cells, however (A. M. Lambert, personal communication), and some serious doubts regarding the meaning of these results have been raised recently (Harper *et al.*, 1989; see also La Claire and Goddard, 1989). Do these data imply the existence of a centrosome in plant cells? We doubt it. There is no evidence for a duplicative cycle of the spindle "poles" and electron microscopy (EM) has failed to demonstrate any particular structure at the ends of the barrel-shaped spindles. Spindle shaping in plant cells, like the organization of their interphase microtubular network, apparently involves specific mechanisms that are linked to the existence of a cell wall (Flanders *et al.*, 1990). If "one of the most remarkable responsibilities of centrosomes is the determination of the plane of cytoplasmic division following mitosis" (Mazia, 1987), a "responsibility" that is apparently strongly supported by the bipolar division of enucleated blastomeres (Nagano *et al.*, 1981; Picard *et al.*, 1988), then there is no centrosome in plant cells, in which the plane of cell division is

determined by the marginal band of microtubules that forms before the establishment of the spindle poles. Obviously, the definition of the centrosome as "the division organ," once proposed by Boveri, does not fit all dividing cells (nor does it exclude the possibility that cells deprived of a centrosome can divide).

We conclude from these examples that the presence of microtubule arrays in a given cell correlates in all likelihood with the presence of MTOCs of various sorts but not necessarily with the presence of a centrosome. It is not necessarily true that "images of arrangements of MTs delineate shapes of centrosomes" (Mazia, 1987) unless one uses "centrosome" and "MTOC" as synonyms. The trouble with the acronym MTOC is that only the "M" and "T" have the same meaning for everybody. The "O" is quite vague and can refer to many types of "clumps, sheets, or strands of densely staining material" (Tucker, 1979) including kinetochores. The "C" can designate structures of any shape and size anywhere in the cell. Whereas the root *centro* in the early definition of centrosome designated "the cell center," the *center* in MTOC simply designates material at one end of a particular array of microtubules. In a taxol-treated epithelial cell, for example, such as the one shown in Fig. 1, double immunofluorescence with anticentrosome and antitubulin sera allows one to identify "MTOCs," but where is the centrosome in this case? There is no functional centrosome in such a cell, since it has been dissociated, or pulled apart, by the taxol-induced assembly of the cellular tubulin. As a result, the centriole pair is in one place and the PCM is in several other places. Interestingly, the centrosome can recover from such an unfortunate experience if taxol treatment is not too long relative to the length of the cell cycle. In addition, the recovery may uncover some principles of centrosome assembly.

Typically, the centrosome insures, at the cellular level, a unique organization and a collective behavior of the microtubular network, both in interphase and in mitotic cells. As such, it acts as the determinant for the organization of the cytoplasm and for the fidelity of cell division. Therefore, the centrosome has been thought, since its discovery, to play a dominant role in those cell functions which, more than any others, require a concerted behavior, that is, in establishing and maintaining cell polarity during cell movement (the kinetic center) and in cell division (the division organ).

In the examples just discussed, we often alluded to the presence of centrioles in the centrosome or to their dissociation from the microtubule-nucleating material of the centrosomes in particular situations. However, we re-emphasize that the presence of centrioles is not a general criterion for identifying centrosomes. Cells without centrioles or basal bodies can possess a centrosome that behaves as described. We know very little about the structure and composition of such centrosomes. When cells do express centrioles or basal bodies, these always form part of the centrosome, except in the case of highly differentiated multiciliated cells.

C. What Are the Functions of Centrosomes?

The centrosome–microtubule complex is involved in cell polarity, cell migration, and cell division. Microtubules determine the cellular distribution and, to some extent, the dynamics of membrane compartments such as the endoplasmic reticulum, the Golgi apparatus, various elements of membrane traffic, and mitochondria, and of other cytoskeletal networks such as the vimentin filaments or the cortical microfilaments. Accordingly, the organization and the dynamics of the microtubular network could exert critical control over cell activity. The centrosome, acting as a unique nucleating center tightly associated with the nucleus, provides the microtubular network with two remarkable properties—an asymmetrical organization and a unique polarity—since the individual microtubules are anchored to the centrosome by their minus end (Euteneuer and McIntosh, 1981).

1. Is the Microtubule-Nucleating Activity of the Centrosome Sufficient to Explain Centrosome Behavior during Cell Growth?

Obviously, a highly controlled mechanism for centrosome duplication has evolved, encompassing the whole cell cycle. There is also a progressive increase in the total nucleating capacity of the centrosome during cell growth. If the microtubular network exerts critical control over cell activity (see previous text), it would reasonably participate in mechanisms that monitor the cell mass during growth since its size would have to increase correspondingly, which also should be reflected by the increased nucleating capacity of the centrosome. The discrete steps in the centrosome duplication process, which starts early in the cell cycle in yeasts as well as in mammalian cells, that is, before S phase, would be a key element of this mechanism. The interaction between the centrosome and cell cycles will be covered in the next section.

2. Is the Microtubule-Nucleating Activity Sufficient to Explain the Centrosome as the Major Determinant for the Organization of the Cytoplasm and for Cell Polarity?

There is no simple answer to this question since we do not know enough about the dynamics of the cytoplasm nor the considerable number of transient interactions between microtubules and the different cellular components, including the nucleus. However, clear evidence exists for the role of the microfilament system in the positioning and motility of centrosomes in cells (Euteneuer and Schliwa, 1985; Buendia *et al.*, 1990).

The structural complexity of the centrosome in mammalian cells suggests that the centrosome is not simply an aggregate of microtubule-nucleating sites: the geometry of the centrosome—that is, the size and the organization of the PCM, the intercentriolar distance, the number of microtubules, and their orientation—can be largely modulated. For example the pericentriolar material is distributed unequally on both centrioles and possesses a complex spatial organization. In human cells, microtubules are anchored *in vivo* at the tips of nine arms surrounding one of the centrioles. This arrangement can be observed after quick-freezing cells at helium vapor temperature, an approach that should provide samples that accurately represent living cells (Paintrand and Bornens, unpublished observations).

3. Is the Centrosome Able to Respond to Distal Events in the Microtubular Network Where Interactions with the Actin System and with Other Organelles Take Place?

Some investigators, including this author, have speculated along these lines (Bornens, 1979; Albrecht-Buehler, 1981; see also Bornens and Karsenti, 1984). In contrast, it has been argued on the basis of the dynamic instability of microtubules that the role of the centrosome is a passive one, since selective stabilization of microtubules in response to a polarized external signal alone could be sufficient to establish cell polarity (Kirschner and Mitchison, 1986). In many cell types, however, overwhelming evidence supports the existence of an intrinsic and permanent cell polarity in the absence of a polarized external signal. Recent experimental data suggest that the centrosome, and especially centrioles, could actively contribute to the establishment of such an endogenous cell polarity.

1. Differentiating systems (mentioned previously) in which significant changes in cell polarity take place show a related change in the structure and the organization of the centrosome. Myotubes produced by the fusion of competent myoblasts have no anterior–posterior polarity and do not migrate. During fusion, the centrioles are eliminated; then the microtubules start to be nucleated on the nuclear periphery, to which the pericentriolar material has migrated (Tassin *et al.*, 1985a). The Golgi apparatus displays a perinuclear distribution, which contrasts with the polarized juxtanuclear position it occupies in myoblasts (Tassin *et al.*, 1985b). Although the association between the Golgi apparatus and the MTOC, a basic feature of cell polarity, is maintained, the two do not display any preferential orientation and cannot change their position. In contrast, migrating precursor cells can change their direction of migration by relocating the centrosome and, consequently, the Golgi apparatus. The stable organization of the myotube and the loss of endogenous cell polarity might be necessary to allow the establishment of a functional polarity of the muscle by external clues such as innervation (Tassin *et al.*, 1985b). This view was recently supported by the demonstration that innervation restricts the Golgi apparatus (Jasmin *et al.*, 1989) and a subset of microtubules (Jasmin *et al.*, 1990) to subsynaptic domains. After denervation, myotube organization is re-established. Thus, muscle cell polarity is apparently established by interaction with a nerve cell, according to the model of selective

stabilization of microtubules by a polarized external signal as proposed by Kirschner and Mitchison (1986). Prerequisites, however, are the loss of intrinsic polarity at the cellular level and the loss of the migrating capacity of the differentiating myoblasts. The simultaneous elimination of the centrioles suggests that they may be involved in defining the endogenous polarity of migrating cells.

The monocyte-derived osteoclast-like cells provide a complementary example from which similar conclusions can be drawn. Unlike myotubes, these syncitial giant cells are capable of active oriented migration. They retain the individual centriole-containing centrosomes of their precursor cells; these are distributed in an organized manner in the giant cytoplasm (Moudjou *et al.*, 1989a).

2. Newt leukocytes that display a highly visible centrosomal area were used to study the effect of laser irradiation of centrosomes on cell migration (Koonce *et al.*, 1984). After irradiation, cells stopped transiently, then resumed migration in a random manner, with a significantly lower average rate. The authors provided evidence for specific damage to the centrioles. The associated centrosomal material was capable of reinitiating microtubule assembly several minutes after irradiation.

3. A dramatic centrosome splitting, that is, separation of the centrosome into two single centrioles, each surrounded by an aster of microtubules, has been observed in human neutrophils when they are activated (Schliwa *et al.*, 1982; see chapter 13). In these cells, microfilaments apparently are capable of affecting the position of the centrosome by interacting with the microtubules radiating from it (Euteneuer and Schliwa, 1985). In cells capable of chemotactic behavior, the cortex plays a decisive role in their response. The centrosome–microtubule complex maintains the motile machinery as an integral part of the cell. This integrative role requires more than a passive relocation of the complex behind the leading lamella. It also requires elaborate interactions with the cell cortex and with other cellular compartments, particularly with the nucleus. The possibility that the actin system interacts directly with the centrosome exists also. Evidence has been provided that basal body/centriole migration during differentiation of ciliated cells in *Beroë* is driven by the directed assembly of microfilaments attached to a centriole-associated rootlet (Tamm and Tamm, 1988).

4. Cells growing in suspension, for example, lymphoblasts, display polarity in cell surface activity in the absence of any polarizing signal. We have recently observed that this polarity is established through a mechanism based on a permanent periodic oscillatory activity of the cortical actomyosin system. Without microtubules, this polarity reverses periodically as one constriction ring moves from one end of the cell to the other (Bornens *et al.*, 1989). Microtubules impose a stable polarity in this system by dramatically reducing the amplitude of the equatorial contraction and of the movement of the constriction ring. Both effects are due to the inhibitory control exerted by the microtubules on the association of myosin II with the cortex. The outcome is that the oscillatory activity of the cortical actomyosin system becomes confined to one side of the cell, resulting in

unidirectional extension of the cytoplasm. The side of confinement is determined internally by the position of the centrosome.

5. Striated rootlets (rhizoplasts) are often associated not only with basal bodies but also with centrosomes, as seen in epithelial PtK_2 cells and many other cell types. Rhizoplasts are calcium-sensitive contractile organelles (Salisbury and Floyd, 1978). Moreover, immunological evidence suggests that homologous proteins are present in the striated rootlets associated with both the basal bodies and the centrosomes of many cells, ranging from unicellular algae to mammalian cells (Klotz *et al.*, 1986; Salisbury *et al.*, 1986; McFadden *et al.*, 1987), and even in the centrosomes of cells without any obvious centrosome-associated rootlets (Moudjou *et al.*, 1989b; 1991). A functional role for these calcium-sensitive proteins in altering the orientation of centrioles or basal bodies in response to calcium fluxes has been proposed (Salisbury *et al.*, 1986; see Chapter 6). Interaction of the rootlets with the actin system in particular could have important functional implications.

These data (see also Rogers *et al.*, 1985; Chapter 11) suggest a specific and important role for the centrosome and its centrioles in the definition of cell polarity associated with cell migration.

D. Toward a Functional Definition of a Centrosome

An apparently objective definition of a centrosome in animal cells was recently proposed (Bornens *et al.*, 1990). The term "centrosome" designates an organelle that, once isolated from cells, would be able to perform, under controlled conditions, two of the basic functions normally associated with centrosomes, that is, the nucleation of microtubules and the establishment of an efficient mitotic spindle. More recent data, obtained with isolates from various animal cell sources, indicate some of the limitations of this definition. Centrosomes may not permanently express all their activities, a possibility for which there is precedence in other cellular organelles, including the nucleus. This concept will be discussed further in the next section.

As for the specific properties common to all centrosomes, from those in diatoms to those in mammals, one may have to wait for a better molecular characterization of the centrosomal components in a few species in order to search for those components in other species.

II. THE ISOLATED CENTROSOME

A. Isolation Procedures

Centrosomes can be isolated as centrosome–nucleus complexes (Bornens, 1977), then further dissociated from the nuclei (Nadezhdina *et al.*, 1978; Maro and

Bornens, 1980). The yield by this approach, however, is low, and the centrosomes are contaminated with perinuclear cytoskeleton. Alternatively, centrosomes can be dissociated from their nuclei by cell lysis at very low ionic strength. Introduced by Blackburn and colleagues (1978), this approach was further developed by Mitchison and Kirschner (1984, 1986). Cells are lysed with nonionic detergents after rapid equilibration in media of decreasing ionic strength. Subsequent purification on a sucrose gradient is monitored by a quantitative immunofluorescent assay (Evans *et al.*, 1985). Adapting this approach, our laboratory has obtained high yields of isolated centrosomes from the human lymphoblastic cultured KE37 cell line (Bornens *et al.*, 1987). KE37 cells grow in suspension and possess a low cytoplasm-to-nucleus ratio, which is favorable for isolating centrosomes. An example of such a preparation is presented in Fig. 2. The most striking feature of these isolates is that the native paired configuration of the centrioles has been maintained during the isolation procedure. Using specific antibodies and double immunofluorescence, centrosomes could be identified without ambiguity. The major potential contaminants of these preparations were pieces of chromatin, which were removed by treating lysate supernatants with DNase II before the gradient purification step. The centrosome yield is directly dependent on the success of the cell lysis step. Optimally, it is about 60% of the starting material, assuming one centrosome per cell, but can drop to 15–20% if chromatin is released in a sticky pseudosoluble state. Recovery of functional centrosomes depends on the care taken to inhibit proteases during cell lysis, since microtubule nucleating activity is shown to be sensitive to proteolysis (Kuriyama, 1984).

The method described is adaptable to other cell types, provided that cell lysis is optimized. Isolation was achieved, for example, with tissue cells instead of cultured cells (Komesli *et al.*, 1989). Calf thymus was chosen as the source because thymocytes have the lowest known cytoplasm-to-nucleus ratio. This favorable starting material provides many highly purified centrosomes from a simplified isolation procedure. For example, drug pretreatment to disassemble microfilaments is unnecessary because microfilaments are not prominent in thymocytes. The method has now been used or adapted to isolate centrosomes from cells in several tissue types as well as from other cultured cells, including human peripheral lymphocytes (Tournier *et al.*, 1989), rat and mouse thymocytes, bovine

---

Fig. 2. Immunofluorescence using an antitubulin serum showing that centrosomes are isolated as pairs. Individual centrioles can be identified in preparations of centrosomes from the human lymphoblast KE37 cell line (a) and from calf thymus (b). Bars: 10 μm. Electron micrograph of a centrosome from the human lymphoblast KE37 cell line (c) showing a distinct link between the two centrioles of a pair that seems to emanate from a network surrounding each centriole. Centrosomes isolated from calf thymus (d) have a collinear orientation. The centrioles are apparently attached by their proximal ends. Subdistal radial arms (arrows) are present only on one of the two centrioles of a pair in both types of centrosome. Bars: 0.2 μm.

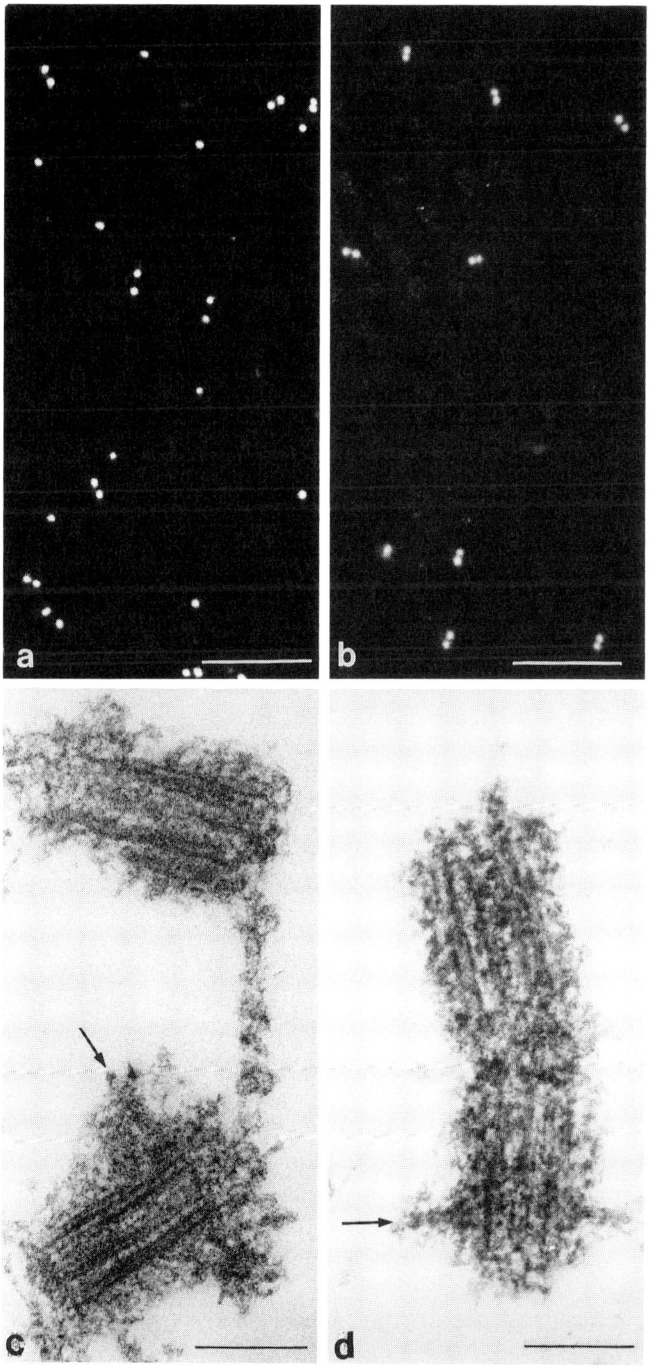
a
b
c
d

peripheral lymphocytes, and cultured bovine cells (Tournier *et al.*, 1991a; 1991b). Centrosomes have also been isolated from mitotic CHO cells and from metaphase sea urchin eggs (Kuriyama, 1984; Kuriyama and Borisy, 1983).

B. Structure of Isolated Centrosomes

For optimal observation of their complex three-dimensional (3-D) organization at the ultrastructural level, isolated centrosomes must be oriented by sedimentation on a coverslip before they are processed for electron microscopy. In this way, the great majority of centrioles is positioned parallel or perpendicular to the coverslip. This orientation is critical, for example, for establishing the structural basis of the paired configuration of the centrioles in each centrosome. A link exists between them that is part of a complex filament network surrounding each centriole (Bornens *et al.*, 1987; Fig. 2). This link often displays a periodic organization in its narrow part. Its chemical nature is unknown. Substantial separation of the two centrioles sometimes occurs *in vivo* (Schliwa *et al.*, 1982; Sherline and Mascardo, 1982; Buendia *et al.*, 1990). Evidence has been provided for a role of the actin system in the positioning and the motility of the centrosome (Euteneuer and Schliwa, 1985; see subsequent text). Actin is consistently present in centrosome preparations in significant amounts (Komesli *et al.*, 1989) and its association with sperm centrioles has been demonstrated in *Hydractinia* (Kleve and Clark, 1980). However, attempts to enhance the link between the centrioles with myosin S_1 fragments have failed.

Both centrioles have an average length of 0.4 μm in human lymphoblasts. The addition of tannic acid to the fixative has greatly improved the ultrastructural preservation of centrioles (Klotz *et al.*, 1990; Moudjou *et al.*, 1991). There is a definite proximal–distal differentiation along the centriolar cylinders. The proximal part displays the typical nine triplets of microtubules and shows a tangential link between microtubule A, the most internal one, and microtubule C, the most external one, of adjacent triplets. Microtubule C in each triplet becomes progressively incomplete and is totally absent at the distal end, as already reported for basal bodies (Wilsman and Farnum, 1983). Moreover, the nine doublets at the distal end are no longer linked tangentially but are oriented in a more tangential manner than the nine proximal triplets (Fig. 3; see also Komesli *et al.*, 1989; Klotz *et al.*, 1990). The lumen of each centriole is filled for about three-fifths of its length on the distal side by an as yet ill-defined material. The lumen at the proximal end of the centrioles appears empty. Each centriole, therefore, has two distinct organizations, one in the proximal part, that is typical of the classical basal body/centriole structure, and another in the distal part, from which a primary cilium can grow in some circumstances (Wheatley, 1982), that is more reminiscent of the axoneme structure. A recent report convincingly demonstrated that centrioles

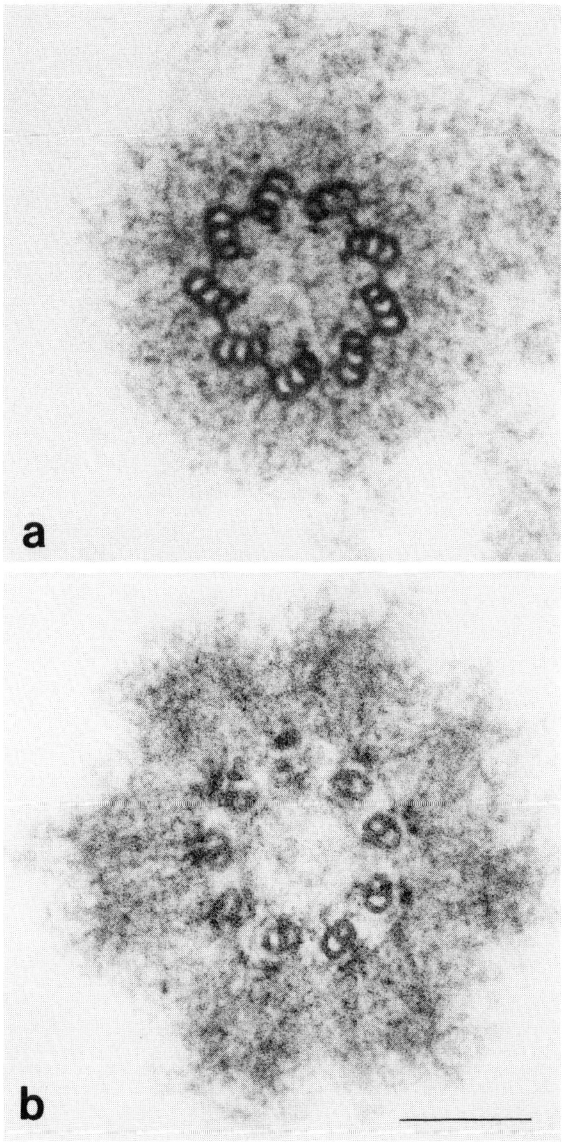

Fig. 3. Transverse section of the subproximal end of either centriole in a centrosome (a) and of the subdistal end of the mother centriole (b). Note that both the centriole and the surrounding noncentriolar material display proximal–distal differentiation. Note also the complexity of the subdistal arms in b. Bar: 0.1 μm.

contain tektins (Steffen and Linck, 1988), suggesting that their structural similarities with axonemes are corroborated by biochemical homologies.

The pericentriolar material appears to be highly complex and also shows variations in proximal–distal organization. The distal end around one of the two centrioles seems incomplete. At the proximal end, the pericentriolar material appears as a sheath about 40 nm thick that covers one-half of its total length. At the distal end of only one of the centrioles, the pericentriolar material displays radial arms distributed more or less according to a ninefold axial symmetry. These arms themselves have a complex structure and are apparently organized from numerous linear threads (Fig. 3; see also Klotz *et al.*, 1990). Cryofixation shows that microtubules are anchored at their tips *in vivo*. These arms are in a subdistal position, whereas another set of nine thin appendages is present in the distal position (Moudjou *et al.*, 1991). The latter appendages are much better preserved or visualized in isolated centrosomes when tannic acid is added to the fixative. They extend outward from the wall of the centriole in an orientation that is slightly tilted with respect to a radial direction. They probably correspond to the transitional fibers of alar sheets observed *in vivo* in some cases (see Wheatley, 1982; see also Vorobjev and Chentsov, 1982). These fibrous appendages usually accompany centrioles that are becoming basal bodies, a transformation which, as far as we know, does not take place in lymphoid cells. The function of these appendages deserves more attention. They may, for example, anchor the centrosome to the actin system.

Centrosomes isolated from calf thymocytes have a peculiar structure: the two centrioles are tightly associated with each other by their proximal ends, giving a linear configuration to the centrosome (Fig. 2); see also Komesli *et al.*, 1989). Granular material filling the narrow space between both centrioles could correspond to some folded form of the intercentriolar link that is observed in centrosomes from human lymphoblasts. The centrioles themselves are similar to those of KE37 cells, with triplets of microtubules at the proximal end and doublets at the distal end. They also have similar dimensions. The pericentriolar material is not as abundant and is differently distributed on each centriole. Radial arms are present at the distal end of only one of the centrioles. The length of the arms is similar to that of those in human cells. There are never more and often less than six of these arms present, compared with the nine seen in human lymphoblasts.

Therefore, some differentiation in the centrosomal structure from one cell type to another can exist, which raises the intriguing possibility that the centrosome has properties specific for differentiated states, despite the similarity in the structure of the centrioles. Moreover, the centrosomes of thymocytes reveal important functional differences from those of lymphoblasts in the two basic assays that are essential for the definition of centrosomes. (1) They nucleate only a few microtubules from the distal ends of the centrioles, as do the basal bodies (Komesli *et al.*, 1989). (2) They do not allow parthenogenetic development of the *Xenopus* egg (Tournier *et al.*, 1991a). The significance of these observations will be discussed subsequently.

An important issue when studying isolated organelles is whether or not the isolation procedure has led to heavy structural or functional modifications. Are centrosomes denatured by the present methods of isolation as inferred by Mazia (1987) from the aster shape of microtubules growing from isolated centrosomes? This criterion might not be the most discriminating one, since an aster can appear to be a hemispindle, depending on the handling of the coverslips at the fixation step (see Fig. 5; see also Komesli *et al.*, 1989). However, apparent structural differences appear between isolated centrosomes and those described *in situ*. Isolated centrosomes have either lost most of their loosely associated material, or satellites, often identified as part of the pericentriolar material *in situ* (see Gosti-Testu *et al.*, 1986; see also Rieder and Borisy, 1982), or this material has been induced to contract around the centrioles by the isolation procedure.

If the loosely associated material is lost, then we can conclude from the functional properties of the isolated centrosomes that this material corresponds to an excess of nucleating material, nonessential for the complete expression of basic centrosomal functions. The amount of this material would be more related, for example, to the size of the cell (and, consequently, to the optimal number of microtubules necessary) than to the actual expression of centrosomal functions. An additional explanation for the large amount of this material present in eggs is that the nucleating material may be stockpiled during oogenesis and later distributed between blastomeres during the rapid segmentation of the embryo. This view is suggested by the work of Sluder and Rieder (1985), who documented by serial sectioning the complete ultrastructure of the large centrosomes in sea urchin blastomeres. The two centrioles that appear embedded in a large cloud of granular material were shown to be markers for the duplicative capacity of spindle poles. From this result, one may infer that these centrosomes, once isolated and free from most of the surrounding cloud of granular material, will retain the essential functions of centrosomes. Sluder and co-workers (1989a) have already demonstrated that the reverse is not true: a sea urchin egg centrosome without centrioles does not reproduce between mitoses (see Chapter 9).

If the material contracts around the centrioles during isolation, one should be able to decondense the contracted pericentriolar material. We are currently trying to distinguish between these two possibilities.

The data just presented provide no evidence to support "the flexible linear centrosome" model put forward by Mazia (1987). This model is based essentially on Boveri's description of mitosis in eggs of *Ascaris* and on studies of mitosis in sea urchin eggs, in which the centrosomes are very large and the cell cycle is very short. The model confines itself to the function of the centrosome as "the division organ of the cell" (Boveri, 1901) in the egg and does not deal with centrosomes at further developmental stages. However, as soon as the segmentation of the egg is complete, the cell cycle lengthens considerably and cells must grow before dividing. Cells begin to move, and spend most of their life in interphase, preparing for the next division. These somatic cells are considerably smaller than the egg from which they come; accordingly, they possess a tiny centrosome, tightly associated with the nucleus, in which a pair of centrioles appears as a prominent feature.

The possibility that the noncentriolar material in the centrosome is fibrillar in nature is certainly not excluded by the ultrastructural data from isolated centrosomes. The two centrioles are held together as a pair by a complex link made of thin fibrils; apparently, this link may exist in some folded state in particular cells (Fig. 2). In other cells, centrosomes are associated with striated rootlets (Salisbury and Floyd, 1978; Klotz *et al.*, 1986; Salisbury *et al.*, 1986). As mentioned earlier, the noncentriolar material, or parts of it, may be able to undergo large reversible conformational changes in response to several effectors, such as Ca^{2+} ions (Salisbury *et al.*, 1986; Moudjou *et al.*, 1989b; 1991). The centrosome is also able to change its overall shape. For example, the two centrioles may split and migrate over large distances, depending on transient activity of the actin system (Euteneuer and Schliwa, 1985). The centrosome may also change when a more permanent state of differentiation is induced (Buendia *et al.*, 1990); change also frequently is observed transiently in unsynchronized epithelial or fibroblastic cells in culture.

The concept of the centrosome that emerges from all these data is at variance with the model of "the flexible linear centrosome." Instead the pair of centrioles appears as the central element on which the whole centrosome is organized. This idea is suggested by two ultrastructural features of the centrosome: (1) The noncentriolar "network" is so highly organized in some parts directly associated with the centriole wall that it also possesses a ninefold symmetry (Fig. 3). (2) This pericentriolar material is unequally distributed on the two centrioles, demonstrating a polarity in centrosome organization (and probably in nucleating capacity) that apparently has its basis in the duplication mechanism of centrioles.

This hypothesis is also suggested by the demonstration that the reproductive capacity of centrosomes in sea urchin eggs depends on the presence of centrioles (Sluder and Rieder, 1985; Sluder *et al.*, 1989a; see Chapter 9).

C. Centrosomal Proteins

The total protein content of our isolated centrosome is $2–4 \times 10^{-2}$ pg, depending on the isolates. The one-dimensional protein profile of isolated human centrosomes is complex but quite distinct from either detergent-soluble or -insoluble cellular proteins (Fig. 4). When preparations near homogeneity, the protein profile is remarkable because of the absence of major components with apparent molecular weight below 35K. One can distinguish about 15 major polypeptides (Fig. 4,

Bornens *et al.*, 1987; Moudjou *et al.*, 1991). High molecular weight proteins (180–350K) and a group of proteins between 40 and 70K are the major features of the centrosomal pattern. Two-dimensional analysis further demonstrates the specificity of the protein pattern for centrosomes (Bornens *et al.*, 1987).

It is, indeed, likely that not all of the proteins identified in the preparations belong to the centrosome proper, and it is necessary to obtain independent evidence for their genuine centrosomal origin. Except for the linkage to the nucleus, the interactions that maintain the centrosome in its cytoplasmic location are not known. A frequent source of contamination is the nuclear material, which can be easily detected by the presence of bands below 35K that correspond to histones. During cell lysis, the centrosome becomes dissociated from its environment in an unknown manner; other contaminating proteins could be present for different reasons. Fodrin, for example, is sometimes abundant in centrosome preparations, but cannot be demonstrated at the centrosome *in situ* (Klotz and Bornens, unpublished results). This protein could be artifactually adsorbed onto centrosomes during the isolation procedure because of its insolubility at low ionic strength. Alternatively, pieces of plasma membrane containing fodrin, actin, and myosin (which are also present in centrosome preparations) could attach to the centrosome during the lysis step.

In addition to the tubulins, two polypeptides have been identified in our laboratory as genuine centrosomal proteins (see Fig. 4). One was identified with the rabbit serum 0013, which has an unprovoked anticentrosome activity essentially restricted to human and monkey cells (Maunoury, 1978; Gosti-Testu *et al.*, 1986). A family of high molecular weight polypeptides was identified at first. More recent work has established the susceptibility of the centrosomal antigen to proteolysis: the native antigen seems to be a doublet with an apparent molecular weight of 350K (Komesli *et al.*, 1989; Klotz *et al.*, 1990; Keryer *et al.*, 1990a). It is localized in the pericentriolar material; indirect evidence supports the involvement of this antigen in microtubule nucleation (see subsequent text). This antigen shares a common epitope with a protein of the nucleolar organizing center (Courvalin *et al.*, 1986) and with the human B isoform of lactate dehydrogenase (Gosti *et al.*, 1987; Gosti and Bornens, submitted). Molecular characterization of the 0013 protein is now underway. Centrosomes isolated from calf thymus were shown to possess a protein with a similar molecular weight that is antigenically related to the 0013 human protein (Komesli *et al.*, 1989). A 350K protein has been shown recently to be the major centrosomal binding protein for the regulatory subunit (RII) of cAMP-dependent protein kinase in human cells, and to be phosphorylated by the catalytic subunit (Keryer *et al.*, 1990a), a property which, with its molecular weight, is reminiscent of MAP2. However, no cross-activity between the human centrosomal doublet p350 and the bovine MAP2 has been detected (Gosti-Testu *et al.*, 1986; Keryer *et al.*, 1990a). The possibility that the p350 doublet contains several unrelated polypeptides is being examined.

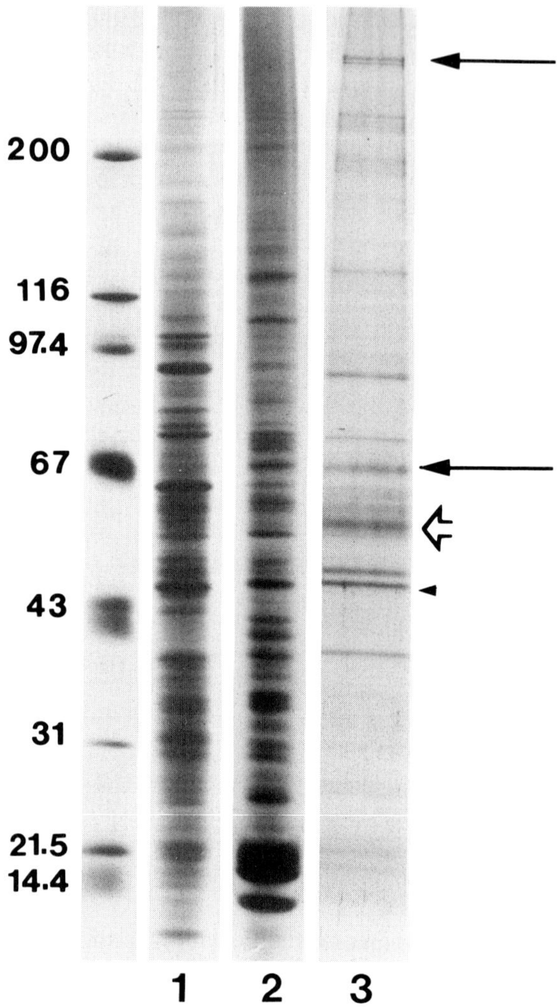

Fig. 4. Centrosome preparations display a specific pattern of proteins. One-dimensional electrophoretic analysis of a centrosome isolate from human KE37 cells on a 6–15% SDS–PAGE gradient, as revealed by silver staining, is shown. Lane 1, Triton X-100 soluble proteins (4 μg); lane 2, Triton X-100 insoluble proteins (4 μg); lane 3, 1×10^7 isolated centrosomes (about 0.3 μg). Tubulins are abundant proteins (open arrow). Actin is also abundant (as judged by two-dimensional analysis and Western blotting), even in the best preparations (arrow head). Two polypeptides (long arrows) have been identified as genuine centrosomal proteins p 350 (Gosti-Testu *et al.*, 1986) and p62/64 (Moudjou *et al.*, 1989b; 1991).

The second centrosomal protein is a protein with an apparent molecular weight of 62–64K, identified by two unrelated antibodies (Moudjou *et al.*, 1989b, 1991). The first antibody was polyclonal and raised against the basal-body-associated protein centrin, a 20K Ca^{2+}-binding protein identified in the green algae *Tetraselmis* (Salisbury *et al.*, 1984) and in *Chlamydomonas* [Huang *et al.*, 1988a (caltractin)] that shows important sequence homology with both calmodulin and the yeast *CDC31* gene product required for spindle pole body (SPB) duplication (Baum *et al.*, 1986; Huang *et al.*, 1988b). The second antibody was raised against a 230K protein that, with a 22K Ca^{2+}-binding protein, forms a basal-body-associated contractile structure in the ciliate protozoan *Polyplastron* (Vigues *et al.*, 1984). Strong evidence that this protein binds Ca^{2+} and is involved in the nucleation of microtubules at the centrosome has been obtained (see subsequent text). The identification of this new calciprotein in human centrosomes suggests that this protein is highly conserved among divergent organisms, as is the structure of the centriole/basal body pair in these organisms.

The anticentrin antibody has also been shown to immunoprecipitate a 165K protein in PtK_2 and to bind specifically the centrosome in these cells (Baron and Salisbury, 1988), a result that suggests that homologous proteins from algae and humans (Moudjou *et al.*, 1989b; 1991) can have apparent molecular weights that are quite different. However, the identity of the 165K protein as a centrosomal protein in PtK_2 cells still must be confirmed directly.

The problem of using antibodies to identify a protein as a rare centrosomal protein deserves a general comment since the technique has pitfalls. When rabbits are used to produce an antibody, caution must be exercised before concluding that the antigen is localized in centrosomes because unprovoked activity against centrosomes or basal bodies is quite frequent in sera from this species and can increase transiently during immunization (Gosti-Testu *et al.*, 1986). Affinity-purified antibodies should be used instead, when the antigen is available or it must be proved independently in some way that the antigen is associated with centrosomes. The latter is an absolute requirement before any conclusion can be drawn from Western blotting or immunoprecipitation experiments with centrosome-reacting sera or monoclonal antibodies. Several proteins reported previously as centrosome-associated still have to meet these criteria.

Monoclonal antibodies have been raised by using purified centrosomes from KE37 cells as the immunogen. As shown by immunofluorescence, some of these antibodies were associated only with centrosomes (Bailly *et al.*, 1989) or centrioles. Others were associated strongly with the centrosome as well as with other structures such as the Golgi apparatus or the nucleus. Most antibodies exhibit cellcycle-dependent changes in the staining of the centrosome, sometimes staining it in an all-or-none manner. These antibodies should allow us to identify new centrosomal proteins. Several already have led to the identification of new proteins in ciliates that associate with cortical or nuclear MTOCs, again suggesting an important conservation of centrosomal proteins among divergent species (Keryer *et al.*, 1989, 1990b).

Biochemical studies on the centrosome are just beginning. As more immunological probes become available, these studies should benefit from a molecular approach and lead to a refined picture of the centrosome (see Chapter 5).

D. Nucleation of Microtubules

Nucleating proteins are still rare, although some were reported years ago in the alga *Polytomella* (Stearns and Brown, 1979). Isolated centrosomes from different sources are efficient in nucleating the assembly of microtubules (Mitchison and Kirschner, 1984; Bornens *et al.*, 1987; Komesli *et al.*, 1989; Tournier *et al.*, 1989; Bré and Karsenti, 1990; Klotz *et al.*, 1990; Verde *et al.*, 1990; Fig. 5) but the mechanism is still unknown. Even the location of sites of microtubule nucleation in the centrosome are not known accurately, nor is it known if nucleation and anchoring of the microtubules are one event or two distinct events.

The human centrosomal calciprotein p62/64 (see previous text) is a candidate for a nucleating protein (Moudjou *et al.*, 1989b, 1991), since the antibodies with which it reacts block centrosome-dependent tubulin assembly but do not inhibit the elongation of microtubules on fragments of axonemes. These data suggest that Ca^{2+} could modulate microtubule nucleation at the centrosome.

From several indirect lines of evidence it can be argued that the 0013 centrosomal antigen p350 also has a role in microtubule nucleation. The protein is localized in the pericentriolar material (Gosti-Testu *et al.*, 1986). The variation of centrosome labeling during the cell cycle coincides with the cyclic changes in the nucleating activity of centrosomes. When the immunoreacting material dissociates from the centrioles under physiological conditions, as in myotubes (Tassin *et al.*, 1985a), or is induced to do so artificially, as in taxol-treated cells (Gosti-Testu *et al.*, 1986), it redistributes as expected for a capping or a nucleating protein (see Fig. 1). Direct demonstration of its involvement in nucleating microtubules has not been obtained, however, and antibodies to p350 do not inhibit microtubule nucleation on isolated centrosomes. The recent demonstration that the 0013 antigen can be phosphorylated by the catalytic subunit of kinase A (Keryer *et al.*, 1990a) further suggests a role for this protein in microtubule nucleation or stabilization. A similar function has been suggested for MAP2, presently the only characterized RII-binding protein (Löhmann *et al.*, 1984).

Two other proteins present in both the centrosphere and the mitotic spindle of sea urchin eggs have been shown recently to be involved in microtubule nucleation or stabilization. A 51K basic protein that was shown to congregate in an insoluble granular material is able to induce aster formation *in vitro* (Toriyama *et al.*, 1988); a 62K protein, when phosphorylated in a Ca^{2+}/calmodulin-dependent

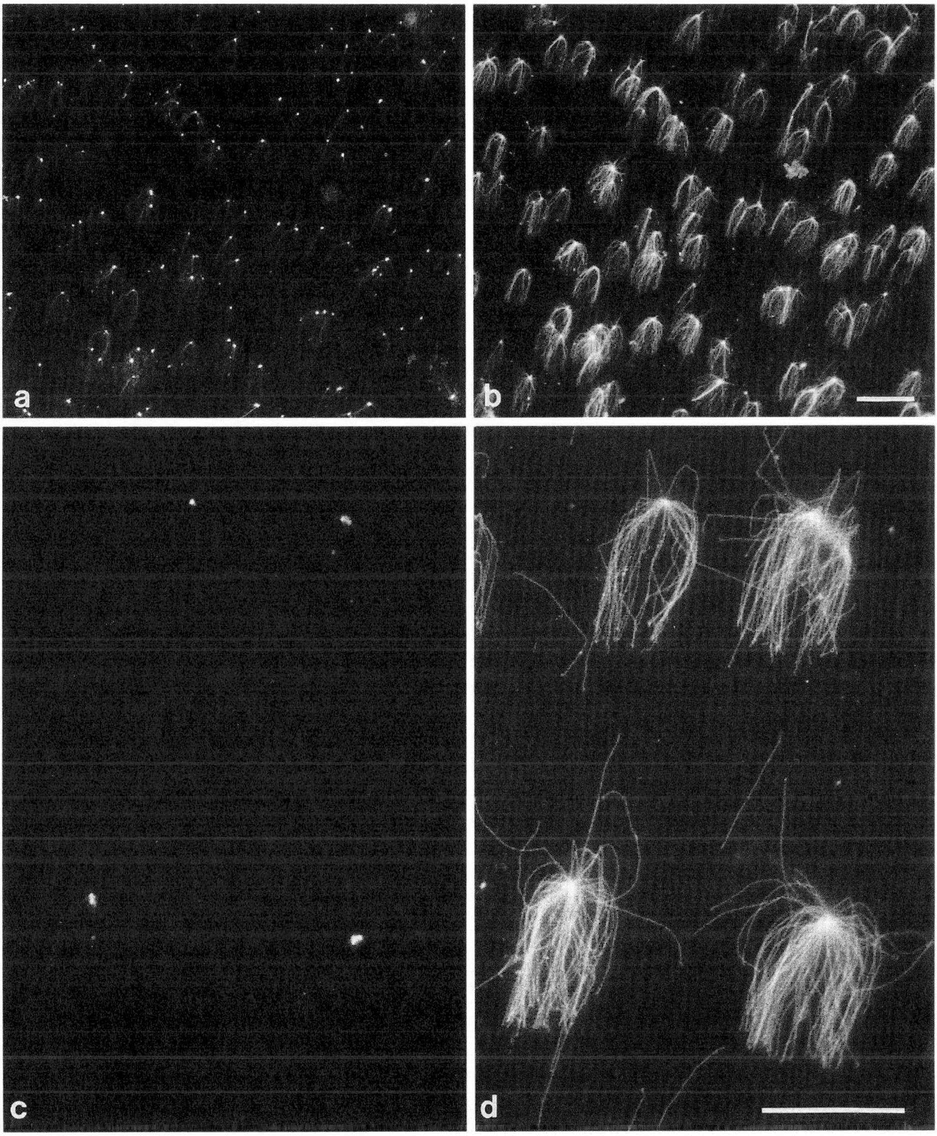

Fig. 5. Isolated centrosomes nucleate microtubules from purified tubulin. Low (a,b) and high (c,d) magnification view of a nucleation experiment in which centrosomes were incubated at 37°C for 8 min with 12 μ*M* tubulin (beef brain) purified on phosphocellulose. Nucleation was monitored by double immunofluorescence. The asters were visualized by an antitubulin antibody (b,d) and the centrosomes by the rabbit serum 0013 (a,c). This experiment shows the presence of a single centrosome in the center of each individual aster. Bars: 10 μm.

manner, induces depolymerization of mitotic spindle microtubules (Dinsmore and Sloboda, 1988).

These recent results should lead to a better understanding of microtubule nucleation at the centrosome. All the data suggest that phosphorylation and Ca^{2+} could be involved in the regulation of this activity.

E. The Centrosome and the Cell Cycle

An essential property of centrosome duplication, as of DNA duplication, is that it happens only once in each cell cycle. The other property, more specific to centrosome duplication, is that the process continues during all four stages of the cell cycle. Two important events occur during the centrosome cycle: (1) the initiation of centriole duplication during G_1 or at the G_1–S border, after the early separation of the two orthogonal centrioles inherited by each daughter cell (Kuriyama and Borisy, 1981a); and (2) the separation of the duplicated centrosomes at the G_2–M border. The molecular mechanisms that govern the different steps of centrosome duplication and separation are unknown.

Our understanding of the cell cycle has increased considerably in recent years. The genetic approach to studies of the growth-limited somatic cell cycle, principally in yeasts, and the biochemical approach to studies of the elementary segmentation cycle in eggs have revealed universal M-phase control mechanisms (Maller, 1990; Nurse, 1990).

What controls the centrosome cycle? Is it entirely driven by the same biochemical clock that governs most of the nuclear and cytoplasmic events or is it governed by independent mechanisms? This is an important issue because, if governed independently, the centrosome would function as an additional pacemaker that would interact with the M-phase clock for the cell cycle to insure correct timing for cell division.

In cells that are not growth-limited, the interphase–metaphase transition in microtubule dynamics is controlled by the mitotic cell division cycle 2 (cdc2) kinase (Verde *et al.*, 1990); conversely, spindle microtubules participate in the timing of the mitosis portion of the cell cycle (Sluder, 1988). In growth-limited cells, cell mass and growth rate must be monitored and coupled to the M-phase control mechanism. As mentioned earlier, the increasing role of the microtubular network in numerous cell functions makes microtubules reasonable candidates for structures that participate in monitoring cell mass. The duplication mechanism of the centrosome, which involves a discontinuous increase in the number of centriole copies from two to four and an increase in the microtubule-nucleating capacity, obviously could be the basis for an appropriate mechanism that could monitor time and mass during the cell cycle. Chromosomes are also candidates for this function (Nurse, 1990), since a correlation between cell mass and ploidy must be maintained. This correlation might require a specific counting mechanism in the cytoplasmic compartment. In fact, some correlation between ploidy and the number of centrioles in cells has often been observed.

1. Centrosome Duplication Cycle

The doubling of centrosomes appears to be largely independent of the nuclear events. In growth-limited cells, DNA synthesis is not required for centrosome duplication (Rattner and Phillips, 1973; Kuriyama and Borisy, 1981b). There is, however, some requirement for the presence of the nucleus. Either some gene product is needed at the initiation step or the physical association between centrosome and nucleus is necessary for procentriole formation (Kuriyama and Borisy, 1981b).

Two genes controlling SPB duplication have been isolated in yeasts. The gene *CDC31* exerts a positive control on SPB duplication (Baum *et al.*, 1986). The product of *CDC31* has significant homology with Ca^{2+}-binding proteins, including calmodulin, and with the basal-body-associated protein centrin or caltractin (Huang *et al.*, 1988a,b). The antigenically related protein p62/64, identified in human centrosomes, apparently is involved in microtubule nucleation (Moudjou *et al.*, 1989b; 1991). The role of Ca^{2+} and this protein in the initiation of centrosome duplication deserves further study.

The other gene, *esp1*, exerts a negative control on SPB duplication (Baum *et al.*, 1988); theoretically, its gene product participates in the mechanism that allows the spindle body to duplicate only once in each cell cycle. The presence of this gene suggests that the mechanism for SPB duplication is largely independent of nuclear events. Interestingly, extra SPBs duplicate asynchronously (see Chapter 7).

In eggs or early embryos, in which cells are not growth-limited, centrosome duplication does not require the nucleus. By removing the nucleus from the eggs of sea urchins at prophase of the first mitosis, Sluder and colleagues (1986) demonstrated that the centrosome reproduced, doubling precisely, with a periodicity slightly slower than in the controls, and that each of the resulting centrosomes contained two centrioles (see chapter 9). Picard and co-workers (1988) further investigated in starfish whether the centrosome could duplicate in the absence of any material from the oocyte germinal vesicle, since this material becomes largely cytoplasmic in the egg. Hormonal induction of meiotic maturation in enucleated oocytes resulted in the growth of the two meiotic asters normally present at the animal pole and in their division at the normal time during the first meiotic cleavage. Moreover, the sperm centrosome also underwent periodic duplications; periodic changes in the organization of asters could be detected that correlated with cell surface reorganization. Uncleaved eggs with 2^9 centrosomes ultimately lysed. This number corresponds to the blastomere number at the mid-blastula transition. When the eggs were enucleated after fertilization by removing both pronuclei, 2^9 blastomeres lacking nuclei were formed, as already demonstrated in this species by Nagano and colleagues (1981) using aphidicolin to inhibit nuclear replication. Therefore, even in the complete absence of nuclear material in the oocyte, the centrosome cycle takes place up to the mid-blastula transition (see Chapter 9). A similar dissociation between centrosomal and nuclear division has been observed in eggs of *Drosophila* carrying the recessive maternal-effect mutation *gnu* (Freeman *et al.*, 1986). Phenotypic copies of the *gnu* mutation have been produced by aphidicolin treatment of the eggs (Raff and Glover, 1988; see Chapter 8). As in starfish, polar cells lacking nuclei but containing centrosomes were observed.

2. *Parthenogenetic Assay in* Xenopus

Amphibian eggs are favorable systems in which to study the initiation of centrosome duplication since they apparently lack a functional seed. Frog eggs are unable to assemble a centrosome from its elements in the absence of a preexisting centrosome, in contrast to mouse eggs, in which this happens as a normal part of the development process (Szollozi *et al.*, 1972), or to sea urchin eggs, in which this can be induced by various treatments (see, for example, Kuriyama and Borisy, 1983). Isolated centrosomes are powerful and exclusive parthenogenetic agents in amphibian eggs (Maller *et al.*, 1976). A striking feature of this system is that apparently no species specificity is required for the centrosomes, that is, heterologous centrosomes isolated from sea urchin (Maller *et al.*, 1976), mouse (Karsenti *et al.*, 1984), human (Bornens *et al.*, 1987), and several other species including rat and calf (Tournier *et al.*, 1991a) are active in these eggs.

In order to induce parthenogenesis, the injected centrosome must be able to assemble microtubules from egg tubulin, to duplicate, and to interact properly with the nuclear compartment of the egg. Centrosomes or related structures that lack one of these properties would be unable to induce parthenogenesis. For example, basal bodies isolated from ciliates or flagellates are capable of assembling egg tubulin into radial asters of microtubules, as is a regular centrosome. They are unable, however, to induce parthenogenesis, suggesting that they are unable to duplicate properly or to interact in an effective way with the nucleus (Heideman and Kirschner, 1975). On the other hand, human centrosomes competent for parthenogenesis become incompetent when preincubated with antibodies that inhibit microtubule-nucleating activity (Tournier, Moudjou, and Bornens, unpublished results).

The parthenogenetic assay can be used in two ways: (1) to address the question of centrosome continuity, that is, to identify the active component in the heterologous centrosome capable of complementing the egg in starting the successive rounds of centrosome doubling; and (2) to study the duplication mechanism of the centrosome.

a. Centrosome Integrity Is Required for Its Activity. Chemical or enzymatic dissection of centrosomes to identify the active subfraction has been undertaken (Klotz *et al.*, 1990). Centrosomes isolated from human KE37 cells and sequentially extracted with salts, urea, or detergents, as well as treated with nucleases or proteases, were monitored by protein analysis, by Western blotting using specific markers for centrioles and pericentriolar material, and by transmission electron microscopy. Structural modifications were correlated with the loss of microtubule-nucleating activity *in vitro* and with the loss of parthenogenetic activity after injection into eggs. Neither the parthenogenetic activity nor the nucleating activity was modified by nuclease treatments (micrococcal nuclease or RNase A) alone or in conjunction with a dissociating agent. No major centrosome-associated RNA resisted the nuclease treatment. We did not suspect the presence of large DNA and concluded that the parthenogenetic activity did not require the presence of nucleic acids in the injected centrosomes. These results do not eliminate the possibility that centrosomes are complemented by some nucleic acid in the egg. It was noted, for example, that the parthenogenetic activity was preserved under conditions that abolished the microtubule nucleating activity *in vitro*, suggesting the possibility that renaturation or complementation could take place in the egg cytoplasm. Both the nucleating activity and the parthenogenetic activity were very sensitive to proteases. We concluded that the parthenogenetic activity was associated with an insoluble proteinaceous structure that was not significantly simpler than the native centrosome, as judged from the protein content and structural organization of centrosomes examined after treatment with urea or salt at concentrations at which they still retained activity. A correlation between the loss of the parthenogenetic activity and the disorganization of the centriolar triplets was observed, but this does not prove that centrioles are the parthenogenetic agent. We were unable to prepare centrioles free of the surrounding material. Overall, the results of these studies are compatible with the idea that the injected centrosome acts as a structural template in the initiation of centrosome duplication.

b. Centrosomes from G_0, G_1, or G_2 Cells Have the Same Activity. Do centrosomes need to be at a particular stage in their duplicative cycle to induce cleavage? That is, is their duplication initiated in response to the egg cytoplasm, or can they only elongate from a preformed bud? Centrosomes isolated from synchronized human lymphoblastic cells and from peripheral human lymphocytes are shown to possess similar activity (Tournier *et al.*, 1989), indicating that duplication of heterologous centrosomes can be triggered in *Xenopus* eggs.

Does this mean that centrosomes isolated from any type of animal cell are competent and allow parthenogenesis? The answer is no. Centrosomes isolated from calf thymus are incapable of inducing cleavage of activated eggs, as discussed subsequently.

c. Is the Centrosome Cycle Independent of the Basic Cell Cycle Oscillator? A precise description of the duplication of the injected centrosome is necessary, since duplication is one of the essential steps in the formation of a functional mitotic spindle. Progress toward this goal recently has been achieved by initiating centriole duplication *in vitro* using *Xenopus* egg extracts (Tournier *et al.*, 1991b). This system should allow us to identify the factors controlling the initiation of procentriole budding. One remarkable result obtained showed that procentriole budding occurs in cycloheximide-treated extracts, suggesting that centrosome duplication is not under translational control in eggs, that is, it does not depend on cyclin synthesis, which regulates progression of the cell cycle. A similar result has been reported recently in *Xenopus* embryos (Gard *et al.*, 1990). Several rounds of centrosome duplication have been observed also in sea urchin eggs in absence of any protein synthesis (Sluder *et al.*, 1990). Centrosome duplication has been shown to be asynchronous in this case, suggesting that under normal conditions it is likely to remain in phase with the cytoplasmic clock (see Chapter 9).

These results might be important for our understanding of the cell cycle. We need to know, in biochemical terms, how the centrosome responds to the cytoplasmic clock in the egg. Some insight has been obtained by the demonstration that centrosomes could interact directly with the mitotic kinase.

3. Centrosomes and the M-Phase Transition

An immunocytochemical study of human cells (HeLa and KE37 cell lines) has demonstrated that, in interphase, the *cdc2*$^{+}$ gene product p34^{cdc2} is distributed in an apparently random manner, both in the cytoplasm and in the nucleus, where it occupies DNA-excluding areas in an irregular manner (Bailly *et al.*, 1989). Some p34^{cdc2} becomes associated with the centrosome during G_2 and at the beginning of M. Cell fractionation corroborates these conclusions from immunochemistry and further demonstrates an enrichment of the slow migrating forms of p34^{cdc2} in isolated centrosomes and other Triton-insoluble fractions at this time.

Mitotic control in *Schizosaccharomyces pombe* involves the products of several other genes, including p13^{suc1} which interacts with p34^{cdc2} (Brizuela *et al.*, 1987). A 13K protein associated with p34^{cdc2} that cross-reacts with an antiserum raised against recombinant yeast p13^{suc1} has been detected in HeLa cells (Draetta *et al.*, 1987). In the same cells, Bailly and co-workers (1989) demonstrated that another affinity-purified antibody directed against the yeast recombinant p13^{suc1} specifically immunoprecipitated a protein with a molecular weight significantly smaller than 13K suggesting that the human homolog could be smaller than the yeast protein. This protein was found to be distributed in HeLa cells in a manner strikingly similar to that observed for p34^{cdc2}, with one exception. Although the presence of both proteins in the centrosomes of cells late in the cell cycle was obvious, even more so for the p13^{suc1} homolog than for the p34^{cdc2} homolog, the centrosomal staining with the anti-p13^{suc1} antibody was definitely seen throughout the cell cycle, that is, in G_1 as well.

One of the roles of the p34^{cdc2} protein kinase at the initiation of mitosis could be to modify the centrosome, probably by protein phosphorylation, in a manner that would lead to the generation of a mitotic spindle. Phosphoproteins are known to be associated with the mitotic poles at the onset of mitosis (Vandre *et al.*, 1986). An increase in centrosome-specific staining with some anticentrosome antibodies occurs during G_2, whereas a dramatic decrease takes place at the onset of anaphase (Gosti-Testu *et al.*, 1986). These centrosomal modifications, which are likely to be correlated with a rise and fall in their microtubule-nucleating activity, might be controlled by phosphorylation.

It is now important to learn if any of the centrosomal proteins are substrates of the p34^{cdc2} protein kinase and, if so, to elucidate the functional implications of such phosphorylation. A more general study of centrosome-associated kinases has been undertaken (Keryer *et al.*, 1990a), because both regulatory and catalytic subunits of the cAMP-dependent protein kinase II have been localized in the centrosome and in the Golgi apparatus (Nigg *et al.*, 1985; De Camilli *et al.*, 1986) and because calmodulin also accumulates at the poles of the mitotic spindle (Andersen *et al.*, 1978; Welsh *et al.*, 1979).

The association of p34^{cdc2} with the centrosome might also regulate its duplicative cycle, that is, control the separation of the two daughter centrosomes. Conversely, these late events of the centrosome cycle may provide signals for the activation of the mitotic process. The release of mature starfish oocytes from arrest in interphase by the injection of centrosomes isolated from KE37 cells (Picard *et al.*, 1987) could be relevant to this issue. Obviously, more work is necessary to understand the precise role of the centrosome in mitotic induction and whether changes in centrosome activity exert a feedback control on the cell cycle oscillator.

The study by Bailly and colleagues (1989) also revealed that, in anaphase, p34^{cdc2} becomes associated with vesicles in the middle of the dividing cell that later surround the reforming nuclei. Although the significance of these observations is unknown, it is possible that they are related to the decrease in kinase activity that is required for leaving the mitotic state.

4. *Centrosomes and the G_0–G_1 Transition*

The involvement of the centrosome, centrioles, or SPBs in the control of the cell cycle has been proposed in the past (Byers and Goetsch, 1975; Brooks *et al.*, 1980). Some experimental support for this proposal was reported for the G_0 to G_1 transition in studies showing a temporal correlation between structural modifications of the centrosome and cell commitment to divide (Tucker *et al.*, 1979; Sherline and Mascardo, 1982). However, no causal link was ever established.

We recently considered the possibility of using the parthenogenetic assay in *Xenopus* eggs to test the potential of centrosomes isolated from various animal cells to start a duplicative cycle. Centrosomes isolated from calf thymocytes are incapable of triggering egg cleavage (Tournier *et al.*, 1991a) and evidence has been obtained that they do not duplicate in the eggs (Tournier *et al.*, 1991b). This result cannot be related simply to the quiescent state of thymocytes, since centrosomes from quiescent human peripheral lymphocytes have been shown to be active (Tournier *et al.*, 1991a). Centrosomes isolated from other bovine cells and from thymocytes of other species have now been tested. This study has established that there is a strict relationship between the linear structure of the calf centrosome and the lack of activity. As a simple working hypothesis, we propose (Tournier *et al.*, 1991a) that the initiation of procentriole budding cannot occur as long as the two centrioles are closely linked to each other. Thus, the structural arrangement of the centrioles could control the duplication of the centrosome. Daughter centrioles do not dissociate from the parent centrioles before the beginning of the next cell cycle, in early G_1, when the so-called disorientation of centrioles occurs (Kuriyama and Borisy, 1981a). As long as the centrioles remain orthogonally associated, and have not terminated their duplication, or linearly associated, a possible intermediate step toward dissociation/disorientation, they cannot reinitiate a new duplicative round. This overlap between the centrosome cycle and two successive cell cycles insures that centrosome duplication happens only once during each cell cycle.

III. FUTURE DIRECTIONS AND PROSPECTS

What are the important aspects for future studies of the centrosome?

The microtubule-nucleating activity will have to be analyzed in molecular terms. We have seen that not only the nucleating activity but the integrity of the centrosome itself depends on environmental conditions. Therefore, unraveling the principles that control centrosome assembly might be important in achieving understanding of centrosome activity.

How is the centrosome, once assembled, maintained as an apparently stable structure? What is, for example, the half-life of microtubules in the centriolar triplets? Good experimental evidence in *Physarum* amebas suggests that centriolar microtubules might be in dynamic equilibrium with the tubulin pool. In this species, centriole length varies during the cell cycle (shorter in mitosis), according to the ploidy of the ameba (longer in diploid than in haploid cells), or after protracted treatment with taxol (Gely and Wright, 1985). Centriolar microtubules are, however, probably considerably more stable than any other microtubules in cultured cells and apparently resist long treatment with disruptive drugs. This could be a basic requirement for maintaining a permanent association between centrioles and the noncentriolar material. The short half-life of cellular microtubules relative to that of centriolar microtubules would cause molecules capable of interacting with microtubular walls to preferentially interact with centrioles. If this material preferentially interacts with, or is transported to, the minus end of microtubules (a property that might be expected in molecules capable of nucleating microtubules), it will accumulate unequally along the centriolar walls. When taxol induces the assembly of cellular tubulin, many stable microtubules are formed, which would result in the progressive displacement of most of the noncentriolar material from the centrioles to these microtubules, as observed. In the absence of centrioles, the abundant noncentriolar material would have a more scattered distribution, as in the mouse egg. In the presence of centrioles, as in sea urchin eggs, this material would accumulate around the centrioles instead.

The possibility exists of a positive correlation between the amount of noncentriolar material and the cell mass. Large cultured cells display large centrosomes when stained with specific anticentrosome antibodies and possess a large microtubular network. However, quantitative data are necessary. We would expect the regulation of the expression of the nucleating material to be linked to the control of tubulin expression.

The recent discovery (Oakley *et al.*, 1990) that the SPB from *Aspergillus nidulans* contains a specific tubulin, the so-called γ-tubulin, is of great interest because it suggests that a tubulin core could be a general feature of centrosomes. Moreover, the demonstration that the γ-tubulin gene is essential for microtubule function suggests the intriguing possibility of an alternative mechanism for microtubule nucleation (Oakley *et al.*, 1990). It will be important to assess the generality of this finding among diverse species.

What is the spatial organization of the centrosome in the cytoplasm? Where are its boundaries? Does the centrosomal material interact with nonmicrotubular structures? There are no simple answers to these questions, since there are probably great variations from one cell type to the other. Epithelial cells, such as Maden–Darby canine kidney (MDCK) cells, forming a polarized epithelium in culture progressively display a microtubular network that is not organized around the centrosome. In these cells, the two centrioles nucleate only a few microtubules and do not have abundant pericentriolar material (Bré *et al.*, 1987). Has the pericentriolar material become widely distributed in the cytoplasm, possibly as a result of interactions with other cytoskeletal structures or with specialized portions of the plasma membrane? A similar modification in the distribution of centrosomal material may occur in cultured cells during the establishment of confluence. It was shown that centrosome-free cytoplasts from confluent L929 cells can assemble microtubules whereas those from nonconfluent cells cannot (Karsenti *et al.*, 1984). Does this finding also indicate that spreading of the centrosomal material away from the centrosome occurs at confluence? In contrast, in lymphoid cells apparently all the microtubules are anchored at the centrosome, even the additional microtubules assembled after taxol treatment (Brown *et al.*, 1985; Bornens *et al.*, 1989; see Chapter 10).

Isolated centrosomes should be useful for experimental studies of centrosome assembly. Another promising approach is the isolation and study of mutants of centrosome assembly in favorable species (Ruiz *et al.*, 1987; Gonzales *et al.*, 1988; Sunkel and Glover, 1988; see Chapter 8).

Understanding the duplication mechanism of the centrosome is another important goal for the future. Growing evidence suggests that duplication can occur to some extent without requiring extrinsic control mechanisms. *In vitro* assays for centrosome duplication appear feasible (Tournier *et al.*, 1991b) and useful. The orthogonal budding of procentrioles from the proximal end of parent centrioles will have to be described in molecular terms. Does the redistribution of the noncentriolar material between the daughter centrosomes also involve a duplication process?

The coupling between the centrosome cycle and the cell cycle is an important issue. Is the initiation of centriole duplication triggered after the cell is committed to divide or is it a critical signal for cell commitment? A similar question can be raised with respect to the onset of mitosis, when some of the mitotic kinase associates with the duplicating centrosomes before they separate (Bailly *et al.*, 1989). Is the end of centrosome duplication controlled by the mitotic clock or is this centrosomal event exerting feedback control on the activation of the mitotic kinase? How are the rise of nucleating activity of the centrosome in G_2–M and the sudden fall at the onset of anaphase controlled?

The presence of several kinases (cyclic AMP-dependent kinase type II, $p34^{cdc2}$) and calcium-binding proteins in the centrosome raises questions about the role of the microtubule-nucleating center in regulation of cell functions, since it could provide amplification mechanisms via the microtubular network.

In animal cells, the paired configuration of the centrosome is a striking feature. This arrangement might be important for the duplication cycle, for centrosomal functions, or for both. Does the acentriolar centrosome in diatoms or in *Heliozoa* also contain two copies of some component?

Regarding centrosome inheritance, obviously, different strategies have evolved among species to restart the centrosome cycle during early development. Uniparental inheritance of the centrosome is generally believed to take place. For example, inheritance is paternal in marine or amphibian eggs and maternal in the mouse. The sea urchin egg is perhaps the best studied system; here the apparently strict paternal inheritance of centrosomes has been studied thoroughly and elegantly by light microscopy and by experimental manipulation (Sluder *et al.*, 1989a; see Chapter 9). The results obtained demonstrate that the ability of the maternal centrosome to double at successive mitoses is lost. However, the possibility that some sort of complementation takes place in the formation of the first mitotic spindle cannot be ruled out. The same holds for *Xenopus* eggs. Injected centrosomes are unique and exclusive parthenogenetic agents in this species, as discussed in the previous section, but this does not rule out the necessity of some complementation for the first cleavage to occur. Maternal inheritance of the centrosome in the mouse is more strongly supported since parthenogenetic development in this species can be experimentally triggered in several ways. In some species, such as *Drosophila*, the first mitotic spindle seems to be formed by the fusion of independent maternal and paternal spindle structures (Sonnenblick, 1950). Grafen (1988), in order to explain short-term evolutionary maintenance of sexual reproduction, has proposed that the centrosome was inherited in a selectively ambiguous way, that is, a sexual offspring receives one centrosome from each parent and selects the better one to pass on to its own offspring. This process would have resulted in the various uniparental inheritance patterns observed in the different species. Perhaps we do not know enough about the centrosomes and their functions to envisage how, according to this original theory, they could compete with each other for presence in the germ line. Complementation in order to re-establish a fully functional centrosome would seem a simpler solution.

What is the framework in which we can develop a comprehensive view of the function of centrosomes? A detailed knowledge of the evolution of the microtubular network and of its organization and activity in various species would allow us to understand the adaptive value of the different types of microtubule nucleating centers that have evolved. In the absence of such knowledge, we can only make some tentative remarks.

In most species, microtubules are apparently quite often nucleated or stabilized near the nuclear periphery. This might be a favorable strategy for cell division and for cell organization in general. This organization is even observed when detectable centrosomes are absent, as, for example, in plant cells. These cells do not locomote nor do they display any obvious anterior–posterior polarity. Nevertheless, the microtubules are apparently involved in determining cell geometry (see, for example, Flanders *et al.*, 1990). When a type of centrosome is present, it introduces a bias in the spatial organization of the microtubular network and the cell displays some polarity. The centrosome can consist of a specialized zone at the nuclear surface, like a plaque or SPB. Such cells are capable of polarized growth, but do not migrate nor show cell surface movement. If, instead, they have a nucleus-associated centrosome, the cells do migrate and exhibit cell surface movements. The centrosome can be with or without internal core. If present, the core can consist of a paracrystalline plaque, an inclusion with a matchbox shape, or a pair of centrioles.

Is there, among the different types of centrosomes, a detectable evolution toward a better adaptation to some centrosome-dependent cell function? For example, does the presence of centrioles, that is, microtubule-containing structures in the centrosome, give the cell better control over the dynamics of the microtubular network? Can we identify a specific type of locomotory activity in cells with each type of centrosome? Is there a difference, for example, in the locomotion of the ameba *Dictyostelium* and the ameba *Physarum*? Indeed, *Physarum* amebas can use their centrosomes transiently to grow flagella and transform themselves into swarm cells that can move considerably faster. Cells with centriole-associated centrosomes are also typical of metazoan cells, which form multicellular aggregates capable of autonomous movements. In contrast, *Dictyostelium* have centrosomes that lack centrioles. When they form slugs, aggregating cells of *Dictyostelium* show very limited movement that cannot compare with the complex forms of locomotion seen in the simplest metazoan.

For locomotion, a cell requires both a motor and a mechanism for integrating the motile activity. It does not seem unreasonable to think of the microtubular network as the integration system. The evolution toward more efficient cell locomotion has been of paramount importance for cell survival—to eat and not be eaten—and in all likelihood has led to changes in both the motor and the integration. In the evolution of the microtubular network there have been apparently more changes in the nucleating centers than in the microtubules themselves, perhaps because this led toward a better integration of cytoplasmic movements. If so, the centrosome would be an integrative device for cell movement, that is, a navigational instrument.

If the primary function of the centrosome is to integrate cell locomotion, the two daughter centrosomes in a dividing cell would be expected to control the establishment of the mitotic spindle and the cleavage furrow. Then the role of the centrosome during cell division and during interphase would not be different after all.

Note Added in Proof

Since the writing of this chapter, our laboratory has carried out a detailed and quantitative study of centrosome organization and centriole architecture, using serial sectioning of centrosomes isolated by two different methods (M. Paintrand, M. Moudjou, H. Delacroix, and M. Bornens, *J. Struct. Biol.*, in press). Several new structural data have been reported which are not reviewed in the present chapter.

ACKNOWLEDGMENTS

This work on isolated centrosomes is a collective endeavor. I warmly thank E. Bailly, J. Bergès, N. Bordes, C. Celati, G. Keryer, F. Gosti, C. Klotz, M. Moudjou, M. Paintrand, and F. Tournier from CGM-CNRS (Gif-sur-Yvette, France), S. Komesli and D. Job from CENG (Grenoble, France), and E. Karsenti from EMBL (Heidelberg, Germany) who actively participated in the research and discussions that are presented in this review. I also thank S. Brown for preparing the English version.

This work was supported by CNRS, by grant M.R.E.S. No. 87-C-0555, and by grants from A.F.M. and from La Ligue Nationale Contre Le Cancer.

REFERENCES

Albrecht-Buehler, G. (1981). Does the geometric design of centrioles imply their function? *Cell Motility* **1**, 237–245.

Andersen, B., Osborn, M., and Weber, R. (1978). Specific visualization of the calcium dependent regulatory protein of cyclic nucleotide phosphodiesterase (modulator protein) in tissue culture cells by immunofluorescence microscopy: Mitosis and intercellular bridges. *Cytobiologie* **17**, 354–364.

Bailly, E., Dorée, M., Nurse, P., and Bornens, M. (1989). p34^{cdc2} is located in both nucleus and cytoplasm; part is centrosomally located at G_2/M and enters vesicles at anaphase. *EMBO J.* **8**, 3985–3995.

Bardele, C. F. (1975). The fine structure of the centrohelidian *Heterophrys marina*. *Cell Tissue Res.* **167**, 85–102.

Baron, A. T., and Salisbury, J. L. (1988). Identification and localization of a novel, cytoskeletal, centrosome-associated protein in PtK_2 cells. *J. Cell Biol.* **107**, 2669–2678.

Baum, P., Furlong, C., and Byers, B. (1986). Yeast gene required for spindle pole body duplication: Homology of its product with Ca^{2+}-binding proteins. *Proc. Natl. Acad. Sci. U.S.A.* **83**, 5512–5516.

Baum, P., Yin, C., Goetsch, L., and Byers, B. (1988). A yeast gene essential for regulation of spindle pole duplication. *Mol. Cell Biol.* **8**, 5386–5397.

Bessis, M., and Breton-Gorius, J. (1967). Rapports entre noyau et centrioles dans les granulocytes étalés. Rôle des microtubules. *Nlle. Rev. Fr. Hematol.* **7**, 601–620.

Blackburn, G. R., Barreau, M. D., and Dewey, W. C. (1978). Partial purification of centrosomes from Chinese hamster ovary cells. *Exp. Cell Res.* **113**, 183–187.

Bornens, M. (1977). Is the centriole bound to the nuclear membrane? *Nature (London)* **270**, 80–82.

Bornens, M. (1979). The centriole as a gyroscopic oscillator. Implications for cell organization and some other consequences. *Biol. Cell* **35**, 115–132.

Bornens, M., and Karsenti, E. (1984). The centrosome. *In* "Membrane, Structure, and Functions" (E. E. Bittar, ed.) pp. 100–171. John Wiley and Sons, New York.

Bornens, M., Paintrand, M., Berges, J., Marty, M. C., and Karsenti, E. (1987). Structural and chemical characterization of isolated centrosomes. *Cell Motil. Cytosk.* **8**, 238–249.

Bornens, M., Paintrand, M., and Celati, C. (1989). The cortical microfilament system of lymphoblasts displays a periodic oscillatory activity in the absence of microtubules: Implications for cell polarity. *J. Cell Biol.* **109**, 1071–1083.

Bornens, M., Bailly, E., Gosti, F., and Keryer, G. (1990). The centrosome: Recent advances on structure and functions. *Prog. Mol. Subcell. Biol.* **11**, 86–114.

Boveri, T. (1901). Über die Natur der Centrosomen. *Jena. Z. Med. Naturw.* **28**, 1–220.

Bré, M. H., Kreis, T. E., and Karsenti, E. (1987). Control of microtubule nucleation and stability in Madin–Darby canine kidney cells: Occurrence of noncentrosomal, stable detyrosinated microtubules. *J. Cell Biol.* **105**, 1283–1296.

Bré, M. H., and Karsenti, E. (1990). Effects of brain microtubule-associated proteins on microtubule dynamics and the nucleating activity of centrosomes. *Cell Motil. Cytosk.* **15**, 88–98.

Brizuela, L., Draetta, G., and Beach, D. (1987). p13^{suc1} acts in the fission yeast cell division cycle as a component of the p34^{cdc2} protein kinase. *EMBO J.* **6**, 3507–3714.

Brooks, R. F., Bennett, D. C., and Smith, J. A. (1980). Mammalian cell cycles need two random transitions. *Cell* **19**, 493–504.

Brown, D. L., Little, J. E., Chaly, N., Schweitzer, I., and Paulin-Levasseur, M. (1985). Effects of taxol on microtubule organization in mouse splenic lymphocytes and on response to mitogenic stimulation. *Eur. J. Cell Biol.* **37**, 130–139.

Buendia, B., Bre, M. H., Griffiths, G., and Karsenti, E. (1990). Cytoskeletal control of centriole movement during the establishment of polarity in MDCK cells. *J. Cell Biol.* **110**, 1123–1135.

Byers, B., and Goetsch, L. (1975). Behavior of spindles and spindle plaques in the cell cycle and conjugation of *Saccharomyces cerevisiae*. *J. Bacteriol.* **124**, 511–523.

Calarco-Gillam, P. D., Siebert, M. C., Hubble, R., Mitchison, T., and Kirschner, M. W. (1983). Centrosome development in early mouse embryos as defined by autoantibody against pericentriolar material. *Cell* **35**, 621–623.

Clayton, L., Black, C. M., and Lloyd, C. W. (1985). Microtubule nucleating sites in higher plant cells identified by an autoantibody against pericentriolar material. *J. Cell Biol.* **101**, 319–324.

Courvalin, J. C., Hernandez-Verdun, D., Gosti-Testu, F., Marty, M. C., Maunoury, R., and Bornens, M. (1986). A protein of M_r 80,000 is associated with the nucleolus organizer of human cell lines. *Chromosoma (Berlin)* **94**, 353–361.

De Brabander, M., Geuens, G., Nuydens, R., Willebrords, R., and De Mey, J. (1981). Taxol induces the assembly of free microtubules in living cells and blocks the organizing capacity of the centrosomes and the kinetochores. *Proc. Natl. Acad. Sci. U.S.A.* **78**, 5608–5612.

De Camilli, P., Moretti, M., Denis Donini, S., and Lohman, S. M. (1986). Heterogeneous distribution of the cAMP receptor protein kinase R-II in the nervous system: Evidence for its intracellular accumulation on microtubules, microtubule organizing centers, and in the area of the Golgi complex. *J. Cell Biol.* **103**, 189–203.

Dinsmore, J. H., and Sloboda, R. D. (1988). Calcium and calmodulin-dependent phosphorylation of a 62 kDa protein induces microtubule depolymerization in sea urchin mitotic apparatuses. *Cell* **53**, 769–780.

Draetta, G., Brizuela, L., Potashkin, J., and Beach, D. (1987). Identification of p34 and p13, human homologs of the cell cycle regulators of fission yeast encoded by *cdc2+* and *suc1+*. *Cell* **50**, 319–325.

Euteneuer, U., and McIntosh, J. R. (1981). Polarity of some motility-related microtubules. *Proc. Natl. Acad. Sci. U.S.A.* **78**, 372–376.

Euteneuer, U., and Schliwa, M. (1985). Evidence for an involvement of actin in the positioning and motility of centrosomes. *J. Cell. Biol.* **101**, 96–103.

Evans, L., Mitchison, T. J., and Kirschner, M. W. (1985). Influence of the centrosome on the structure of nucleated microtubules. *J. Cell Biol.* **100**, 1185–1191.

Fais, D., Nadezhdina, E. S., and Chentsov, Y. S. (1984). Evidence for the nucleus–centriole association in bovine cells obtained by ultracentrifugation. *Eur. J. Cell Biol.* **133**, 190–196.

Fèvre-Chevalier, C., and Fèvre, J. (1984). Axonemal microtubular pattern of *Cienkowskya mereschkovskyi* and a revision of heliozoan taxonomy. *Origins of Life* **13**, 315–338.

Flanders, D. J., Rawlins, D. J., Shaw, P. J., and Lloyd, C. W. (1990). Nucleus-associated microtubules help determine the division plane of plant epidermal cells: Avoidance of four-way junctions and the role of cell geometry. *J. Cell Biol.* **110**, 1111–1122.

Freeman, M., Nüsslein-Volhard, C., and Glover, D. M. (1986). The dissociation of nuclear and centrosomal division in *gnu*, a mutation causing giant nuclei in *Drosophila*. *Cell* **46**, 457–468.

Gard, D. L., Hafezi, S., Zhang, T., and Doxsey, S. T. (1990). Centrosome duplication continues in cycloheximide-treated *Xenopus* blastulae in the absence of a detectable cell cycle. *J. Cell Biol.* **110**, 2033–2042.

Gely, C., and Wright, M. (1985). Centriole size modifications during the cell cycle of the amoeba of the myxomycete *Physarum polycephalum*. *J. Ultrastr. Res.* **91**, 127–137.

Gonzalez, C., Casal, J., and Ripoll, P. (1988). Functional monopolar spindles caused by mutation in *mgr*, a cell division gene of *Drosophila melanogaster*. *J. Cell Sci.* **89**, 39–48.

Gosti, F., Marty, M. C., Courvalin, J. C., Maunoury, R., and Bornens, M. (1987). Centrosomal proteins and lactate dehydrogenase possess a common epitope in human cell lines. *Proc. Natl. Acad. Sci. U.S.A.* **84**, 1000–1004.

Gosti-Testu, F., Marty, M. C., Berges, J., Maunoury, R., and Bornens, M. (1986). Identification of centrosomal proteins in a human lymphoblastic cell line. *EMBO J.* **5**, 2545–2550.

Grafen, A. (1988). A centrosomal theory of the short term evolutionary maintenance of sexual reproduction. *J. Theor. Biol.* **131**, 163–173.

Hall, J. L., Ramanis, Z., and Luck, D. J. L. (1989). Basal body/centriolar DNA: Molecular genetic studies in *Chlamydomonas. Cell* **59**, 121–132.

Harper, J. D. I., Mitchison, J. M., Williamson, R. E., John, P. C. L. (1989). Does the autoimmune serum 5051 specifically recognise microtubule organising centres in plant cells? *Cell Biol. Intern. Reports* **13**, 471–475.

Hauser, M. (1986). Taxol affects both the microtubular arrays of heliozoan axonemes and their microtubule-organizing centers. *Eur. J. Cell Biol.* **42**, 295–304.

Heath, I. B. (1981). Variant mitoses in lower eukaryotes: Indicators of the evolution of mitosis? *Intern. Rev. Cytol.* **64**, 1–79.

Heidemann, S. R., and Kirschner, M. W. (1975). Aster formation in eggs of *Xenopus laevis*. Induction by isolated basal bodies. *J. Cell Biol.* **67**, 105–117.

Huang, B., Watterson, D. M., Lee, V. D., and Schibler, M. J. (1988a). Purification and characterization of a basal body-associated Ca^{2+}-binding protein. *J. Cell Biol.* **107**, 121–131.

Huang, B., Mangersen, A., and Lee, V. D. (1988b). Molecular cloning of cDNA for caltractin, a basal body-associated Ca^{2+}-binding protein: Homology in its protein sequence with calmodulin and the yeast *cdc31* gene product. *J. Cell Biol.* **107**, 133–140.

Jasmin, B., Cartaud, J., Bornens, M., and Changeux, J. P. (1989). Golgi apparatus in chick skeletal muscle: Changes in its distribution during end plate development and after denervation. *Proc. Natl. Acad. Sci. U.S.A.* **86**, 7218–7222.

Jasmin, B., Changeux, J. P., and Cartaud, J. (1990). Compartmentalization of cold-stable and acetylated microtubules in the subsynaptic domain of chick skeletal muscle fibre. *Nature (London)* **344**, 673–675.

Johnson, K. A., and Rosenbaum, J. L. (1990). The basal bodies of *Chlamydomonas reinhardtii* do not contain immunologically detectable DNA. *Cell* **62**, 615–619.

Karsenti, E., Newport, J., Hubble, R., and Kirschner, M. W. (1984). Interconversion of metaphase and interphase microtubule arrays as studied by the injection of centrosomes and nuclei into *Xenopus* eggs. *J. Cell Biol.* **98**, 1730–1745.

Karsenti, E., and Maro, B. (1986). Centrosomes and the spatial distribution of microtubules in animal cells. *TIBS* **11**, 460–463.

Karsenti, E., Bravo, R., and Kirschner, M. (1987). Phosphorylation changes associated with the early cell cycle in *Xenopus* eggs. *Dev. Biol.* **119**, 442–453.

Keryer, G., Marty, M. C., Garreau de Loubresse, N., and Bornens, M. (1989). Identification of a spindle-associated protein in ciliate micronuclei. *J. Cell Sci.* **93**, 287–298.

Keryer, G., Rios, R. M., Lohmann, S. M., and Bornens, M. (1990a). The association of the regulatory subunit of cAMP-dependent protein kinase type II to the centrosome of human cells is mediated by a 350 kDa RII-binding proteins *Cell Biol. Intern. Reports* **14**, Abs. Supplt., 149.

Keryer, G., Iftode, F., and Bornens, M. (1990b). Identification of proteins associated with microtubule-organizing centres in the oral apparatus of the ciliate *Paramecium tetraurelia. J. Cell Sci.* **97**, 553–563.

Kirschner, M. W., and Mitchison, T. (1986). Beyond self-assembly: From microtubules to morphogenesis. *Cell* **45**, 329–342.

Kitanishi-Yamura, T., Blose, S. H., and Fukui, Y. (1985). Role of the MT-MTOC complex in determination of the cellular locomotory unit in *Dictyostelium. Protoplasma* **127**, 133–146.

Kleve, M. G., and Clark. W. H. (1980). Association of actin with sperm centrioles: Isolation of centriolar complexes and immunofluorescent localisation of actin. *J. Cell Biol.* **86**, 87–95.

Klotz, C., Bordes, N., Laine, M. C., Sandoz, D., and Bornens, M. (1986). A protein of 175,000 daltons associated with striated rootlets in ciliated epithelia as revealed by a monoclonal antibody. *Cell Motil. Cytosk.* **6**, 56–67.

Klotz, C., Dabauvalle, M. C., Paintrand, M., Weber, H., Bornens, M., and Karsenti, E. (1990). Parthenogenesis in *Xenopus* eggs requires centrosomal integrity. *J. Cell Biol.* **110**, 405–415.

Komesli, S., Job, D., Paintrand, M., Margolis, R., and Bornens, M. (1989). Mass isolation of calf thymus centrosomes: The identification of a specific configuration. *J. Cell Biol.* **109**, 2869–2878.

Koonce, M. P., Cloney, R. A., and Berns, M. W. (1984). Laser irradiation of centrosomes in newt eosinophils: Evidence of a centriole role in motility. *J. Cell Biol.* **98**, 1999–2010.

Kuriyama, R. (1984). Activity and stability of centrosomes of Chinese hamster ovary cells in nucleation of microtubules *in vitro*. *J. Cell Sci.* **66**, 277–295.

Kuriyama, R., and Borisy, G. G. (1981a). Centriole cycle in Chinese hamster ovary cells as determined by whole mount electron microscopy. *J. Cell Biol.* **91**, 814–821.

Kuriyama, R., and Borisy, G. G. (1981b). Microtubule-nucleating activity of centrosomes in Chinese hamster ovary cells is independent of the centriole cycle but coupled to the mitotic cycle. *J. Cell Biol.* **91**, 822–826.

Kuriyama, R., and Borisy, G. G. (1983). Cytasters induced within unfertilized sea urchin eggs. *J. Cell Sci.* **61**, 175–189.

La Claire, II, J. W., and Goddard, R. H. (1989). Immunolocalization of pericentriolar material in algal mitotic spindles. *Cell Motil. Cytosk.* **13**, 225–238.

Löhmann, S. M., De Camilli, P., Einig, I., and Walter, U. (1984). High-affinity binding of the regulatory subunit (RII) of cAMP-dependent protein kinase to microtubule-associated and other cellular proteins. *Proc. Natl. Acad. Sci. U.S.A.* **81**, 6723–6727.

McFadden, G. I., Schulze, D., Surek, B., Salisbury, J. L., and Melkonian, M. (1987). Basal body reorientation mediated by a Ca^{2+}-modulated contractile protein. *J. Cell Biol.* **105**, 903–912.

Maller, J. (1990). *Xenopus* oocyte and the biochemistry of cell division. *Biochemistry* **29**, 3157–3166.

Maller, J., Poccia, D., Nishioka, D., Kido, P., Gerhart, J., and Hartman, H. (1976). Spindle formation and cleavage in *Xenopus* eggs injected with centriole containing fractions from sperm. *Exp. Cell Res.* **99**, 285–294.

Maro, B., and Bornens, M. (1980). The centriole–nucleus association: Effects of cytochalasin B and nocodazole. *Biol. Cell* **39**, 287–290.

Maro, B., Paintrand, M., Sauron, M. E., Paulin, D., and Bornens, M. (1984). Vimentin filaments and centrosomes: Are they associated? *Exp. Cell Res.* **150**, 452–458.

Maro, B., Howlett, S. K., and Webb, M. (1985). Nonspindle microtubule organizing centers in metaphase II arrested mouse oocytes. *J. Cell Biol.* **101**, 1665–1672.

Maunoury, R. (1978). Localization immunocytochimique de la centrosphere de cellules tumorales humaines par utilisation d'anticorps naturels de lapin. *C. R. Acad. Sci. (Paris). Ser. 3* **286**, 503–506.

Mazia, D. (1987). The chromosome cycle and the centrosome cycle in the mitotic cycle. *Intern. Rev. Cytol.* **100**, 49–92.

Mitchison, T., and Kirschner, M. (1984). Microtubule assembly nucleated by isolated centrosomes. *Nature (London)* **312**, 232–237.

Mitchison, T. J., and Kirschner, M. (1986). Isolation of mammalian centrosomes. *Meth. Enzymol.* **134**, 261–268.

Mogensen, M. M., and Tucker, J. B. (1987). Evidence for microtubule nucleation at plasma membrane-associated sites in *Drosophila*. *J. Cell Sci.* **88,** 95–107.

Moudjou, M., Lanotte, M., and Bornens, M. (1989a). The fate of the centrosome–microtubule network in monocyte-derived giant cells. *J. Cell Sci.* **94**, 237–244.

Moudjou, M., Vigues, B., and Bornens, M. (1989b). Centrosome and calcium: Identification of a 62–64 centrin homologue protein in human centrosomes. *J. Cell Biol.* **109**, 345a (*Abs*).

Moudjou, M., Paintrand, M., Vigues, B., and Bornens, M. (1991). A human centrosomal protein is immunologically related to basal-body associated proteins from lower eucaryotes and is involved in the nucleation of microtubules. *J. Cell Biol.* **115**(1), 129–140.

Nadezhdina, E. S., Fais, D., and Chentsov, Y. S. (1978). Partial purification of centrioles from spleen cells. *Cell Biol. Int. Rep.* **2**, 601–606.

Nagano, H., Hirai, S., Okana, K., and Ikegami, S. (1981). A chromosomal cleavage of fertilized starfish eggs in the presence of Aphidicolin. *Dev. Biol.* **85**, 409–415.

Nigg, E. A., Schafer, G., Hiltz, H., and Eppenberger, H. M. (1985). Cyclic-AMP-dependent protein kinase type II is associated with the Golgi complex and with centrosomes. *Cell* **41**, 1039–1051.

Nurse, P. (1990). Universal control mechanism regulating onset of M phase. *Nature (London)* **344**, 508.

Oakley, B. R., Oakley, C. E., Zoon, Y., and Jung, M. K. (1990). γ-Tubulin is a component of the spindle pole body that is essential for microtubule function in *Aspergillus nidulans*. *Cell* **61**, 1289–1301.

Omura, F., and Fukui, Y. (1985). *Dictyostelium* MTOC: Structure and linkage to the nucleus. *Protoplasma* **127**, 212–221.

Picard, A., Karsenti, E., Dabauvalle, M. C., and Doree, M. (1987). Release of mature starfish ovocytes from interphase arrest by microinjection of human centrosomes. *Nature (London)* **327**, 170–172.

Picard, A., Harricane, M. C., Labbe, J. C., and Doree, M. (1988). Germinal vesicle components are not required for the cell cycle oscillator of the early starfish embryo. *Dev. Biol.* **128**, 121–128.

Pickett-Heaps, J. D., Schmid, A. M. M., and Tippit, D. H. (1984). Cell division in diatoms. *Protoplasma* **120**, 132–154.

Policard, A., and Bessis, M. (1953). Etude microcinematographique et microscopie electronique du centre cellulaire des leucocytes de vertébrés. *Exp. Cell Res.* **4**, 202–207.

Porter, K. R., and McNiven, M. A. (1982). The cytoplast: A unit structure in chromatophores. *Cell*, **29**, 23–32.

Raff, J. W., and Glover, D. M. (1988). Nuclear and cytoplasmic mitotic cycles continue in *Drosophila* embryos in which DNA synthesis is inhibited with Aphidicolin. *J. Cell Biol.* **107**, 2009–2019.

Rattner, J. B., and Phillips, S. G. (1973). Independence of centriole formation and DNA synthesis. *J. Cell Biol.* **57,** 359–372.

Rieder, C. R., and Borisy, G. G. (1982). The centrosome cycle in PtK_2 cells: Asymmetric distribution and structural changes in the pericentriolar material. *Biol. Cell* **44**, 117–132.

Robbins, E., and Gonatas, N. K. (1964). The ultrastructure of a mammalian cell during the mitotic cycle. *J. Cell Biol.* **21**, 429–458.

Rogers, K. A., McKee, N. H., and Kalnins, V. I. (1985). Preferential orientation of centrioles towards the heart in endothelial cells of major blood vessels is reestablished after reversal of a segment. *Proc. Natl. Acad. Sci. U.S.A.* **82**, 3272–3276.

Rondanelli, E. G., Carosi, G., Gerna, G., and Magliulo, E. (1968). Aspects morphologiques et dynamiques de l'appareil de la sphère dans les hystiocytes et les macrophages cultives *in vitro*. *Acta Anat.* **70**, 85–98.

Roobol, A., Havercroft, J. C., and Gull, K. (1982). Microtubule nucleation by the isolated microtubule-organizing centre of Physarum polycephalum myxamoeba. *J. Cell Sci.* **55,** 365–381.

Roos, U. P. (1975). Fine structure of an organelle associated with the nucleus and cytoplasmic microtubules in the cellular slime mould *Polyshondylium violaceum*. *J. Cell Sci.* **18**, 315–326.

Ruiz, F., Garreau de Loubresse, N., and Beisson, J. (1987). A mutation affecting basal body duplication and cell shape in *Paramecium*. *J. Cell Biol.* **104**, 417–430.

Salisbury, S. L., and Floyd, G. L. (1978). Calcium-induced contraction of the rhizoplast of a quadriflagellate green alga. *Science (Washington, D.C.)* **202**, 975–978.

Salisbury, J. L., Baron, A. T., Surek, B. and Melkonian, M. (1984). Striated flagellar roots-isolation and partial characterization of a calcium-modulated contractile organelle. *J. Cell Biol.* **99,** 962–970.

Salisbury, J. L., Baron, A. T., Coliny, B. E., Marlindale, V. E., and Sanders, M. A. (1986). Calcium modulated contractile proteins associated with the eucaryotic centrosome. *Cell Motil. Cytosk.* **6**, 193–197.

Salisbury, J. L., Sanders, M. A., and Haraps, L. (1987). Flagellar root contraction and nuclear movement during flagellar regeneration in chlamydomonas reinhardtii. *J. Cell Biol.* **105,** 1799–1805.

Schatten, H., and Schatten, G. (1986). Motility and centrosomal organization during sea urchin and mouse fertilization. *Cell Motil. Cytosk.* **6**, 163–175.

Schliwa, M., Pryzwansky, K. B., and Euteneuer, U. (1982). Centrosome splitting in neutrophils: An unusual phenomenon related to cell activation and motility. *Cell* **81**, 705–717.

Sellito, C., and Kuriyama, R. (1988). Distribution of pericentriolar material in multipolar spindles induced by Colcemid treatment in Chinese hamster ovary cells. *J. Cell Sci.* **89**, 57–65.

Sherline, P., and Mascardo, R. J. (1982). EGF-induced centrosomal separation: Mechanism and relationship to mitogenesis. *J. Cell Biol.* **95**, 316–322.

Sluder, G. (1988). Control mechanism of mitosis: The role of spindle microtubules in the timing of mitotic events. *Zoological Science* **5**, 653–665.

Sluder, G., and Rieder, C. L. (1985). Centriole number and the reproductive capacity of spindle poles. *J. Cell Biol.* **100**, 887–896.

Sluder, G., Miller, F. J., and Rieder, C. L. (1986). The reproduction of centrosomes. Nuclear versus cytoplasmic controls. *J. Cell Biol.* **103**, 1873–1881.

Sluder, G., Miller, F. J., and Rieder, C. L. (1989a). Reproductive capacity of sea urchin centrosomes without centrioles. *Cell Motil. Cytosk.* **13**, 264–273.

Sluder, G., Miller, F. J., Lewis, K., Davison, E. D., and Rieder, C. L. (1989b). Centrosome inheritance in starfish zygotes: Selective loss of the maternal centrosome after fertilization. *Dev. Biol.* **131**, 567–579.

Sluder, G., Miller, F. J., Cole, R., and Rieder, C. L. (1990). Protein synthesis and the cell cycle: Centrosome reproduction in sea-urchin eggs is not under translational control. *J. Cell Biol.* **110**, 2025–2032.

Sonnenblick, B. P. (1950). The early embryology of *Drosophila melanogaster*. *In* “Biology of *Drosophila*” (M. Demerec, ed.) pp. 62–163. Hafner.

Stearns, M. E., and Brown, D. L. (1979). Purification of cytoplasmic tubulin and microtubule organizing center proteins functioning in microtubule initiation from the alga *Polytomella. Proc. Natl. Acad. Sci. U.S.A.* **76**, 5745–5749.

Steffen, W., and Linck, R. (1988). Evidence for tektins in centrioles and axonemal microtubules. *Proc. Natl. Acad. Sci. U.S.A.* **85**, 2643–2647.

Sunkel, C. E., and Glover, D. M. (1988). Polo, a mitotic mutant of *Drosophila* displaying abnormal spindle poles. *J. Cell Sci.* **89**, 25–3.

Szollozi, D., Callarco, P., and Donahue, R. P. (1972). Absence of centrioles in the first and second meiotic spindles of mouse ovocytes. *J. Cell Biol.* **11**, 521–541.

Tamm, S. L., and Tamm, S. (1988). Development of macrociliary cells in *Beroë*, I. Actin bundles and centriole migration. *J. Cell Sci.* **89**, 67–80.

Tassin, A. M., Baro, B., and Bornens, M. (1985a). Fate of microtubule-organizing centers during myogenesis *in vitro. J. Cell Biol.* **100**, 35–46.

Tassin, A. M., Paintrand, M., Berger, E. G., and Bornens, M. (1985b). The Golgi apparatus remains associated with microtubule organizing centers during myogenesis. *J. Cell Biol.* **101**, 630–638.

Toriyama, M., Ohta, K., Endo, S., and Sakai, H. (1988). 51-kDa protein, a component of microtubule-organizing granules in the mitotic apparatus involved in aster formation *in vitro. Cell Motil. Cytosk.* **9**, 117–128.

Tournier, F., Karsenti, E., and Bornens, M. (1989). Parthenogenesis in *Xenopus* eggs injected with centrosomes from synchronized human lymphoid cells. *Dev. Biol.* **136**, 321–329.

Tournier, F., Komesli, S., Paintrand, M., Job, D., and Bornens, M. (1991a). The intercentriolar linkage is critical for the ability of heterologous centrosomes to induce parthenogenesis in *Xenopus. J. Cell Biol.* **113**(6), 1361–1369.

Tournier, F., Cyrklaff, M., Karsenti, E., and Bornens, M. (1991b). Centrosomes competent for patheno-genesis in *Xenopus* eggs support procentriole budding in cell-free extracts. *Proc. Natl. Acad. Sci. U.S.A.*, in press.

Tucker, R. W. (1979). Spatial organization of microtubules. *In* "Microtubules" (K. Roberts and J. S. Hyams, eds.) pp 347–357. Academic Press, London.

Tucker, J. B., Pardee, R., and Fujiwara, K. (1979). Centriole ciliation is related to quiescence and DNA synthesis in 3T3 cells. *Cell* **17**, 527–535.

Tucker, J. B., Milner, M. J., Currie, D. A., Muir, J. W., Forrest, D. A., and Spencer, M. J. (1986). Centrosomal microtubule organizing centres and a switch in the control of the protofilament number for cell surface-associated microtubules during *Drosophila* wing morphogenesis. *Eur. J. Cell Biol.* **41**, 279–289.

Vandre, D. D., Davis, F. M., Rao, P. N., and Borisy, G. G. (1986). Distribution of cytoskeletal proteins sharing a conserved phosphorylated epitope. *Eur. J. Cell Biol.* **41**, 72–81.

Verde, F., Labbé, J. C., Dorée, M., and Karsenti, E. (1990). Regulation of microtubule dynamics by cdc2 protein kinase in cell-free extracts of *Xenopus* eggs. *Nature (London)* **343**, 233–238.

Vigues, B., Metenier, G., and Senaud, J. (1984). The sub-surface cytoskeleton of the ciliate Polyplas-tron multi vesiculatum-isolation and major proteins components. *Eur. J. Cell Biol.* **35,** 336–342.

Vigues, B., and Grolière, C. A. (1985). Evidence for a Ca^{2+}-binding protein associated to non-actin microfilamentous systems in two ciliated protozoans. *Exp. Cell Res.* **159**, 366–376.

Vorobjev, I. A., and Chentsov, Y. S. (1982). Centrioles in the cell cycle. I. Epithelial cells. *J. Cell Biol.* **93**, 938–949.

Vorobjev, I. A., and Nadehzdina, E. S. (1987). The centrosome and its role in the organization of microtubules. *Int. Rev. Cytol.* **106**, 227–284.

Watt, F. M., and Harris, H. (1980). Microtubule organizing centres in mammalian cells in culture. *J. Cell Sci.* **44**, 103–121.

Welsh, M. J., Dedman, J. R., Brinkley, B. R., and Means, A. R. (1979). Tubulin and calmodulin. Effects of microtubule and microfilament inhibitors and localization in the mitotic apparatus. *J. Cell Biol.* **81**, 624–634.

Wheatley, D. N. (1982). The centriole: A central enigma of cell biology. Elsevier Biomedical, New York.

Wise, G. E., and Prescott, D. M. (1973). Ultrastructure of enucleated mammalian cells in culture. *Exp. Cell Res.* **81**, 63–72.

Wick, S. M. (1985). Immunofluorescence microscopy of tubulin and microtubule arrays in plant cells. III. Transition between mitotic/cytokinetic and interphase microtubules arrays. *Cell Biol. Int. Rep.* **9**, 357–371.

Wilsman, N. J., and Farnum, C. E. (1983). Arrangement of C tubule protofilaments in mammalian basal bodies. *J. Ultrast. Res.* **84**, 205–212.

Wright, M., Mir, L., and Moisand, A. (1980). The structure of the pro-flagellar apparatus of the amoeba of *Physarum polycephalum*: Relationship to the flagellar apparatus. *Protoplasma* **103**, 69–81.

Wright, M., and Moisand, A. (1982). Spatial relationships between centrioles and the centrosphere in monoasters induced by taxol in *Physarum polycephalum* amoeba. *Protoplasma* **113**, 69–79.

Wright, R. L., Salisbury, J., and Jarvik, J. W. (1985). A nucleus–basal body connector in *Chlamydomonas reinhardtii* that may function in basal body localization or segregation. *J. Cell Biol.* **101**, 1903–1912.

Zickler, D. (1970). Division spindle and centrosomal plaques during mitosis and meiosis in some Ascomycetes. *Chromosoma (Berlin)* **30**, 287–304.

2

Ultrastructure of Centrosome Domains and Identification of Their Protein Components

J. B. RATTNER

Departments of Anatomy and Medical Biochemistry
University of Calgary
Calgary, Alberta T2N 4N1 Canada

I. INTRODUCTION

The study of the ultrastructure of the centrosome would at first appear to be a straightforward goal. However, this task is complicated by the fact that the centrosome is, to a large degree, a cellular focus rather than a discrete organelle such as a mitochondrion or chromosome. Further, major architectural landmarks in the centrosome may vary; for example, the presence or absence of centrioles and its perimeter may fluctuate considerably. Despite these problems, ultrastructural imaging of the centrosome has provided a wealth of information about this cellular domain. The relevance of this information to the function of the centrosome is only now becoming clear in light of recent advances in the study of the composition of this microtubule organizing center (MTOC).

Copyright © 1992 Academic Press, Inc.
All rights of reproduction in any form reserved.

A. Ultrastructure of Centrosomal Domains

In most animal cells, the centrosome can be identified in electron micrographs as the region containing at least one centriole and appearing as the preferred focus for microtubules. Most microtubules do not originate at the centriole but in the region around it. This morphology suggests that the animal centrosome has two distinct domains, one containing the centriole and the other, the pericentriolar material (PCM). Both domains show cell-cycle-specific changes, in both ultrastructure and composition. In addition, it is possible to find ultrastructural variation in each of these domains among animal species (see Vorobjev and Nadezhdina, 1987, and references therein). If one considers the centrosome as a collection of functionally independent but spatially related domains, then one can understand the structural variability of the eukaryotic centrosome. In this view, it is the PCM domain that is the MTOC necessary for the organization of interphase and, most particularly, the mitotic spindle microtubule arrays (see, for example, Tilney and Goddard, 1970; Berns *et al.*, 1977; Gould and Borisy, 1977; Kuriyama and Borisy, 1981). This domain may exist on its own or in concert with the centriole domain as seen, for example, in multipolar mitosis found in cells recovering from mitotic arrest with nocodazole (Alieva and Vorobjev, 1991). The function of the centriole domain, when it is present, may be to act as the site of nucleation of cilia and flagella, for orientation of the centrosome in the cytoplasm, as a nucleus for the collection of the PCM, as a determinant for the number of PCM foci, and/or for orienting the spindle axis (Vorobjev and Nadezhdina, 1987; Karsenti and Maro, 1986).

The centriole domain is perhaps the most pronounced and most studied region of the centrosome. In animal cells, the number of centrioles found in this domain, as well as their arrangement and architecture, has been found to be cell cycle specific (for reviews, see Peterson and Berns, 1980; Vorobjev and Nadezhdina, 1987). Since the daughter centriole is short and immature throughout most of the cell cycle, the parent is most closely associated with the PCM. Careful examination indicates that there may be several subdomains in the centrosome region. One subdomain, the centriole lumen, is delineated along its sides by the nine triplet blades of the centriole and contains an amorphous hub at the distal end (Fig. 1A) and a pinwheel structure at the proximal end. This subdomain may be the site of centriolar DNA (Hall *et al.*, 1989); microtubules, vesicles, and electron-dense material have been sighted in this region (see, for example, Brinkley and

---

Fig. 1. (A) Interphase centrosome from a mouse L929 cell. Large arrow indicates site of direct centriole–microtubule interaction. Small arrow denotes electron-dense aggregates within the PCM. Note placement of vesicles, microtubules, filaments, and PCM components. Bar: 0.5 μm. (B) Centrosomal region from an L929 cell in S phase of the cell cycle. Three types of electron-dense aggregates are present: (1) unassociated with the cytoskeleton, (2) associated with microtubules, and (3) associated with filaments (F) and microtubules (MT). Note the vesicles associated with the centriole (*lower left*). Bar: 0.5 μm. Reproduced from Rattner and Phillips (1973).

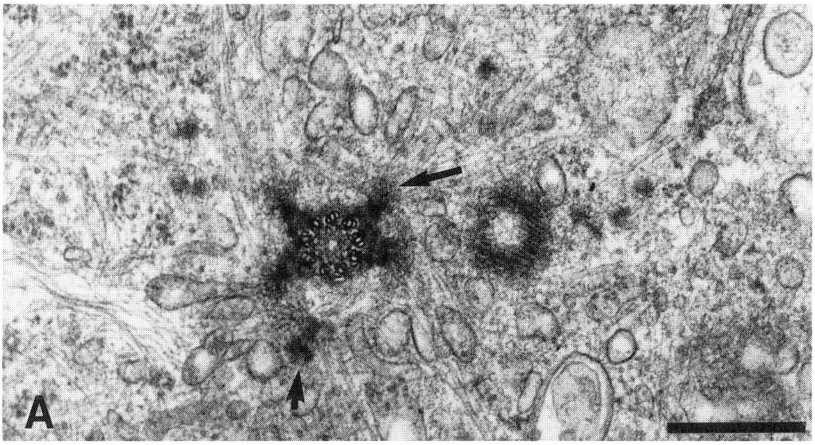
A

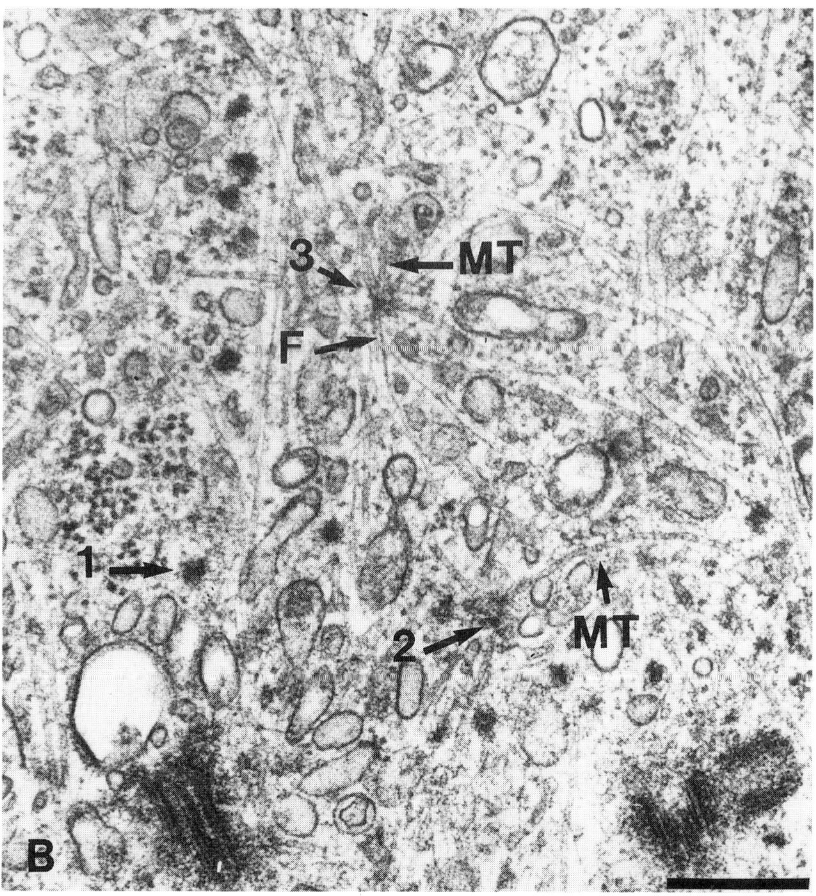
3
MT
F
1
2
MT
B

Stubblefield, 1970; Baron and Salisbury, 1988; Rattner and Bazett-Jones, 1988a). Another subdomain surrounds the centriole and is delineated by the walls of the centriole cylinder on one side and the PCM on the other. This subdomain is occupied by a large amount of electron-dense material. Some of this material adheres to the microtubule blades, and highly structured appendages (called feet or satellites by some authors) extend from the walls of the centriole (Fig. 1A). The electron-dense nature of this subdomain is most prominent during interphase when it is closely associated with the origins of some interphase microtubules. It is apparent that several regions of the centrosome have the ability to organize microtubules. The microtubules from each of these regions may be involved in different functions. For example, microtubules that associate with the highly structured appendages that extend from the outer surface of the centriole blades may function to anchor the centrosome in the cytoplasm, whereas microtubules originating from the surface of the centriole may play a role in orienting and positioning the centriole duplex. During mitosis, the amount of electron-dense material around the centrioles is either greatly reduced or completely absent and the centrosomal region is characterized by abundant microtubules that frequently originate in the PCM (Fig. 2B).

The structure of the PCM has been studied with a wide range of techniques in samples prepared in a variety of ways. These include analysis by conventional transmission electron microscopy and high-voltage electron microscopy of thin sections of intact cells or tissues, detergent extracted cells subjected to brief cold treatment to destroy microtubules, and glutaraldehyde/tannic acid-fixed material (Robbins *et al.*, 1968; Zeligs and Wollman, 1979; Reider and Borisy, 1982; Vorobjev and Chentsov, 1982; Paweletz and Mazia, 1989). In conventional electron micrographs, the PCM does not appear highly structured and is frequently referred to as a cloud or halo around the centriole domain. During interphase the PCM is not very extensive and appears to contain filaments (3–8 nm in diameter) as well as aggregates of material that have been termed "satellites" or electron-dense granules (see, for example, Brinkley and Stubblefield, 1970; Reider and Borisy, 1982; Vorobjev and Nadezhdina, 1987). These electron-dense aggregates appear to be of at least two types: those that are directly associated with cytoskeletal elements and those that are not (Fig. 1B). It is interesting to note that both microfilaments and microtubules can associate with the same electron-dense granule (Fig. 1B). Thus, these structures can act to integrate different elements of the

Fig. 2. (A) Separating centrosome from a late prophase L929 cell. Filaments (F) are oriented parallel to the axis of separation. Most microtubule (MT) profiles are found at the periphery of the centrosomes. Note electron-dense granules (small arrow). Bar: 0.75 μm. (B) Metaphase centrosome from an L919 cell. Numerous microtubule profiles extend through the PCM (bracket). Bar: 0.5 μm. (C) Late telophase–early interphase centrosome from an L929 cell. The centrosome has migrated to a position near the midbody (M) and is now on the opposite side of the cell with respect to its metaphase position. Electron-dense material (small arrows), interphase microtubules (MT), and filaments (F) have reappeared. Bar: 0.75 μm. Reproduced from Rattner and Phillips (1973).

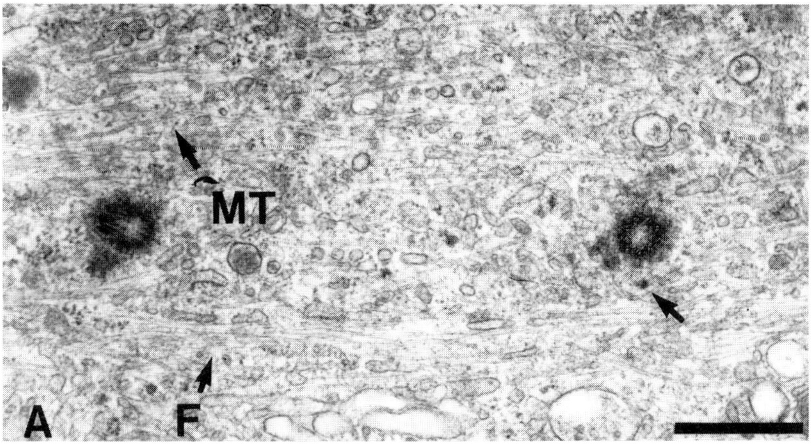
MT
A
F

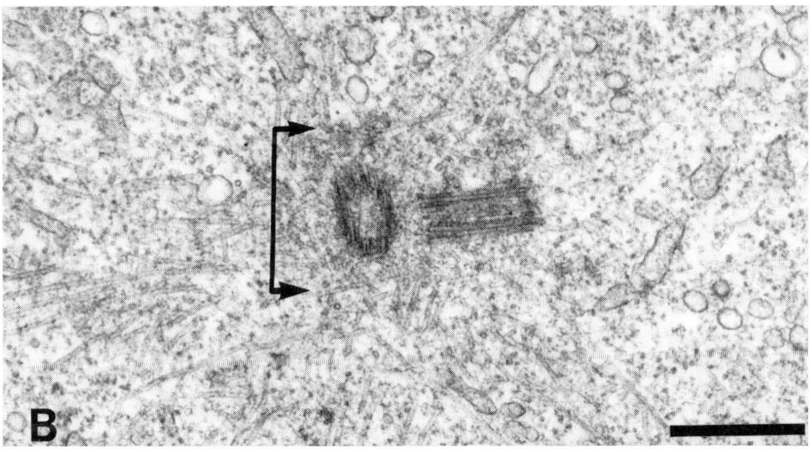
B

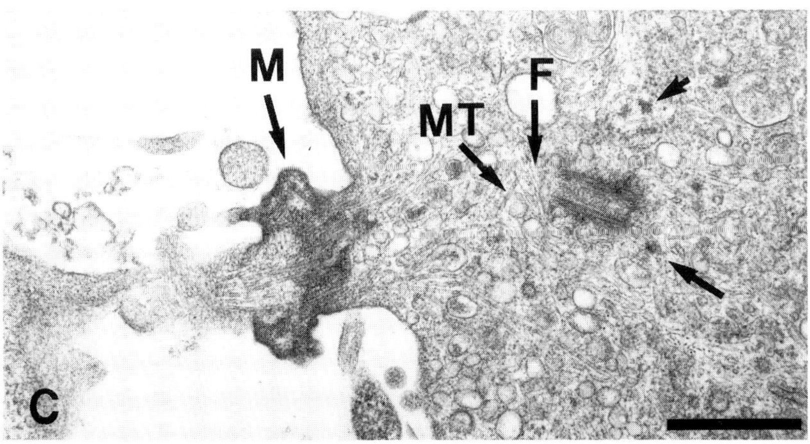
M
F
MT
C

cytoskeleton, a function that may be required for positioning and motility of centrosomes (Euteneuer and Schliwa, 1985).

The PCM is infiltrated by microtubules and filaments throughout the cell cycle. These elements tend to extend radially from the interphase centrosome. At the interphase–mitosis transition, as the centrosomes separate, the interphase microtubules depolymerize and a new array of microtubules is formed (compare Fig. 1A,B with Fig. 2A). At this stage, the filaments tend to align along the axis of separation of the centrosome while the majority of the microtubules extends in bundles angled approximately 30° on either side of this axis (Rattner and Berns, 1976). The microtubules that extend from each centrosome overlap, forming a diamond-shaped array. This type of arrangement may participate in the coordination of the complex linear and arc movements that often accompany centrosome separation and the establishment of the spindle poles. The region occupied by the PCM expands as the cell enters mitosis; this is correlated with an increase in the microtubule-nucleating activity of the centrosome (Kuriyama and Borisy, 1981). The number of large electron-dense aggregates found in the PCM during mitosis decreases, but these structures reappear as the cell returns to interphase (Fig. 2C). The reappearance of this interphase morphology coincides with the migration of the centrosome to a new position in the cytoplasm after the completion of cell division (Fig. 2C) (Rattner and Phillips, 1973).

The three-dimensional configuration of centrosome-associated microtubules is both dynamic and complex. Organization can be astral or anastral, and the distribution of spindle microtubules into spherical or barrel-shaped configurations can be species and developmental-stage specific. This implies that there is a network of microtubule nucleating centers in the PCM and that the spatial distribution of these sites, while quite flexible (Mazia, 1984), is under strict control. The location of these nucleating sites may best be visualized using immunoelectron microscopy. For example, antibodies to a 225K thiophosphoprotein map to small granules from which centrosome microtubules emanate (Kuriyama, 1989), perhaps representing nucleating protein complexes of the PCM. It is possible that nucleating sites in the PCM are not equivalent since they are able to nucleate at least two types of microtubule that differ in their associated proteins. Thus, PCM nucleating sites may not only nucleate microtubules but may determine their properties by determining, at least in part, their protein composition (McIntosh, 1989).

In addition to fibrillar and granular components, the PCM often contains a large number of vesicles, elements of the endoplasmic reticulum (Fig. 1B). These components often have been ignored, and their function and content has not been determined. It is possible that they play a role in sequestering ions in the PCM. For example, they may contain Ca^{2+} that, when bound by calmodulin, may directly affect a number of mitotic processes at the centrosome (Hepler, 1989). It is not uncommon to see the centrioles abut larger vesicles (Fig. 1B). Many ubiquitous cytoplasmic components are also found in the PCM domain; however, larger structures such as mitochondria are excluded from this region (Figs. 1, 2).

B. Centrosome Composition

The composition of the centrosome region has been of interest since its discovery. There are reports that the centrosome stains preferentially for ribonucleoprotein (Reider, 1979). The fibrillar material and the satellites have been shown to be digestible with RNase in lysed cell models and in thin sections. Both RNase T1 and RNase A affect the ability of the PCM to nucleate microtubules (Brinkley and Stubblefield, 1970; Pepper and Brinkley, 1980). Reports of DNA at the centrosome also have been made (Goodenough, 1989; Hall *et al.*, 1989).

In addition to containing nucleic acids, the centrosome, particularly the PCM domain, is rich in a variety of proteins. The identity of some centrosomal components are known; these include centrin, kinesin, tubulin, calmodulin, tektin, a microtubule-activated Mg^{2+}-ATPase, RNA polymerase, cAMP-dependent protein kinase, transferrin receptor, MAP 1, p34^{cdc2}, purine nucleotide phosphorylase, intermediate filament proteins, centrophilin, HSP-70, and several proteins detected by antibody binding (Pepper and Brinkley, 1976; Dedman *et al.*, 1980; Brinkley *et al.*, 1981; Stearns *et al.*, 1991: Oliver *et al.*, 1981; Sherline and Mascaro, 1982; Turksen and Kalnins, 1982; Moroi *et al.*, 1983; Tuffanelli *et al.*, 1983; Nigg *et al.*, 1985; Willingham and Pastan, 1985; Osborn *et al.*, 1986; Baron and Salisbury, 1988; Neighbors *et al.*, 1988; Steffen and Linck, 1988; Schatten *et al.*, 1987; Tousson *et al.*, 1991; Rattner, 1991; Kuriyama, 1989; Sager *et al.*, 1986). Recently, we have detected glycolytic enzyme and creatine phosphokinase epitopes at the centrosome. In addition, we have documented that many centrosomal proteins such as p34^{cdc2} are also present at other sites in the mitotic apparatus. These observations are discussed in the following sections.

II. GLYCOLYTIC ENZYME AND CREATINE PHOSPHOKINASE EPITOPES AT MICROTUBULE ORGANIZING CENTERS

Gosti and co-workers (1987) have identified proteins that share a common epitope with lactate dehydrogenase and are localized to the centrosome. Intrigued by the possibility that glycolytic enzymes or evolutionarily related proteins may be present at the centrosome, we investigated the localization of enolase in cells of the Indian muntjac during various phases of the cell cycle.

Enolase (2-phospho-D-glycerate hydrolase) is a glycolytic enzyme that is found as three isozymes in eukaryotes (Rider and Taylor, 1974). Isozyme $\alpha\alpha$ is found in various tissues and in glial cells of adult nervous tissue [nonneuronal enolase

(NNE)]; ββ is found in muscle tissue; γγ, originally found in nervous tissue [neuron-specific enolase (NSE)] is now known to be present in various tissues. We used NSE from bovine brain to generate a polyclonal antibody. This antibody was affinity purified against NSE and shown on Western blots to react against the γ isotope (NSE) and to cross-react weakly against the αα and ββ isotypes. This antibody also recognized a single 48K protein on Western blots of whole-cell extracts from Indian muntjac mitotic cells. Indirect immunofluorescence (IIF) using logarithmically growing Indian muntjac cells revealed a site-specific distribution of NSE reactivity. At interphase, the antibody reacted with material distributed throughout the cytoplasm. The staining was especially prominent around the nucleus. Each centrosome appeared as a well-defined strongly staining dot adjacent to the nucleus (Fig. 3A). Cytoplasmic NSE reactivity appeared to aggregate around the chromosomes during nuclear breakdown at prometaphase and the centrosomes appeared as strongly stained regions larger than those detected at interphase. At metaphase, staining was seen at the two centrosomes and at each of the paired kinetochores of the chromosomes (Fig. 3B). At telophase, NSE staining persisted at the centrosome and could also be seen at the midbody of the intercellular bridge joining the two daughter cells (Fig. 3C).

Antienolase antibody staining of the kinetochores was particularly prominent in muntjac mitotic cells but was difficult to detect in other cell types (HeLa, L929, and HEp-2). The karyotype of the Indian muntjac is unique among mammals, containing a small number ($2n$=6♀,7♂) of large chromosomes with extensive centromeres and large kinetochores. It has been suggested that the karyotype of the Indian muntjac has evolved through a series of Robertsonian fusions from an ancestral species, the Chinese muntjac ($2n$=46) (Shi and Ye, 1980). In thin sections, the kinetochore appears similar to that found in other mammals. However, several investigations have upheld the view that these structures are compound in nature, composed of several unit kinetochores that are visible only at interphase with antikinetochore antibodies or by whole-mount electron microscopy (Brinkley *et al.*, 1984; Rattner, 1986). The large size of the Indian muntjac centromere therefore makes it ideal for detecting and evaluating the distribution of putative kinetochore-associated proteins as well as those proteins that may be common to the kinetochore and the centrosome. Using this rationale, we investigated whether calmodulin, found at the centrosome and kinetochore-to-pole microtubules, may

Fig. 3. (A,B,C) Confocal images of Indian muntjac cells stained with anti-NSE antibodies. (A) Interphase; c, centrosomes; bar: 10 μm. (B) Metaphase; c, centrosome; k, kinetochore; bar: 10 μm. (C) Telophase; c, centrosome; m, midbody; bar: 20 μm. (D) Affinity-purified antienolase antibodies from serum of a patient with autoimmune diseases reacted with HEp-2 cells. The antibodies react strongly with the centrosomal region (arrows); Bar: 40 μm. (E,F) Metaphase Indian muntjac cell stained with DAPI (E) and IIF (F) with antibodies to antilactate dehydrogenase. The centrosome (c) and kinetochores (small arrows) are strongly stained; bar: 10 μm. (G,H) Metaphase Indian muntjac cell stained with DAPI (G) and IIF (H) with antibodies to anticreatine phosphokinase. The centrosome (c) and kinetochores (arrows) are strongly stained; bar: 7 μm.

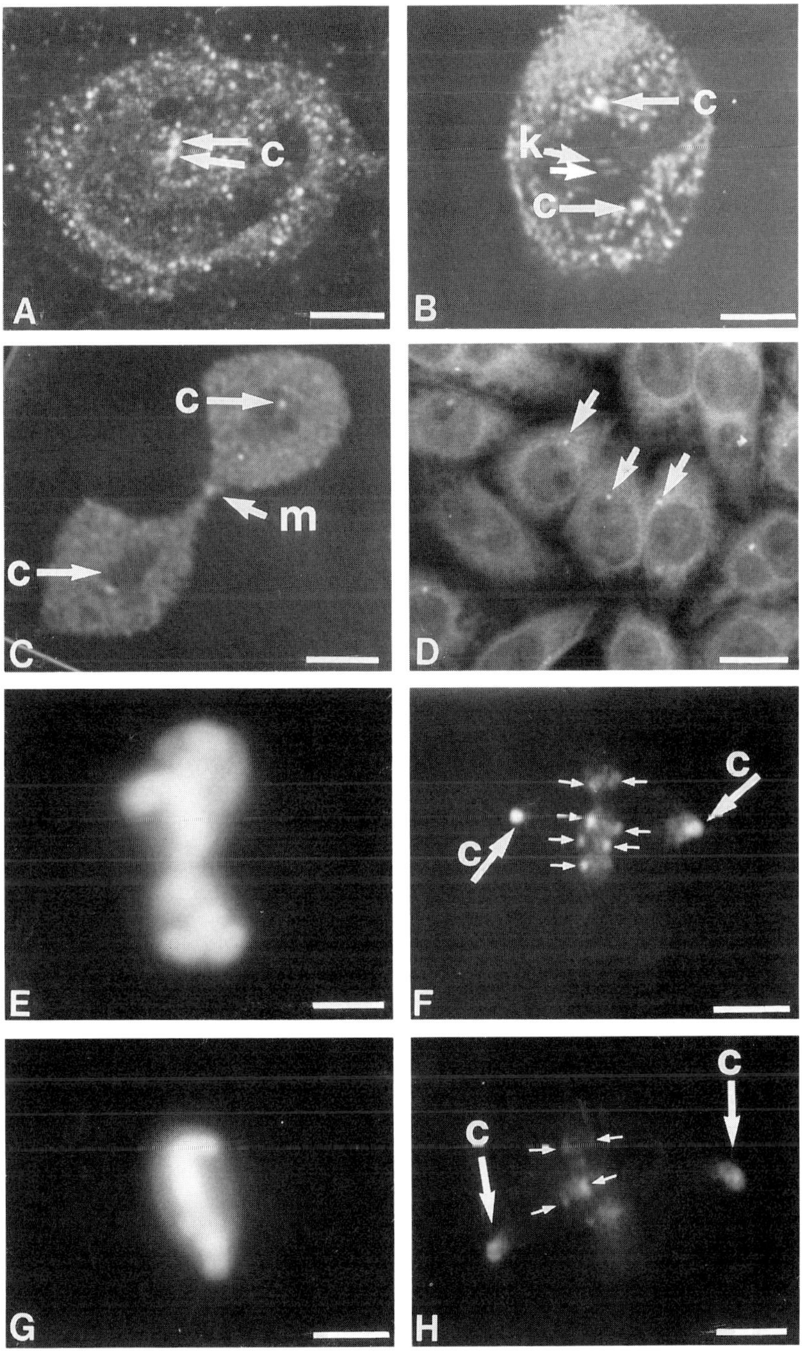
c
A
k
c
c
B
c
m
c
C
D
E
c
c
F
G
c
c
H

also reside at the kinetochore. Observations of intact cells stained with an anti-calmodulin antibody confirmed that the kinetochores of the Indian muntjac contain calmodulin.

Because tubulin is a major component of MTOCs including the kinetochore, we investigated whether calmodulin and the enolase epitope were bound to the kinetochore through an interaction with tubulin. We grew Indian muntjac cells in the presence of vinblastine (40 μg/ml) for 6 hr. Chromosomes isolated from these cells did not react with antitubulin antibodies. Calmodulin reactivity was also abolished by this treatment. However, this was not the case for enolase reactivity, suggesting that some proteins may be associated with the kinetochore in a tubulin-dependent manner, whereas others are present independent of tubulin.

To determine if other glycolytic enzymes, or their epitopes, and associated proteins are also present at the centrosome and other MTOCs, we obtained commercially available antibodies to creatine phosphokinase and lactate dehydrogenase for IIF studies with Indian muntjac cells. We did not detect creatine phosphokinase staining at the interphase centrosome. However, this region was stained with antibodies to lactate dehydrogenase. Both antibodies stained the centrosome and the kinetochore in dividing cells (Fig. 3E,F,G,H) as well as the midbody region at telophase.

Because autoimmune sera have been used to detect both centrosomes and kinetochores (Moroi *et al.*, 1980, 1983), we screened sera from a group of 29 patients that did not fall into the CREST category of autoimmune disease but contained antibodies that produced centrosome or spindle staining in HEp-2 cells for reactivity with glycolytic enzymes and centrosome-associated proteins. Of those tested, 2 patients were found to contain antibodies to pyruvate kinase, 4 to triose-phosphate isomerase, 5 to lactate dehydrogenase, 3 to creatine phosphokinase, 1 to glyceraldehyde 3-phosphate, 1 to aldolase, and 4 to calmodulin. Several patients showed reactivity to more than one of these proteins. We were also able to identify 4 patients with antibodies to NSE. When these antibodies were affinity purified using NSE and applied to a HEp-2 cells, they produced a characteristic staining pattern that included reactivity at the centrosome (Fig. 3D) (Rattner *et al.*, 1991).

The results just described suggest two possibilities: either glycolytic enzymes and their associated proteins are localized to MTOCs or a family of MTOC and centrosomal proteins shares epitopes with a number of different glycolytic enzymes. Although it was initially assumed that glycolytic enzymes were "free floating" in the cytoplasm, evidence for compartmentalization of these enzymes has been accumulating (Knull, 1980; Vinores *et al.*, 1984, 1986) Clark and colleagues (1985) have suggested that glycolytic enzymes are capable of dual functions, playing both a structural and a catalytic role. Structurally, some glycolytic enzymes have been shown to function *in vitro* in the bundling of microfilaments (Clark *et al.*, 1985) and microtubules (Huitorel and Pantaloni,

1985); it is possible that they play such a role at MTOCs. Catalytically these enzymes may function at MTOCs in glycolysis, specifying the site-specific production of ATP and GTP, both of which have been shown to play a role in microtubule growth and function. ATP in particular has been shown to be required for increased microtubule turnover in mitosis (De Brabander *et al.*, 1981). Creatine phosphokinase might play a role in the local storage or accumulation of high-energy phosphates and calmodulin may participate in the regulation of site-directed glycolysis. If our antibody studies simply detect shared epitopes between glycolytic enzymes and centrosomal proteins, then there must be an evolutionary relationship between these two classes of proteins. It is possible that functional domains have been conserved between these two groups. In this case, there may be centrosomal proteins that function in a manner similar to glycolytic enzymes. Although these results are preliminary, they are intriguing and indicate an avenue for future work.

III. p34^{cdc2} PRESENT AT CENTROSOME, KINETOCHORE, AND ALONG KINETOCHORE-TO-POLE MICROTUBULES

All eukaryotes appear to have a protein kinase homologous to the p34^{cdc2} gene product of the fission yeast *Schizosaccharomyces pombe*. p34^{cdc2} is a component of M-phase promoting factor (MPF), an inducer of mitosis in a wide variety of species (Maller, 1985). MPF induces the breakdown of the nuclear membrane, chromosome condensation, nuclear lamin breakdown, and the formation of the mitotic spindle (Lamb *et al.*, 1990). In human cells, p34^{cdc2} reaches its maximum activity at mitosis. Injection of anti-p34^{cdc2} antibodies prevents cell division (Riabowal *et al.*, 1989). p34^{cdc2} has been localized to the nucleus during interphase and a fraction has been shown to be associated with the centrosome only at mitosis (Bailly *et al.*, 1989; Riabowal *et al.*, 1989). Microtubule dynamics have been shown to be affected by p34^{cdc2}, and it has been suggested that this kinase plays a major role in the transition between interphase and metaphase microtubule arrays (Verde *et al.*, 1989; Lamb *et al.*, 1990). The introduction of an isolated centrosome into interphase *Xenopus* egg extracts results in the polymerization of long microtubules (Verde *et al.*, 1989). Addition of p34^{cdc2} kinase results in shorter centrosome-associated microtubules and the elimination of free cytoplasmic microtubules. The level of phosphorylation associated with the centrosomes increases after the addition of p34^{cdc2} kinase. This increased phosphorylation is accompanied by an increase in the nucleating activity of the centrosome. Interestingly, although p34^{cdc2} kinase is found in association with cyclin in these eggs, cyclin is not necessary to obtain an effect on microtubule dynamics (see Chapter 1).

Using polyclonal antibodies to either the C-terminal or N-terminal portions of p34^{cdc2}, we carried out an IIF study on African green monkey and Indian muntjac cells (Rattner *et al.*, 1990). Using antibodies to either of these domains, we were able to confirm that the centrosome showed pronounced staining only during mitosis. Thus the distribution of this protein is different from that of other centrosomal components, such as those detected by antibodies to enolase that are also found in association with the interphase centrosome. However, in contrast to previous studies, both antibodies showed staining at the kinetochores and along the kinetochore-to-pole microtubules in several organisms (Fig. 4A–D) (Verde *et al.*, 1989, Riabowal *et al.*, 1989). We were unable to detect staining of the pole-to-pole microtubules at metaphase or as the chromosomes separated during anaphase.

The appearance of $p34^{cdc2}$ in specific spindle domains suggests that, in addition to a role in the passage of cells from G_2 to the mitotic phase of the cell cycle, this protein plays a role in spindle dynamics. Since $p34^{cdc2}$ has the ability to modulate microtubule dynamics and stability, it may be an important component in the regulation of microtubule polymerization–depolymerization during chromosome movement (anaphase A). Further, its presence in the intercellular bridge suggests that it may also have a role in spindle elongation (anaphase B). The finding that many centrosomal components are also present at other MTOCs probably reflects their common function. In addition, some of these components may be instrumental in coordinating events between the centrosome and other MTOCs. The emerging pattern of protein distribution in the mitotic apparatus is beginning to provide important insights into the manner in which spindle dynamics are controlled both spatially and temporally.

One general feature of the centrosome, derived from ultrastructural information, is that the organization and size of the centrosome are cell cycle specific. This concept is confirmed by IIF data. For example, the extent of staining of several centrosomal proteins, such as those detected by antibodies to enolase and lactate dehydrogenase, indicates that they are confined to a small well-defined area during interphase and show a more diffuse expanded localization as the centrosome approaches and proceeds into mitosis. In some instances, as is the case with creatine phosphokinase and $p34^{cdc2}$, a protein may be present at mitosis but undetectable during the interphase period. The lack of staining in this instance may be the result of a masked epitope or the complete absence of the protein. Considered simultaneously, these observations suggest that the ultrastructural alterations of the centrosome reflect changes in the types, amounts, and functional states

Fig. 4. Staining of African green monkey (A,B) and Indian muntjac (C,D) mitotic cell components with antibody to $p34^{cdc2}$. DAPI image (*left*), IIF image (*right*). $p34^{cdc2}$ is present at the centrosome (c), kinetochore-to-pole microtubules (mt), and kinetochores (small arrows, k) in both species. A and D, metaphase; B, telophase, C, isolated metaphase chromosome of the Indian muntjac cell in which the anti-$p34^{cdc2}$ staining is confined to the kinetochore domain. Bar: (A) 8 μm; (B) 10 μm; (C) 5 μm; (D) 10 μm.

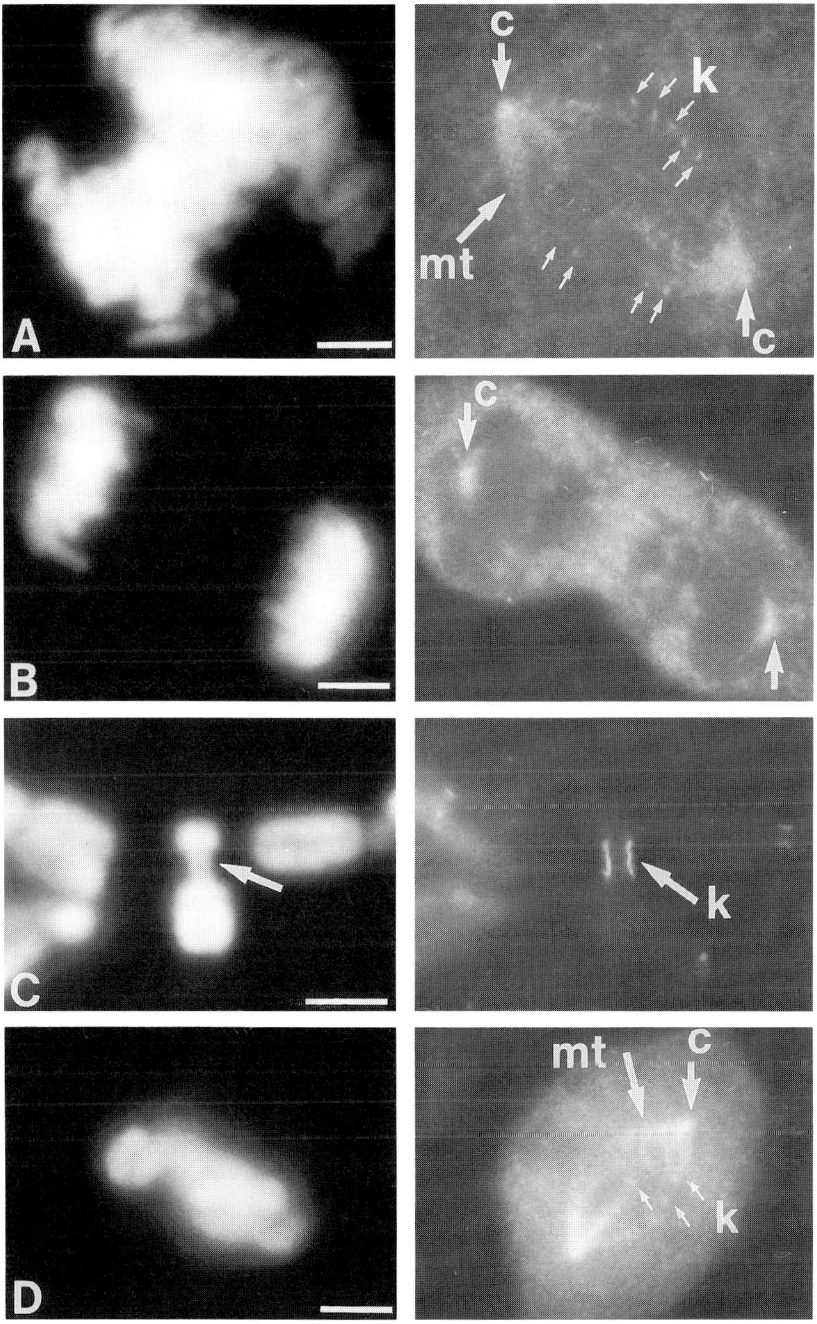
c
k
mt
c
A
c
B
k
C
mt
c
k
D

of centrosome-associated proteins. It is interesting to note that most of the centrosomal proteins identified to date are ubiquitous cellular constituents that are also found throughout the cytoplasm. A major function of the centrosome may be to act as a sink for a variety of ubiquitous cellular proteins. Changes in the concentration of these cellular components may be a key to the control of microtubule dynamics by the centrosome. This may also be true for MTOCs in general, since many of the proteins found at the centrosome are also present at the midbody and kinetochore. The presence of these proteins in the cytoplasm may explain why microtubule assembly is also possible outside the major MTOCs (Bré *et al.*, 1987; McNiven and Porter, 1988).

The cycling of cells from interphase to mitosis is generally characterized by phosphorylation events. Antibodies to phosphorylated epitopes of mitotic spindle proteins have been used to show that these proteins are associated with MTOCs, including the centrosome (Vandre *et al.*, 1984, 1986). A general dephosphorylation occurs at the onset of anaphase (Vandre and Borisy, 1987, 1989). One 225K thiophosphoprotein has been identified in the centrosome and other MTOCs (Kuriyama, 1989), and the calcium- and calmodulin-dependent phosphorylation of a 62K protein has been found to induce microtubule depolymerization (Dinsmore and Sloboda, 1988). Further, phosphorylation of a 205K protein may be coupled to spindle elongation *in vitro* (Wordman and Cande, 1987). The colocalization of $p34^{cdc2}$ and phosphoproteins in the mitotic spindle is consistent with the idea that the kinase activity of $p34^{cdc2}$ may control phosphorylation events that are essential for spatial and temporal changes in spindle dynamics.

IV. ELECTRON SPECTROSCOPIC IMAGING OF CENTROSOMES

The precise localization of phosphorus-containing components at the centrosome can be determined by employing a relatively new electron microscopic procedure, electron spectroscopic imaging (ESI). In the electron microscope, incident electrons pass through the specimen and cause excitations and ionizations of the specimen's atoms. It is possible to analyze the resulting X-ray spectra or the energy-loss spectra of electrons that have passed through the sample. The energy-loss spectrum reflects the chemical composition of the sample. An electron microscope has been developed in which an electron spectrometer has been installed, so spectra can be obtained and images can be recorded at any particular energy loss. This microscope can produce a two-dimensional distribution map of a particular chemical element characterized by both high spatial resolution and sensitivity. This technique has been applied to the detection of phosphorus in the basic 10-nm chromatin fiber. The resolution of the technique is great enough to allow the detection of the path of the DNA double helix as it passes from nucleosome to nucleosome (Bazett-Jones and Ottensmeyer, 1981). ESI has been applied to the examination of physical specimens in both imaging and diffraction modes (Reimer *et al.*, 1988) and also has been employed to investigate the composition and ultrastructural localization of heavy metal deposits in embedded and sectioned *Escherichia coli* cells (Taylor *et al.*, 1988). Taking advantage of the ability of the microscope to detect phosphorus in both whole-mount preparations and thin sections, this instrument also has been used to study the elemental distribution of phosphorus at MTOCs including the centrosome, kinetochore, and midbody (Rattner and Bazett-Jones, 1988a,b, 1989).

To prepare specimens for ESI analysis, samples are fixed by standard procedures. However, buffers containing phosphorus are eliminated. The material is then embedded by standard protocols and sectioned for electron microscopy. The section thickness is about one-third that of a conventional thin section and does not exceed 30 nm. This thickness is essential for proper imaging and also affords better resolution of structure since there is less overlap of structures within the section. All material is viewed in the absence of stain. Since the phosphorus $L_{ii,iii}$ ionization edge is superimposed on a plasmon-like background, it is necessary to eliminate its contribution by recording two images. These images include a reference image at an energy loss below the ionization edge (110 eV) and a phosphorus-enhanced image near the maximum of the ionization spectral peak (150 eV). Areas containing structures of interest, such as the centrosome, are digitized using a linear high-resolution camera. An image-analysis computer is then used to calculate a normalization factor to compensate for differing exposures calculated over regions of plastic embedding matrix between cells that are known to be devoid of phosphorus-containing material. The image pairs are then aligned to within one pixel and subtracted to produce a net phosphorus-distribution map.

Although the focus of this volume is the centrosome, it may be of interest to briefly review ESI findings at the intercellular bridge and kinetochore in order to understand the distribution of phosphorus at the centrosome in the context of other MTOCs. Unlike the centrosome, the kinetochore has a well-defined ultrastructure. In the Indian muntjac, as in most mammals, the kinetochore appears along the surface of the primary constriction at metaphase as a tripartite structure of alternating electron-dense and electron–translucent layers. ESI of the phosphorus distribution in these layers reveals that both the inner and outer layers are phosphorus rich whereas the middle layer is phosphorus poor. This distribution does not change with kinetochore activity or phase of mitosis. In the inner and outer layers, the phosphorus signal appears to reside predominantly in fibers that have a cross sectional diameter of approximately 30 nm (Rattner and Bazett-Jones, 1989). Fibers in the kinetochore have been found to be DNase susceptible and have a morphology similar to that of bulk chromatin (Rattner, 1986). This has led to the suggestion that the phosphorus-rich fibers contain DNA. It is also possible that the plates are sites of phosphorylated proteins. The cytoplasm adjacent to active and inactive kinetochores appears rich in phosphorus-containing material; however, the signal from this region does not appear to be significantly greater than that from the cytoplasm in general. Interestingly, the fibrous corona associated with the outer surface of the outer plate of microtubule-free kinetochores appears phosphorus poor. Microtubules associated with the metaphase kinetochore also demonstrate a phosphorus signal; this signal persists until late telophase.

The mitotic spindle consists in part of two overlapping half-spindles. The microtubules of each half-spindle are arranged so their plus ends are located at the spindle equator at metaphase. Several lines of evidence suggest that elongation of microtubules at their plus ends, along with antiparallel sliding, contributes to the elongation of the spindle during anaphase B and the separation of the spindle poles (McDonald *et al.*, 1979, Tippet *et al.*, 1984; Saxton and McIntosh, 1987). Ultrastructural studies indicate that the interaction between the two half-spindles first becomes apparent at anaphase, when several bundles of overlapping microtubules, called stem bodies, become visible. The stem bodies are characterized by a matrix of electron-dense material that coats the outer surface of the microtubules of the bundle. As mitosis proceeds, the stem bodies join one another and coalesce to form a single structure. This portion of the spindle comes to lie in a membrane-bound tube, the intercellular bridge that joins the two daughter cells at the completion of mitosis. The region in which microtubules are associated with electron-dense material, that is, the midbody, comes to lie at the equator of the bridge.

ESI of the stem bodies and intercellular bridge have revealed a pronounced phosphorylation–dephosphorylation cycle in this region (Rattner and Bazett-Jones, 1988b). At anaphase, the electron-dense matrix surrounding the stem bodies displays a strong phosphorus signal. The microtubule profiles associated with this material also display a signal. After the microtubule bundles have coalesced and been positioned in the intercellular bridge, the midbody region containing the amorphous material of the anaphase stem body continues to display a strong phosphorus signal. Microtubules in the midbody also show a strong signal. However, microtubules extending from either side of the midbody show a low diffuse signal. Thus, at this stage, the midbody is rich in phosphorus whereas the region on either side of it is comparatively phosphorus poor. ESI of older, more mature intercellular bridges reveals a phosphorus distribution pattern that is opposite that seen at earlier stages. That is, the midbody is devoid of a major phosphorus signal whereas the regions on either side of this area now show a strong signal. Therefore, the midbody and intercellular bridge show a pronounced temporal change in phosphorus distribution. It is interesting to note that some proteins that are positioned in other parts of the cell come to lie at the midbody at the completion of mitosis (Cook *et al.*, 1987; Earnshaw and Bernat, 1991). Dephosphorylation may be associated with the inactivation of these proteins. In this regard, the intercellular bridge may serve as a vehicle for the removal of some mitotic constituents, since it is eliminated from both daughter cells at G_1.

Using Indian muntjac cells, the phosphorus distribution at the centrosome has also been studied during different phases of the cell cycle. At interphase, the centrosome of this species is associated with microtubules, which can be seen inserting into the lumen of the parent centriole, radiating from the centriolar triplets, and radiating from the PCM (Fig. 5A). The matrix associated with the walls of the muntjac centriole is extremely electron dense at interphase. When centrosomes of interphase cells are analyzed for phosphorus, a signal is detected originating from the cytoplasmic microtubules as well as from the triplet microtubules that form the blades of the centrioles. Many inner structures of the centriole also show a positive signal, including the proximal and distal ends of the centriole and the arms that interconnect adjacent blades. The lumen of the centriole and the matrix coating the microtubule blades are devoid of signal whereas the PCM shows a diffuse phosphorus signal (Fig. 5B). In phosphorus enhanced micrographs, the PCM appears to be composed of a fibrogranular matrix network. It should be noted that the distribution of phosphorus does not follow the distribution of electron-dense material. Thus, electron density cannot be equated with the distribution of phosphoproteins or nucleic acids.

Although the strong phosphorus signal originating from the microtubule blades of the centriole and their interconnecting structures persists as the cell enters mitosis, the focus of phosphorus signal in the PCM becomes more concentrated (Fig. 5C). Longitudinal profiles of the centrioles reveal a thin plate of phosphorus signal at both the proximal and distal ends of the lumen (Fig. 5C, short arrow). The frequent appearance of microtubules that insert into the lumen of the centrioles suggests that they are able to penetrate these layers. Microtubule profiles of the mitotic spindle show a phosphorus signal throughout mitosis (Rattner and Bazett-Jones, 1989). As the cell completes mitosis and the nucleus begins to reform, the centrosome region begins to assume its interphase conformation. In early telophase, the PCM shows a weaker signal than in interphase. The short microtubules that are remnants of the mitotic spindle are now devoid of phosphorus signal (Rattner and Bazett-Jones, 1989).

The distribution of phosphorus signal detected by ESI suggests that phosphorus-containing substances (nucleic acids/proteins) are present in the centriole, PCM, and microtubules that extend from the centrosome. These components are, however, largely excluded from the centriole lumen and matrix coating the outer walls of the microtubule blades. The localization of purine nucleoside phosphorylase to the outer and inner surface of the centriole blades (Oliver *et al.*, 1981) has led to the suggestion that this protein may provide a substrate for the replication of the RNA that has been reported in association with the centrioles. Phosphorus signal at the microtubule blades may reflect the presence of this RNA.

One interesting finding of our ESI study is that there is relatively little phosphorus signal in the lumen of the centriole. Recent evidence has suggested that the centriole lumen, at least in *Chlamydomonas* basal bodies, may be the site of a DNA subset. In this case, the DNA appears to be a 6–9 megabase linear molecule

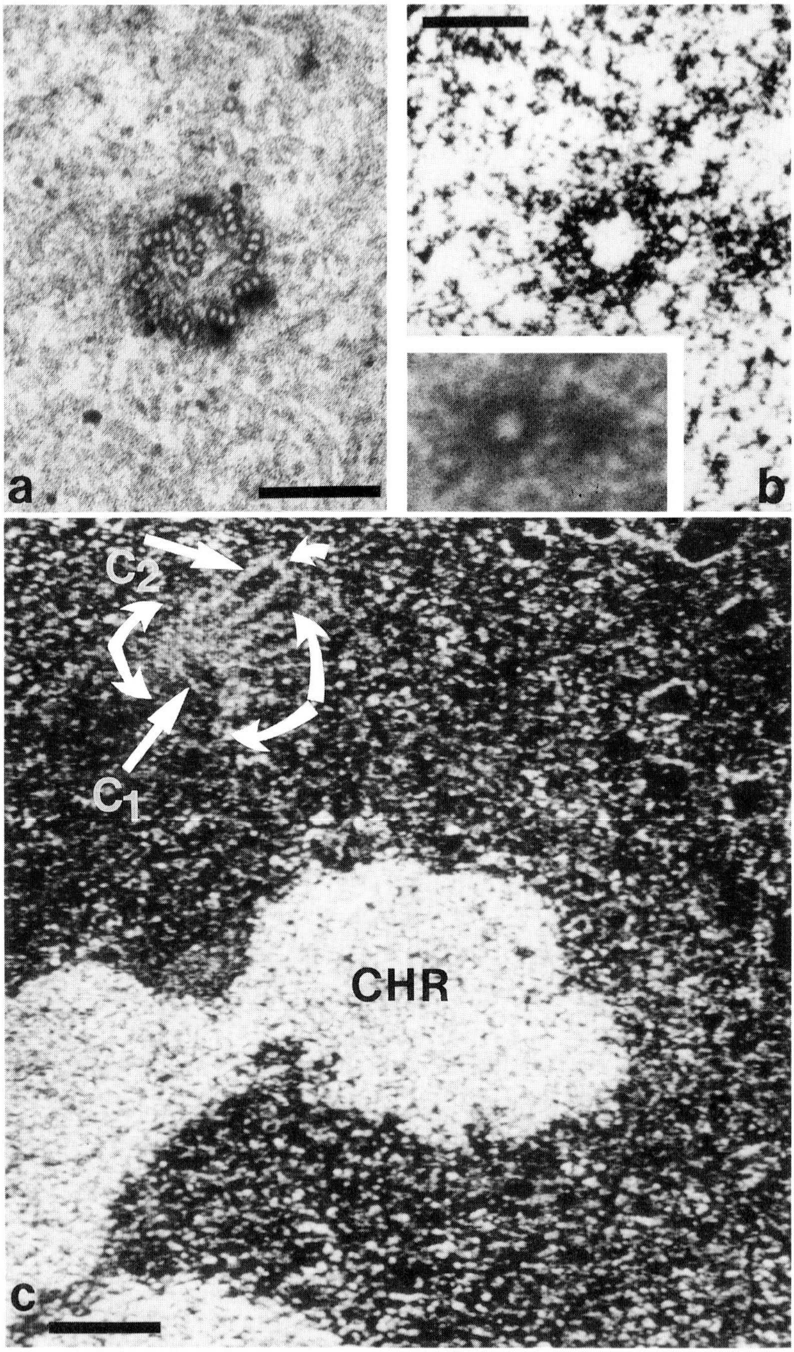
a
b
C2
C1
CHR
c

(Hall *et al.*, 1989). Clearly an aggregate accommodating DNA of this size is not detectable in the centrioles of the Indian muntjac nor is it possible to detect a prominent mass of phosphorus-rich material outside the centriole at any time during the cell cycle. Thus it seems unlikely that a DNA mass of the size found in *Chlamydomonas* is associated with the muntjac centrosome. However, it is possible that the amount of centriole-associated DNA is species specific. In this case, there may be only a small amount present in the muntjac. Whether this small amount of DNA is represented by the phosphorus concentrations already visualized, for example, at the ends of the lumen remains to be investigated.

Initial studies using ESI have demonstrated the usefulness of this technique in establishing elemental distributions in precise regions of the cell. ESI also has the potential to address many questions that remain unanswered about the centrosome. For example, it should be possible to establish the precise location of DNA in cells, particularly in the centrosome, by labeling the DNA with elements such as mercury or bromine that are detectable by ESI. Similarly, it should be possible to map the distribution of specific protein components of the centrosome using antibodies tagged with elements such as boron or gold that are detectable by ESI. Since this technique does not require large tags that affect reactivity and accessibility, it should be possible to obtain high-resolution distribution maps using this approach.

V. SUMMARY

This review suggests several features of the centrosome. (1) Many protein components of the centrosome are also found at other microtubule organizing sites. (2) Microtubule organizing sites including the centrosome may be foci for ubiquitous cytoplasmic proteins. (3) The ultrastructural changes in the size and morphology of the centrosome are correlated with a change in the composition, activity, and/or amount of centrosomal proteins. (4) Proteins with epitopes similar to those in glycolytic enzymes, or the enzymes themselves, are components of

---

Fig. 5. (A) Cross section of an Indian muntjac interphase parent centriole. Microtubules insert through the lumen and are attached to the microtubule blades. Note matrix material associated with the outer surface of the blades. Bar: 0.5 μm. (B) Net phosphorus image of an Indian muntjac interphase centriole, cut slightly oblique. Phosphorus, indicated by black material, is absent from the lumen and matrix but present in the microtubules of the blades of the centrioles and in structures that interconnect the blades. The reference image (see text) is presented in the inset. Bar: 0.5 μm. (C) Net phorphorus image of a mitotic Indian muntjac cell. Phosphorus is indicated by white material. Longitudinal sections of the two centrioles (c_1,c_2) at one spindle pole are illustrated. Note the aggregation of phosphorus-rich PCM in this region (large arrows). Phosphorus signal is also seen at the end of the centriole lumen (small arrow). CHR, chromosome. Bar: 0.25 μm. [Fig. 5A,B reproduced from Rattner and Bazett-Jones, *Journal of Cell Biology* (1988), **91**, 5–11, by copyright permission of the Rockefeller University Press.]

MTOCs, including the centrosome. (5) MTOCs, including the centrosome, contain protein complexes that may act as immunogens in autoimmune disease. (6) p34^{cdc2} is localized to three domains in the mitotic apparatus. This placement suggests that it may play a role in cell division. (7) The distribution of p34^{cdc2} is correlated with the distribution of phosphoproteins and sites of phosphorylation–dephosphorylation. (8) ESI provides a high-resolution map of phosphorus distribution in the centrosome.

ACKNOWLEDGMENTS

The original work presented in this chapter was done in collaboration with David Bazett-Jones, Marvin Fritzler, Liam Martin, Sharon Johnstone, John Lew, David Waisman, and Jerry Wang. I thank Teresa Wang for her excellent technical assistance and Pamela Leco and Alex Wong for help in preparing this chapter. J.B.R. is supported by grants from the National Cancer Institute of Canada and National Science and Engineering Research Council of Canada.

REFERENCES

Alieva, I. B., and Vorobjev, I. A. (1991). Induction of multipolar mitoses in cultured cells: Decay and restructuring of the miotic apparatus and distribution of centrioles. *Chromosoma (Berl)* **100,** 532–542.

Bailly, E., Dorée, M., Nurse, P., and Bornens, M. (1989). p34^{cdc2} is located in both nucleus and cytoplasm: Part is centrosomally associated at G_2/M and enters vesicles at anaphase. *EMBO J.* **8**, 3985–3995.

Baron, A. T., and Salisbury, J. L. (1988). Identification and localization of a novel, cytoskeletal, centrosome-associated protein in PtK_2 cells. *J. Cell Biol.* **107**, 2669–2678.

Bazett-Jones, D. P., and Ottensmeyer, F. P. (1981). Phosphorus distribution in the nucleosome. *Science* **211**, 169–170.

Berns, M. W., Rattner, J. B., and Brenner, S. (1977). The role of the centriolar region in animal cell mitosis: A laser microbeam study. *J. Cell Biol.* **72**, 351–367.

Bré, M. H., Kreis, T. E., and Karsenti, E. (1987). Control of microtubule nucleation and stability in Maden–Darby canine kidney cells: The occurrence of noncentrosomal stable detyrosinated microtubules. *J. Cell Biol.* **105**, 1283–1296.

Brinkley, B. R., and Stubblefield, E. (1970). Ultrastructure and interaction of the kinetochore and the centriole in mitosis and meiosis. *Adv. Cell Biol.* **1**, 119–185.

Brinkley, B. R., Cox, S. M., Pepper, D. A., Wible, L., Brenner, S. L., and Pardue, R. L. (1981). Tubulin assembly sites and the organization of cytoplasmic microtubules in cultured mammalian cells. *J. Cell Biol.* **90**, 554–562.

Brinkley, B. R., Valdovia, M. M., Tousson, A., and Brenner, S. L. (1984). Compound kinetochores of the Indian muntjac. Evolution by linear fusion of unit kinetochore. *Chromosoma (Berlin)* **91**, 1–11.

Clark, F. M., Morton, D. J., Stephan, P., and Wiedemann, J. (1985). The functional duality of glycolytic enzymes: Potential integrators of cytoplasmic structure and function. *In* "Cell Motility. Mechanism and Regulation (10th Yamada Conference)" (H. Ishikawa, S. Hatano, and H. Sato, eds.) pp. 235–250. Tokyo University Press, Tokyo.

Cook, C. A., Heck, M. M. S., and Earnshaw, W. C. (1987). The inner centromere protein (INCENP) antigens. Movement from inner centromere to midbody during mitosis. *J. Cell Biol.* **105**, 2053–2067.

De Brabander, M., Geuens, G., DeMey, F., and Jomoau, M. (1981). Nucleated assembly of mitotic microtubules in living PtK_2 cells after release from nocodazole treatment. *Cell Motil.* **1**, 469–483.

Dedman, J. R., Lin, T., Marcum, J. M., Brinkley, B. R., and Means, A. R. (1980). Calmodulin: Its role in the mitotic apparatus. *In* "Calcium-Binding Proteins: Structure and Function" (F. L. Siegel, E. Carafoli, R. H. Kretsinger, D. H. MacLennan, and R. H. Wasserman, eds.) pp. 181–188. Elsevier-North Holland, Amsterdam.

Dinsmore, J. H., and Sloboda, R. D. (1988). Calcium- and calmodulin-dependent phosphorylation of a 62 kD protein induces microtubule depolymerization in sea urchin mitotic apparatuses. *Cell* **53**, 769–780.

Earnshaw, W. C., and Bernat, R. L. (1991). Chromosomal passenger proteins towards an integrated view of mitosis. *Chromosoma (Berl)* **100,** 139–146.

Euteneuer, U., and Schliwa, M. (1985). Evidence for an involvement of actin in the positioning and motility of centrosomes. *J. Cell Biol.* **101**, 96–103.

Goodenough, U. W. (1989). Basal body chromosomes? *Cell* **59**, 1–3.

Gosti, F., Marty, M-C., Courvalin, J. C., Maunoury, R., and Bornens, M. (1987). Centrosomal proteins and lactate dehydrogenase possess a common epitope in human cell lines. *Proc. Natl. Acad. Sci. U.S.A.* **84**, 1000–1004.

Gould, R. R., and Borisy, G. G. (1977). The pericentriolar material in Chinese hamster ovary cells nucleate microtubule formation. *J. Cell Biol.* **90**, 554–562.

Hall, J. L., Ramanis, Z., and Luck, D. J. L. (1989). Basal body/centriolar DNA: Molecular genetic studies in *Chlamydomonas*. *Cell* **59**, 121–132.

Hepler, P. K. (1989). Membranes in the mitotic apparatus. *In* "Mitosis, Molecules and Mechanisms" (J. S. Hyams and B. R. Brinkley, eds.) pp. 241–271. Academic Press, San Diego.

Huitorel, P., and Pantaloni, D. (1985). Bundling of microtubules by glyceraldehyde-3-phosphate dehydrogenase and its modulation by ATP. *J. Biol. Chem.* **150**, 265–269.

Karsenti, E., and Maro, B. (1986) Centrosomes and the spatial distribution of microtubules in animal cells. *TIBS* **11**, 460–463.

Knull, H. R. (1980). Compartmentalization of glycolytic enzymes in nerve endings as determined by glutaraldehyde fixation. *J. Biol. Chem.* **255**, 6439–6444.

Kuriyama, R. (1989). 225-Kilodalton phosphoprotein associated with mitotic centrosomes in sea urchin eggs. *Cell Motil. Cytosk.* **12**, 90–103.

Kuriyama, R., and Borisy, G. G. (1981). Microtubule-nucleating activity of centrosomes in Chinese hamster ovary cells is independent of the centriole cycle but coupled to the mitotic cycle. *J. Cell Biol.* **91**, 822–826.

Lamb, N. J. C., Fernandez, A., Watrin, A. Labbé, J., and Cavadore, J. (1990). Microinjection of $p34^{cdc2}$kinase induces marked changes in cell shape, cytoskeletal organization, and chromatin structure in mammalian fibroblasts. *Cell* **60**, 151–165.

McDonald, K. L., Edwards, M. K., and McIntosh, J. R. (1979). Cross-sectional structure of the central mitotic spindle in *Diatoma vulgare*: Evidence for the specific interaction between antiparallel microtubules. *J. Cell Biol.* **83**, 443–461.

McIntosh, J. R. (1989). Mitosis in perspective. *In* "Mitosis, Molecules and Mechanisms" (J. S. Hyams and B. R. Brinkley, eds.), pp. 327–340. Academic Press, San Diego.

McNiven, M. A., and Porter, K. R. (1988). Organization of microtubules in centrosome-free cytoplasm. *J. Cell Biol.* **106**, 1593–1605.

Maller, J. L. (1985). Regulation of amphibian oocyte maturation. *Cell Differ.* **16**, 211–221.

Mazia, D. (1984). Centrosomes and mitotic poles. *Exp. Cell Res.* **153**, 1–15.

Moroi, Y., Pickles, C., Fritzler, M. J., Stergerwald, J., and Tan, E. M. (1980). Autoantibody to centromere (kinetochore) in scleroderma sera. *Proc. Natl. Acad. Sci. U.S.A.* **77**, 1627–1631.

Moroi, Y., Murata, I., Sakeuchi, A., Kamatani, N., Tanimoto, K., and Yokohari, R. (1983). Human anticentriole autoantibodies in patients with scleroderma and Raynauds' phenomena. *Clin. Immunol. Immunopathol.* **29**, 381–392.

Neighbors, B. W., Williams, R. C., Jr., and McIntosh, R. J. (1988). Localization of kinesin in cultured cells. *J. Cell Biol.* **106**, 1193–1204.

Nigg, E. A., Schafer, G., Hilz, H., and Eppenberger, H. M. (1985). Cyclic-AMP-dependent protein kinase type II is associated with the Golgi complex and with centrosomes. *Cell* **41**, 1039–1051.

Oliver, J. B., Osborne, W. R. A., Pfeiffer, J. R., Child, F. M., and Berlin, R. D. (1981). Purine nucleoside phosphorylase is associated with centrioles and basal bodies. *J. Cell Biol.* **91**, 837–847.

Osborn, T. G., Ryerse, J. S., Bauer, N. E., Urhahn, J. M., Blair, D., and Moore, T. L. (1986). Anticentriole antibody in a patient with progressive systemic sclerosis. *Arthritis Rheum.* **29**, 142–146.

Paweletz, N., and Mazia, D. (1989). Fine structure of the mitotic cycle of unfertilized sea urchin eggs activated by ammoniacal sea water. *Eur. J. Cell Biol.* **20**, 37–44.

Pepper, D. A., and Brinkley, B. R. (1976). Localization of tubulin in the mitotic apparatus of mammalian cells by immunofluorescence and immunoelectron microscopy. *Chromosoma (Berlin)* **60**, 223–235.

Pepper, D. A., and Brinkley, B. R. (1980). Tubulin nucleation and assembly in mitotic cells: Evidence for nucleic acids in kinetochores and centrosomes. *Cell Motil. Cytosk.* **1**, 1–15.

Peterson, S. P., and Berns, M. W. (1980). The centriolar complex. *Int. Rev. Cytol.* **64**, 81–106.

Rattner, J. B. (1986). Organization within the mammalian kinetochore. *Chromosoma (Berlin)* **93**, 515–520.

Rattner, J. B. (1991). HSP-70 is localized to the centrosome of dividing HeLa cells. *Exptl. Cell Res.* **195,** 110–113.

Rattner, J. B., and Phillips, S. G. (1973). Independence of centriole formation and DNA synthesis. *J. Cell Biol.* **57**, 359–372.

Rattner, J. B., and Berns, M. W. (1976). Distribution of microtubules during centriole separation in rat kangaroo cells. *Cytobiol.* **15**, 37–43.

Rattner, J. B., and Bazett-Jones, D. P. (1988a). Electron spectroscopic imaging of the centrosome in cells of the Indian muntjac. *J. Cell Sci.* **91**, 5–11.

Rattner, J. B., and Bazett-Jones, D. P. (1988b). Electron spectroscopic imaging of the intercellular bridge. *Cell Motil. Cytosk.* **11**, 106–116.

Rattner, J. B., and Bazett-Jones, D. P. (1989). Kinetochore structure: Electron spectroscopic imaging of the kinetochore. *J. Cell Biol.* **108**, 1209–1219.

Rattner, J. B., Lew, J., and Wang, J. H. (1990). $p34^{cdc2}$ kinase is localized to four domains within the mitotic apparatus. *Cell Motil. Cytosk.* **17**, 227–236.

Rattner, J. B., Martin, L., Waisman, D. M., Johnstone, S. A., and Fritzler, W. J. (1991). Autoantibodies to the centrosome (centriole) react with determinants present in the glycolytic enzyme enolase. *J. Immunol.* **146,** 2341–2344.

Reider, C. L. (1979). Ribonucleoprotein staining of centrioles and kinetochores in newt lung cell spindles. *J. Cell Biol.* **80**, 1–9.

Reider, C. L., and Borisy, G. G. (1982). The centrosome in PtK_2 cells: Asymmetric distribution and structural changes in the pericentriolar material. *Biol. Cell* **44**, 117–132.

Reimer, L., Fromm, I., and Rennekamp, R. (1988). Operation modes of electron spectroscopic imaging and energy loss spectroscopy in a transmission electron microscope. *Ultramicroscopy* **24**, 339–354.

Riabowal, K., Draetta, G., Bizuela, L., Vandre, D., and Beach, D. (1989). The cdc2 kinase is a nuclear protein that is essential for mitosis in mammalian cells. *Cell* **57**, 393–401.

Rider, C. C., and Taylor, C. B. (1974). Enolase isozymes in rat tissue's electrophoretic, chromatographic, immunological, and kinetic properties. *Biochim. Biophys. Acta.* **365**, 285–300.

Robbins, E., Jentzsch, G., and Micali, A. (1968). The centriole cycle in synchronized HeLa cells. *J. Cell Biol.* **36**, 329–339.

Sager, P. R., Rothfield, N. I., Oliver, J. M., and Berlin, R. D. (1986). A novel miotic spindle pole component that ordinates from the cytoplasm during prophase. *J. Cell Biol.* **103,** 1863–1872.

Saxton, M. W., and McIntosh, J. R. (1987). Interzone microtubule behaviour in late anaphase and telophase spindles. *J. Cell Biol.* **105**, 875–886.

Schatten, H., Walter, M., Mazia, D., Biessmann, H., Paweletz, N., Coffe, G., and Schatten, G. (1987). Centrosome detection in sea urchin eggs with a monoclonal antibody against *Drosophila* intermediate filament proteins: Characterization of stages of division cycle of centrosomes. *Proc. Natl. Acad. Sci.* **84,** 8488–8492.

Sherline, P., and Mascaro, R. N. (1982). Epidermal growth factor induces rapid centrosomal separation in HeLa and 3T3 cells. *J. Cell Biol.* **93**, 507–511.

Shi, L., and Ye, Y. D. (1980). Comparative cytogenic studies on the red muntjac, Chinese muntjac, and their F_1 hybrids. *Cytogenetics Cell Genetics* **26**, 22–27.

Stearns, T., Evans, L., and Kirschner, M. (1991). y Tubulin is a highly conserved component of the centrosome. *Cell* **65,** 825–836.

Steffen, W., and Linck, R. W. (1988). Evidence for tektins in centrioles and axonemal microtubules. *Proc. Natl. Acad. Sci. U.S.A.* **85**, 2643–2647.

Taylor, D. E., Walter, E. G., Sherburne, R., and Bazett-Jones, D. P. (1988). Structure and localization of tellurium deposited in *Escherichia coli* cells harboring tellurite resistance plasmids. *J. Ultrastruc. Mol. Struc. Res.* **99**, 18–26.

Tilney, L. G., and Goddard, J. (1970). Nucleating sites for the assembly of cytoplasmic microtubules in the ectoderm cells of blastula of *Arabcie punctulata. J. Cell Biol.* **46**, 564–575.

Tippet, D. H., Field, C. T., O'Donald, K. I., Pickett-Heaps, J. D., and McLaughin, D. J. (1984). The organization of microtubules during anaphase and telophase spindle elongation in the rust fungus *Puccinia. Eur. J. Cell Biol.* **30**, 9–17.

Tousson, A., Zeng, C., Brinkley, B. R., and Valdivia, M. M. (1991). Centrophilin. A novel Miotic Spindle Protein Involved in Microtubule nucleation. *J. Cell Biol.* **112:** 427–440.

Tuffanelli, D. L., McKeon, F., Kleinsmith, D. M., Burnham, T. K., and Kirschner, M. (1983). Anticentromere and anticentriole antibodies in the scleroderma spectrum. *Arch. Dermatol.* **19**, 560–566.

Turksen, K., and Kalnins, V. I. (1982). Identification of a centriole-associated protein by antibodies present in normal rabbit sera. *Nature (London)* **298**, 763–765.

Vandre, D. D., Davis, F. M., Rao, P. N., and Borisy, G. G. (1984). Phosphoproteins are components of mitotic microtubule organizing centers. *Proc. Natl. Acad. Sci. U.S.A.* **81**, 4431–4443.

Vandre, D. D., Davis, F. M., Rao, P. N., and Borisy, G. G. (1986). Distribution of cytoskeletal proteins sharing a conserved phosphorylated epitope. *Eur. J. Cell Biol.* **41**, 72–81.

Vandre, D. D., and Borisy, G. G. (1987). Anaphase onset and dephosphorylation of mitotic phosphoproteins occurs consistently. *J. Cell Biol.* **105**, 175a

Vandre, D. D., and Borisy, G. G. (1989). The centrosome cycle in animal cells. *In* "Mitosis, Molecules, and Mechanisms" (J. S. Hyams and B. R. Brinkley, eds.), pp. 39–75. Academic Press, San Diego.

Verde, F., Labbé, J. C., Dorée, M., and Karsenti, E. (1989). Regulation of microtubule dynamics by cdc2 protein kinase in cell-free extracts of *Xenopus* eggs. *Nature (London)* **343**, 233–238.

Vinores, S. A., Bonnin, J. M., Rubinstein, L. J., and Marangos, P. J. (1984). Electron microscopic localization of neuron-specific enolase in rat and mouse brain. *Acta Pathol. Lab. Med.* **108**, 536–540.

Vinores, S. A., Herman, M. M., and Rubinstein, L. J. (1986). Electron-immunocytochemical localization of neuron-specific enolase in cytoplasm and on membranes of primary and metastatic cerebral tumours and on glial filaments of glioma cells. *Histopathology* **10**, 891–908.

Vorobjev, I. A., and Chentsov, Y. S. (1982). Centrioles in the cell cycle I epithelial cells. *J. Cell Biol.* **93**, 938–949.

Vorobjev, I. A., and Nadezhdina, E. S. (1987). The centrosome and its role in the organization of microtubules. *Int. Rev. Cytol.* **106**, 227–293.

Willingham, M. C., and Pastan, I. (1985). Ultrastructural immunocytochemical localization of transferin receptors using monoclonal antibodies in human KB cells. *J. Histochem. Cytochem.* **33**, 59–64.

Wordman, L., and Cande, W. Z. (1987). Reactivation of spindle elongation *in vitro* is correlated with phosphorylation of a 250 kD spindle associated protein. *Cell* **50**, 535–543.

Zeligs, J. D., and Wollman, S. H. (1979). Mitosis in rat thyroid epithelial cells *in vivo*. II. Centrioles and pericentriolar material. *J. Ultrastruc. Res.* **66**, 97–108.

3

Function and Formation of Centrioles and Basal Bodies[1]

GUENTER ALBRECHT-BUEHLER

Department of Cellular, Molecular, and Structural Biology
Northwestern University Medical School
Chicago, Illinois 60611

[1]This chapter is based on and uses material from "The iris diaphragm model of centriole and basal body formation" by Geunter Albrecht-Beuhler in *Cell Motility and the Cytoskeleton*, volume 17, pp. 197–215. Copyright © 1991 by Wiley-Liss, a division of John Wiley & Sons, Inc.

Copyright © 1992 Academic Press, Inc.
All rights of reproduction in any form reserved.

I. INTRODUCTION

A. Structure

More than 30 years have passed since it was recognized that centrioles and basal bodies share the same spectacular ultrastructural organization (Burgos and Fawcett, 1955; Porter, 1956). Soon it also was discovered that they belong to nature's most conserved cellular structures. In different organisms and cell types, they may differ regarding the presence and morphology of rootlets, alar sheets, basal feet, basal plates, or cartwheels, but with very few exceptions they all share the "skeleton" structure of microtubules that is shown schematically in Fig. 1. Exceptional centriole structures are found only in a number of sperm cells (Phillips, 1966; Bacetti *et al.*, 1973; Phillips, 1974; Shay, 1974; Goldstein, 1977; Prensier *et al.*, 1980).

The majority of cases that were studied over the past 30 years by electron microscopy (e.g., Brinkley and Stubblefield, 1970) show centrioles or basal bodies as cylindrical structures 0.2–0.3 μm in diameter and 0.4–0.7 μm in length, that taper toward one end at an angle of 3°–7° (see Fig. 1). The narrower end is called distal, the wider end, proximal. The mantle of the cylinder is divided laterally by nine identical and equally spaced microtubular arrays, called blades. Each such array consists of three microtubules that form an angle (pitch) of 10°–15° with respect to the centriole axis (Anderson, 1972). The cross section of each blade shows the three microtubules lined up side by side and forming an arc that points its convex face away from the centriole axis. Each blade contains one complete microtubule (The A tubule with a circular cross section) and two incomplete microtubules (B and C tubules with c-shaped cross sections). The nine arc-shaped triplet blades slant outward from the perimeter of the centriole in a clockwise rotation if viewed from the distal end. In cross section, the angle between the blades and the perimeter of the centriole decreases toward the distal end in mature centrioles (cf. Fig. 2c,d). For the sake of simplicity, the exact cross section and the

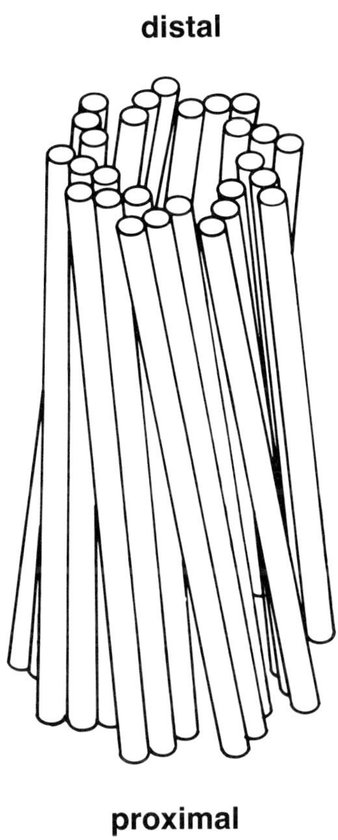

Fig. 1. Simplified schematic of the microtubular skeleton of centrioles and basal bodies. Note that, with the exception of Figs. 2c,e and 16k, all cross-sectional views of centrioles illustrated in this chapter are shown from the distal to the proximal end.

changing slant of the triplet blades were omitted from Fig. 1, but will be used in subsequent text to support the model presented here.

Condensing this description into a list of key words, one may say that the microtubular skeleton of centrioles and basal bodies is characterized by the following features: (1) cylindrical structure, (2) taper of the cylinder, (3) nine-fold rotational symmetry, (4) composition of the blades from one complete and two incomplete microtubules, (5) convex shape of the blades, (6) pitch of the blades, (7) clockwise rotational slant of the blades, and (8) the increasingly tangential slant of the blades toward the distal end.

B. Function

Considering that cells that express centrioles and basal bodies appear to use identical geometric designs for these complex and enigmatic organelles regardless of the taxonomic or evolutionary placement of the organism, one may hypothesize that centriole and basal body geometry is not the result of an evolutionary accident, but a necessary consequence of their function(s). In view of their complexity, the geometric properties of centrioles and basal bodies may thus allow us to guess the particular function that is related to their geometry with little ambiguity. As reviewed in an earlier article (Albrecht-Buehler, 1981), if one keeps in mind the physical conditions of the microscopic world of cells and tries to design the best possible cellular device to detect and locate ambient signal sources around the cells, then the geometric features of centrioles and basal bodies arise immediately as necessary consequences of this function. In this chapter, we will repeat these arguments in a condensed form and extend them to suggest that the physical nature of the detected signals is infrared radiation at a wavelength of about 800 nm.

C. Formation

Not only the function, but the formation and replication of centrioles and basal bodies are still a mystery. The literature, however, provides many consistent observations, such as the following, which may serve as clues:

1. Fibrous coils of about 1200-Å diameter have frequently been seen to precede centriole and basal body formation (Kalnins and Porter, 1969; Manton *et al.*, 1970; Perkins, 1970; Anderson and Brenner, 1971; Picket-Heaps, 1973).
2. The inner and outer diameters grow during maturation of centrioles and basal bodies (cf. Fig. 2a–d).
3. The initially radial slant of the blades turns more tangential as centrioles and

Fig. 2. The formation of centrioles of 3T3 cells in cross section. Each column of panels represents a series of sections through a centriole with the distal end in the top panel. (a) Procentriole. (*Top*) Fibrous coil surrounding the beginning blade structures. Bar: 0.2 μm. (*Middle*) The arrow points to one of the blades whose C tubule is still missing. (*Bottom*) The arrow points to one of the vesicles surrounding the forming structure. (b–d) Centrioles in increasingly mature stages. (b) The arrows point to the central hub of the proximally located cartwheel, which is disassembled in mature centrioles (cf. column d). (c,d) The arrows point to dense plaques that are attached to some but not all blades. (e) Final stage of many centrioles is the formation of a primary cilium, characterized by the absence of a central microtubular doublet (*top*).

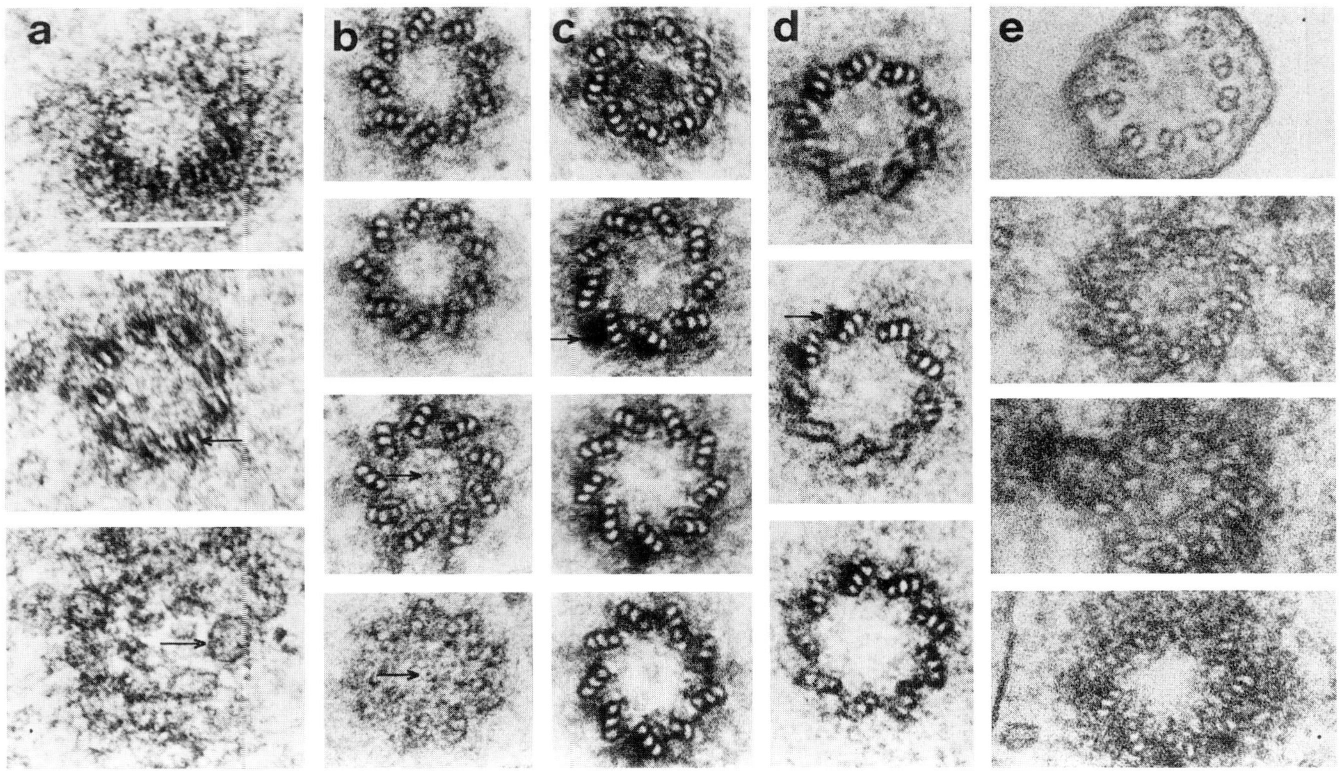

a
b
c
d
e

basal bodies mature (Dippell, 1968; Turner, 1968; Brinkley and Stubblefield, 1970; Moser and Kreitner, 1970; Grimes, 1973; Tamm and Tamm, 1980; Alvey, 1985; cf. Fig. 2b,d).

4. The slant of the blades is more tangential at the distal than at the proximal end. In many cases the mature centriole becomes the basal body of a primary cilium (Sorokin, 1962; Wheatley, 1966; Albrecht-Buehler and Bushnell, 1980; see Fig. 2e). Consequently, the slant of the blades at the distal end of the centriole must be almost tangential (see Fig. 2e, second panel from top) in order to accommodate the tangential orientation of the ciliary doublets (Fig. 2e, upper panel).
5. Connecting fibers appear to span the distance between adjacent blades in centrioles and basal bodies (Gibbons and Grimstone, 1960; Dippell, 1968; Turner, 1968; Allen, 1969; Anderson and Brenner, 1971; cf. Fig. 2d).
6. The A tubule of the forming blades is the first of the three blade microtubules to grow in length, followed by the B and C tubule (Dippell, 1968; Kalnins and Porter, 1969; Anderson and Brenner, 1971; Grimes, 1973).

Some well-established observations about centriole and basal body formation may be ignored by the model, however, because they are not generally true. For example, it is well demonstrated that centrioles and basal bodies usually arise from existing centrioles or basal bodies. However, they can also form *de novo* (Turner, 1968; Manton *et al.*, 1970; Moser and Kreitner, 1970; Perkins, 1970; Pickett-Heaps, 1973).

Another such well-established observation is the usual location of the newly forming so-called procentrioles or probasal bodies (Gall, 1961) facing the wall of the parent centriole or parent basal body. Nevertheless, there are exceptions to this rule as well. In several cases, the new centrioles or basal bodies do not form at right angles but rather end-to-end to an existing one (Berlin and Bowen, 1964; Moser and Kreitner, 1970; Perkins, 1970) or in a “cloud” of up to 10 precursors during ciliogenesis (Kalnins and Porter, 1969; Anderson and Brenner, 1971; Komatsu and Fujita, 1978).

Therefore, we may regard the usual method of centriole and basal body replication orthogonally to the axis of an existing parent as incidental to the mechanism of replication, although we consider it vital for their function (Albrecht-Buehler and Bushnell, 1979; Albrecht-Buehler, 1981).

The features of the structure and formation of centrioles and basal bodies listed previously seem sufficiently informative to allow one to reduce the observations to a very simply geometric principle that may explain the cellular method of centriole and basal body formation. This chapter suggests such a principle.

II. SPECULATION ABOUT THE FUNCTION OF CENTRIOLES AND BASAL BODIES

Assume that animal cells during their morphogenetic movements require an instrument to locate distant signal sources in order to orient the direction of their migration. Assume further that nature would build the best possible design for such an instrument. What would it look like? Based on the following arguments we submit that it would look precisely like a pair of centrioles.

A. Definition of a Cellular Eye

An instrument to locate the direction of distant signal sources may be called an eye. As illustrated in Fig. 3, an eye mediates the one-to-one mapping of different directions of an object space (objects outside the eye) to certain points of an image space, which in the case of the human eye is the retina. For example, the two different directions toward the object points O and O′ are mapped onto the two

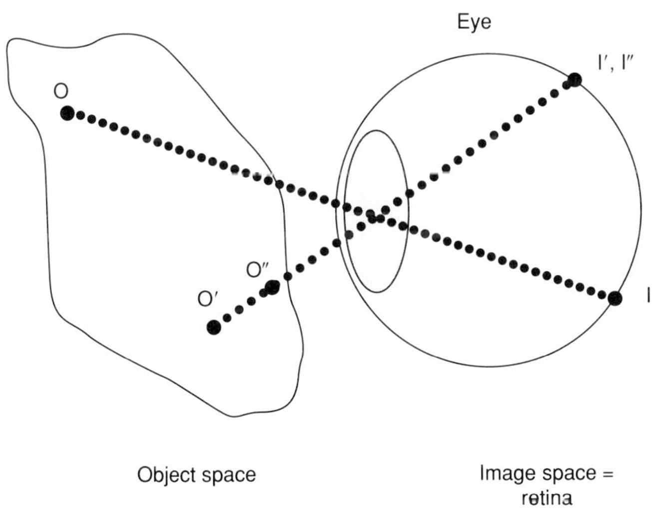

Fig. 3. Schematic illustration of the function of a generalized eye to map different directions of the object space (extraocular space) onto different image points of the image space (retina) in a one-to-one fashion. The different directions are represented by dotted lines connecting object points O, O′, O″ with the center of the lens (shaded area). If object points lie in the same direction (e.g., O′ and O″), they are mapped to the same image point, I′ = I″.

different image points I and I′. If different object points lie in the same direction, such as O′ and O″, then they are mapped onto the same image point (I′ = I″), albeit with some modification. In the case of the human eye, the image points I′ and I″may be blurred to different degrees due to differences in focus.

B. Impossibility of Using Cellular Optical Lenses

Could cells have eyes with lenses like those of large organisms? As shown in Fig. 4, the use of lenses in cellular dimensions is entirely impossible. Given a normal cell size of 10 μm, the cellular lens can be expected to be considerably smaller (e.g., D = 1 μm). Likewise, its focal length must remain inside the cell body (f ≈ 10 μm). The wavelength of the signals focused by this lens is denoted as λ. Consequently, the diameter of the image point (Airy disk) behind the lens is approximately $d = \lambda f/D$. In order to provide a useful optical image, each image point must be much smaller than the lens diameter (e.g., $d = D/1000$). As a result, $\lambda = D^2/1000f \approx 10^{-10}$ m = 1Å. In other words, cellular lenses would require X-rays as the form of radiation that they focus. The essential result remains the same if we change the numbers used in this estimate, as long as they remain compatible with cellular dimensions. Obviously, no materials for lens construction are available with a refractive index large enough for X-rays; therefore, cellular lenses are not possible.

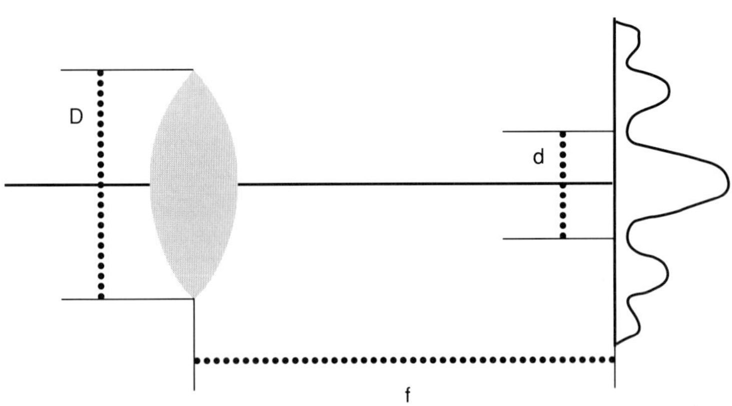

Fig. 4. Diagram relating the diameter D and focal length f of a lens (shaded area) and the wavelength L of a light wave with the diameter d of the Airy disk of the image point. The profile on the right indicates the intensity distribution of the focused light.

C. Impossibility of Using Signal Intensities for Mapping

If radiation cannot be focused by a cellular eye, perhaps an instrument could be designed that compares the various intensities of the signals from different sources in space, similar to bacterial chemotaxis. Unfortunately, the large thermal noise in cellular dimensions prohibits this imaging method also. Comparing intensities at different points is a feasible method to navigate toward a signal source, but the large thermal noise in cellular dimensions would make it impossible to generate a one-to-one map of the locations of multiple sources in a reasonable amount of time required by the function of an eye.

D. Principle of Design of a Cellular Eye

The arguments just given do not mean that cellular eyes are impossible, but they must use principles other than geometric optics or intensity comparisons. In order to identify the geometric characteristics of a cellular eye, we will begin with principles and postulate that nature used the optimal design.

As mentioned earlier, an eye may be defined as an instrument that maps, in a one-to-one fashion, different directions toward an object space onto different points of an image space (see Fig. 5). If the different points of the object space (extra ocular space) are sources of a certain radiation, then the task of providing a one-to-one map of their direction can be formulated as the task of locating the direction of the sources.

The general situation of an eye locating a source is illustrated in Fig. 6. The

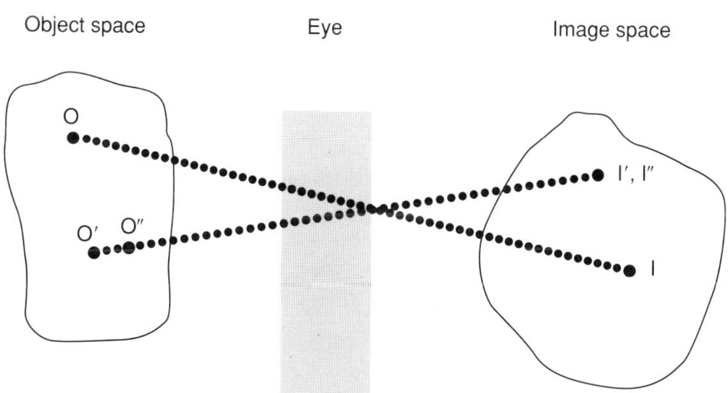

Fig. 5. Generalized definition of an eye to achieve a one-to-one mapping of different directions in an object space onto different points in an image space.

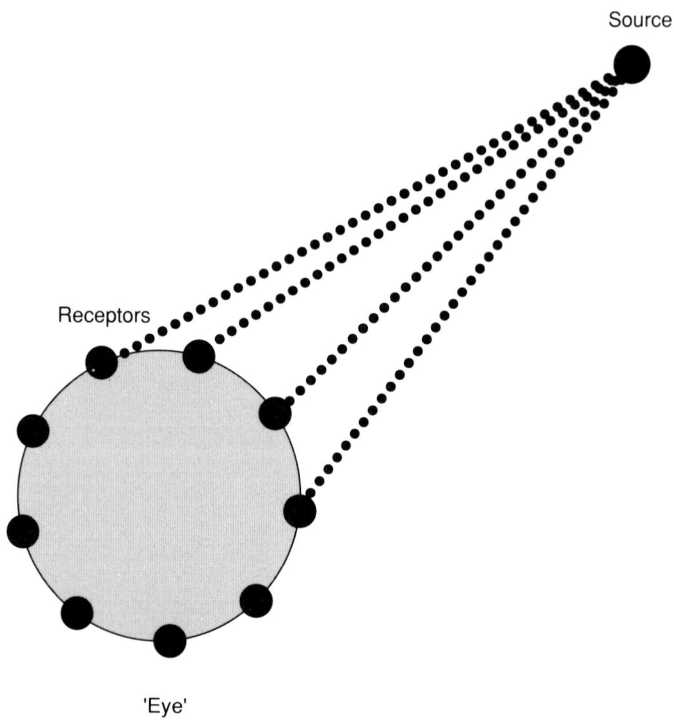

Fig. 6. The most general design of a cellular eye to locate signal sources in its environment. The dotted lines indicate some of the directions of propagation of the signal toward signal receptors around the periphery of the eye. Even with an opaque interior (shaded area), the design does not yield a one-to-one mapping of source directions (object space) onto signal receptors (image space).

object space is considered as a number of signal sources outside the cellular eye. Only one of them is depicted in Fig. 6. The ideal eye has to be rotationally symmetrical or it would be biased and prefer the detection of certain directions over others. It must contain receptors for the type of radiation that the sources emit (black dots in Fig. 6); these receptors must be spaced evenly to avoid directional bias. The receptors provide the points of the image space onto which the different directions of the signal sources are to be mapped. The design of Fig. 6. however, does not provide a one-to-one mapping because more than one receptor is exposed to the signals of the source. If the inside of the eye were transparent for the signal, the situation would be worse because every receptor would receive the signal. In order to minimize the ambiguity, we take the first step toward the final design of the cellular eye by filling the inside of the cellular eye with a material that is designed to absorb the signal (indicated by the gray area in Fig. 6).

E. Use of Blinds

Because optical methods fail to provide the one-to-one mapping between directions of signal sources and the set of receptors, the ambiguity of the signal reception must be reduced by other means. Perhaps the simplest and most effective block of the signals would be by blinds that are attached to one side of each receptor as indicated in Fig. 7. The blinds are supposed to absorb the signal by their thickness. Obviously, the use of the blinds reduces the ambiguity, although a one-to-one mapping is still not achieved. However, if we close the blinds in the manner shown in Fig. 8, only one receptor is exposed to the source, thus yielding the desired one-to-one mapping function.

Optimally, the angle of the blinds is such that they extrapolate to the next receptor in the counterclockwise direction (see Fig. 8). If the angle is more radial,

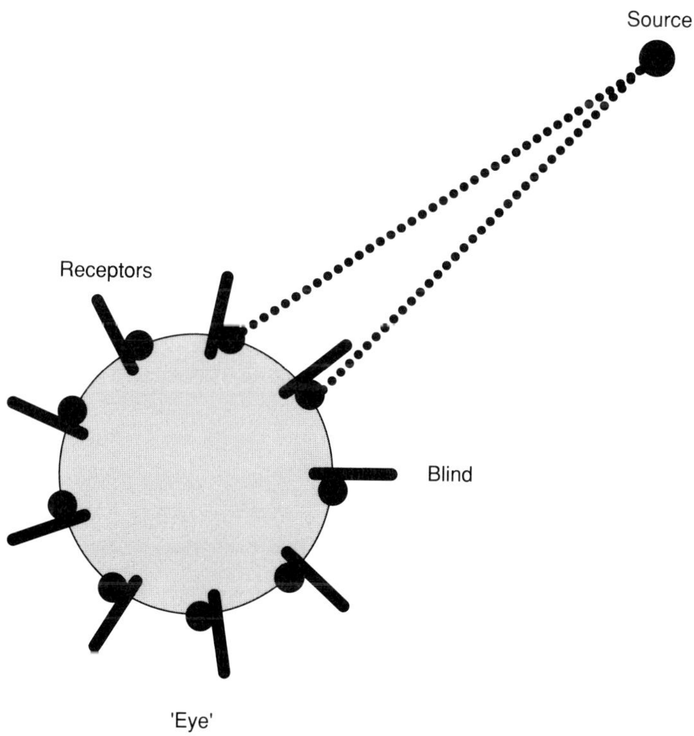

Fig. 7. Reduction of the ambiguity of the mapping function of the design shown in Fig. 6 by attaching blinds to the side of each receptor. The blinds are supposed to absorb the signal in their thickness. The improved design reduces the ambiguity, but does not eliminate it, because more than one receptor is exposed to the signal.

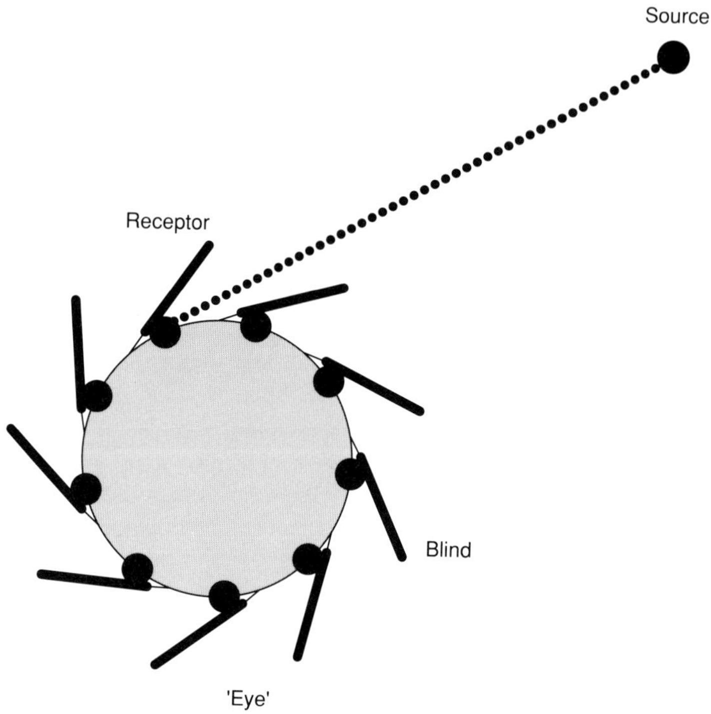

Fig. 8. Final design of the geometry of a two-dimensional cellular eye. After closing the blinds in the depicted way, one and only one receptor is exposed to the signal source, insuring a one-to-one mapping.

the ambiguity of Fig. 7 reappears. If it is more tangential, the design generates blind spots where signal sources cannot reach any of the receptors.

F. Three-Dimensional Shape of the Eye

The design as developed thus far works as long as the signal source remains within the plane of the ring of receptors. However, we must expect that the signal sources around the real cell are also located above and below this plane, as shown in Fig. 9. In this case, the full extent of the ambiguity of signal reception would reappear, since all receptors are exposed to the signal. Therefore, the next step in the construction of a cellular eye is the extension of the planar design of Fig. 8 into a three-dimensional one. Considering that the optimal design must remain rotationally symmetrical, there are only two possibilities for its spatial structure: a

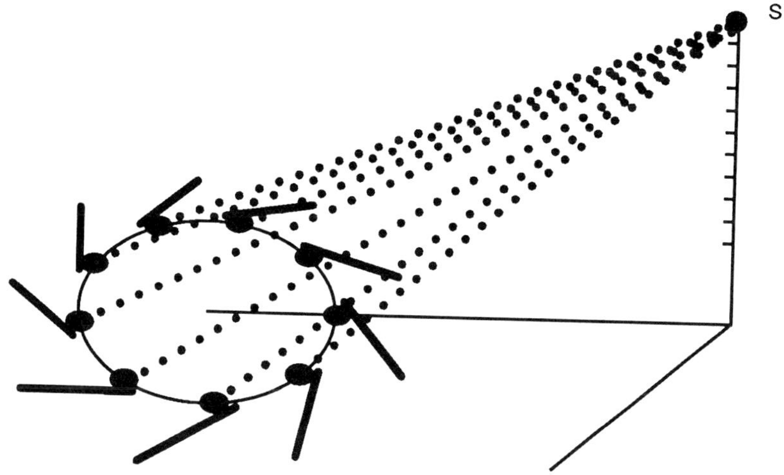

Fig. 9. Demonstration that the design of Fig. 8 is ineffective if the signal source moves out of the plane of the two-dimensional cellular eye. From the position of the source above the plane, all receptors are exposed to its signal, excluding a one-to-one mapping.

cylinder and a sphere. In the case of a sphere, it is apparent that it would not allow the attachment of blinds in a slanting fashion all over its surface. At least at the poles of the sphere the slanting directions would contradict each other. (This fact is known to mathematicians as the Theorem of the Well-Combed Hedgehog.) Therefore, the sphere has to be capped as shown in Fig. 10. In other words, the three-dimensional structure of the cellular eye has to be a straight or bulging cylinder, but a cylinder.

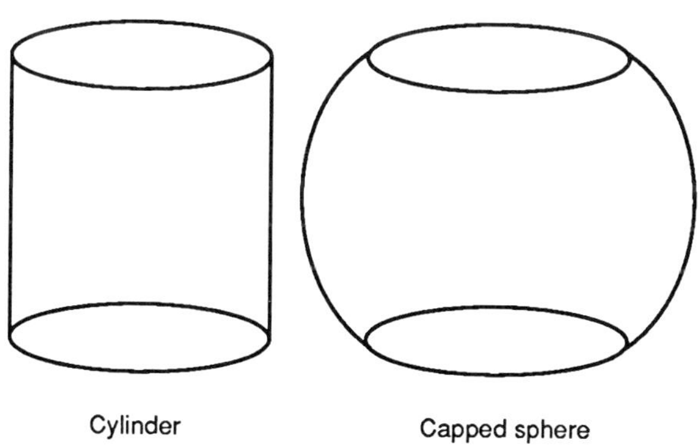

Fig. 10. The only two possibilities for extending the two-dimensional cellular eye into the third dimension are a straight cylinder or a capped sphere (a bulging cylinder).

G. Requirement of Two Orthogonal Cylinders

The design as developed so far permits a one-to-one mapping between the set of signal receptors and the angle of the signal source as measured around the axis of the cylinder. Calling this angle the longitude of the signal source, it is obvious that we need to map its latitude as well if we wish to identify the direction of the signal source. Since the longitude and latitude of a point are angles measured around axes orthogonal to each other, the design requires a second cylinder at right angles to the first, as shown in Fig. 11.

Therefore, we conclude that the cellular eye would be a pair of cylinders at right angles to each other, each having the cross section shown in Fig. 8. It hardly needs mentioning that this design bears striking resemblance to a pair of centrioles. If the arguments as presented are correct, then the structure of centrioles and basal bodies is the expression of a physical necessity and not the result of an accident

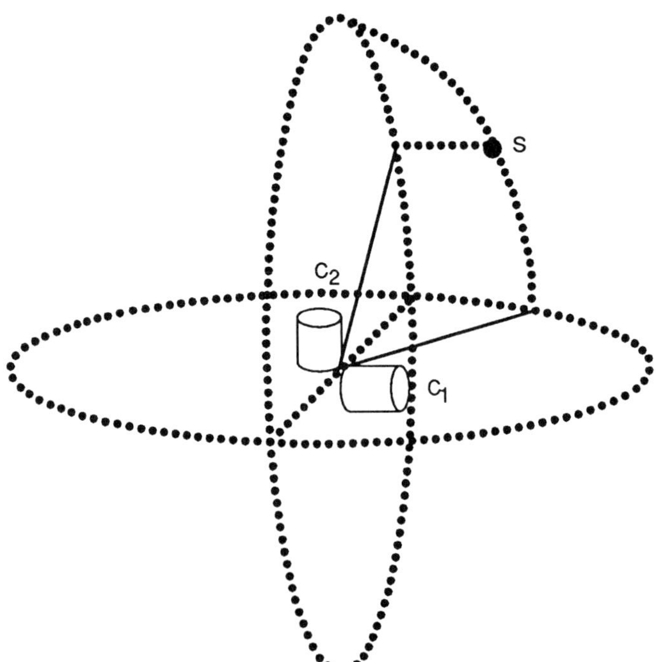

Fig. 11. Necessary use of a pair of orthogonally oriented cylinders as a cellular eye to determine simultaneously the longtitude and latitude of the signal source (see text).

of evolution, which may explain the high degree of conservation of centriole structure throughout nature.

The right-angle orientation can be expected to be found whenever the signal sources are distributed in three-dimensional space. If under special circumstances the relevant signal sources are distributed in a plane, only one cylinder is required that is oriented perpendicular to this plane. We speculate, therefore, that the array of basal bodies that are oriented perpendicular to the surface of a ciliated cell not only form nucleating structures for the cilia but also act as monitoring devices for signal sources at the cell surface.

H. Refinement of Angular Resolution

The similarities between a hypothetical cellular eye and the structure of centrioles and basal bodies extend even further if we try to optimize the angular resolution of the device. The design developed so far limits the angular resolution by the number of receptors and blinds around the perimeter of the cylinder. If a signal source is located anywhere in the sector defined by the angle between adjacent blinds, it will be detected by one and the same receptor. In other words, if the cross section of the cylinder has *N* blinds, then the limit of angular resolution is $\theta = 360°/N$. Therefore, it would be desirable to improve the design in such a way that it can distinguish various directions in each sector.

The most direct method to improve the angular resolution would be to increase the number of receptors and blades (*N*). However, each blade must have a certain thickness in order to absorb the signals. Therefore, the number *N* cannot be increased at will.

A much more elegant way to achieve better angular resolution is to attach the blinds in such a way that they lean across one sector as they run from the top of the cylinder to the bottom (see Fig. 12). The blades of actual centrioles and basal bodies are, indeed, pitched in this way (cf. Fig. 1).

Assume that the viewer is positioned at the location of the signal source *S*. As shown in the perspective drawing of Fig. 12, the viewer would see the outside of some blades (gray shading), the inside of other blades (white areas), the opaqueness of the inside of the cylinder (black areas), and the rows of receptors exposed to the direction of the source (dots lining the blades). The main difference from the earlier design is the exposure of some receptors along blade *n* and some receptors along blade *n*+1 to the source. Therefore, at first sight, the pitched blades may seem to defeat our initial intention of providing a one-to-one mapping between receptors and directions of signal sources. However, the exposed receptors on adjacent blades differ by their position along the blades. More importantly, the apparent ambiguity offers the possibility to determine the location of the direction of the signal source in the sector of blade n. By determining the ratio

$$r = \frac{\text{number of receptors exposed on blade } n}{\text{number of all receptors on blade } n}$$

(see Fig. 12), the angle ϕ of the signal source reaching receptors on blade n becomes

$$\phi = (n + r)\theta$$

This angle is no longer an integral multiple of the former limit of resolution $\theta = 360°/N$, but contains fractions of θ, thus improving the angular resolution. It is still limited, though, because of the finite number of receptors lining the inner edge of each blade.

The explanation of the calculation just presented is obvious if one considers a series of cross sections along the cylinder axis, as shown in Fig. 12. As we move up the axis, the cross sections rotate by the angle of one sector. Depending on the direction of the source within the sector, the blade will cut off the signal source at some distance from the bottom of the cylinder, giving rise to the ratio calculated.

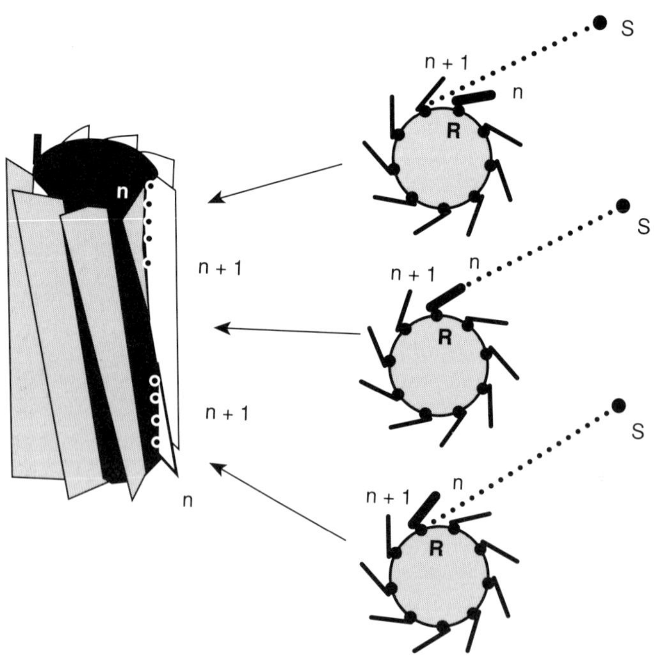

Fig. 12. Pitch of the blades as a means to improve the angular resolution of the design. The percentage of receptors on blade n exposed to the signal source S is a measure of the location of the signal source in the sector between two adjacent blades (see text).

I. Likely Wavelength of Signal Sources

The next important question about our speculation concerns the physical nature of the postulated signal sources. Considering that the function of the cellular eye is to locate objects (signal sources) in the environment of the cell, one may argue that the wavelength emitted by the sources must be linked to the size of these objects. If the objects were smaller than the wavelength, then the cellular eye would not be able to resolve two objects closer than the wavelength. It seems reasonable to assume that the interesting objects around probably include other cells; therefore, we can expect that the wavelength of the sources is smaller than a cell diameter or $\lambda < 10$ μm.

This wavelength excludes sound as a signal, because the frequency of a sound wave with 10-μm wavelength in water is $n = (1200 \text{ m/sec})/(10 \text{ μm}) = 1.2 \times 10^8$ Hz. At this frequency, water is a completely absorbing medium for ultrasound and, therefore, unsuitable as medium of transmission of signals. Consequently, we are left essentially with electromagnetic radiation as the suspected signal. The range of the wavelength places the radiation in the range of visible or infrared light.

There are various reasons to exclude visible light as potential signal carrier: (1) Most animal cells are located deep inside organisms and are thus cut off from visible light. (2) Centrioles and basal bodies are surrounded by cytoplasmic organelles such as mitochondria and lysosomes that are up to 1 μm in size. In order to reach the cellular eye without much scattering by such obstacles in the line of sight, the signal sources should emit wavelengths that are larger than these objects. In other words, this argument places the wavelength outside the visible range.

Thus we are left with infrared light of wavelengths shorter than 10 μm as the most likely signal radiation to be detected and located by centrioles. To be sure, it is quite possible that cells or cell parts emit such radiation, because at room temperature these wavelengths belong to the higher harmonics of molecular vibrations and rotations. Further, it is conceivable that cells can receive such signals because certain animals, including snakes, are able to see infrared sources, suggesting that infrared-sensitive receptors are part of the genome of some animals; the postulate of infrared-sensitive receptors in centrioles cannot be dismissed.

The following argument seems to narrow the range of wavelengths even more. As shown in Fig. 13, the main wavelengths of black body radiation at 27°C range from 3 μm to > 10 μm. A very similar curve applies to room temperature or 37°C. A cellular eye sensitive to these wavelengths would detect a constant bright glow around the cell, stemming from the ambient temperature, that might bury the signals from surrounding sources in high background noise. Therefore, the wavelength of the sources would be placed best above the visible range and below 3 μm (Fig. 13). Unfortunately, in this range of infrared radiation, water is already rather opaque and one would suspect that the relevant wavelengths are located in relative "water windows." As shown in the water absorption curve in Fig. 13, there

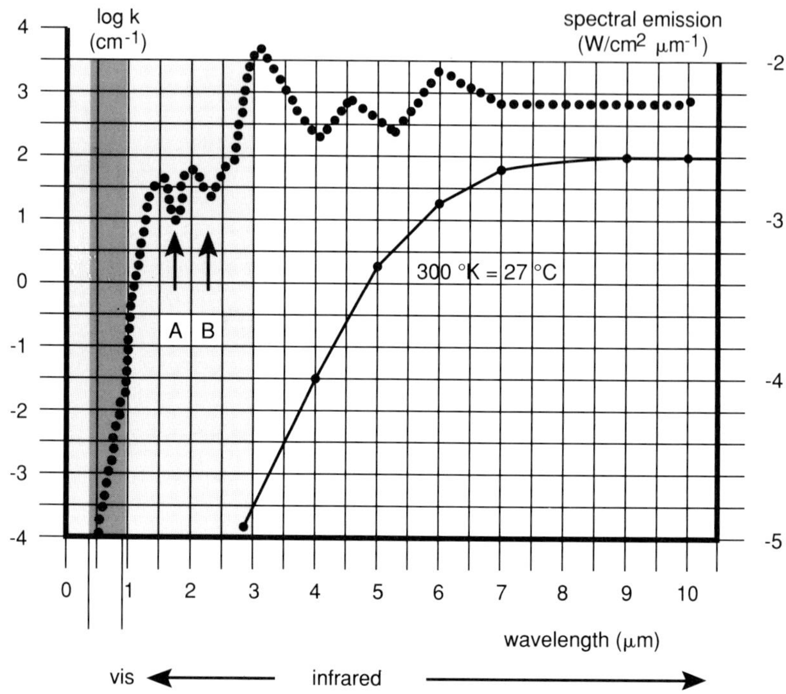

Fig. 13. Black body radiation at and around room temperature (continuous curve; right ordinate shows the logarithm of the spectral emission; 300 K = 27°C) and extinction coefficient of water (dotted curve; left ordinate) as functions of wavelength in the visible and infrared range (redrawn after Wolfe and Zissis, 1989).

are three areas of relative transparency: 0.8–1.3 μm, 1.6–1.7 μm (marked A), and 2.2–2.4 μm (marked B). Using a specially designed microscope, we tested the ability of 3T3 cells to locate distant infrared light sources. The experiments, indeed, identify the window between 0.8–0.9 μm as the most active range (Albrecht-Buehler, 1991).

III. SPECULATION ABOUT THE FORMATION OF CENTRIOLES AND BASAL BODIES

So far we have considered only the optimal design for cellular eyes if, indeed, they exist. We have paid no attention to the practical problem of actually building them. Obviously, centrioles exist, and the way to build them seems to be highly conserved throughout nature. In the following discussion we will propose that the practical construction of the centriole is quite simple, at least in principle, and that this principle agrees quantitatively with the observed structures.

A. Bud Formation

We assume that the formation of a centriole (basal body) begins with a ring of nine microtubular precursors spaced in such a way that nine microtubules could be accommodated in a wall-to-wall configuration (Fig. 14a). This structure will be called the bud. It is likely to resemble a fibrous coil because the connections between the different precursors are likely to involve fibrous polymers. As mentioned in the introduction, fibrous coils having diameters in the range of 1000 Å have been observed to precede centriole and basal body formation. The ring structure of nine wall-to-wall microtubules has been observed in connection with centrioles, too. The mature centriole of the *Ascaris* sperm in cross section appears identical to Fig. 14a (Goldstein, 1977). Therefore, from the point of view of this model, the *Ascaris* sperm appears as the minimal centriole structure.

In order to generate the more complex structure of common centrioles, the bud must add 18 more microtubular precursors. We assume that the additional precursors are packed together in the densest possible way (see Fig. 14b,c). In consequence, the diameter of the bud grows to about 1400 Å. As shown in Fig. 14, the densest packing automatically generates a curved arrangement of microtubular cross sections similar to the known cross section of centriolar blades (cf. Fig. 2).

B. Formation of Swivel Arms

A major postulate of the model is the formation of fibrous arms that extend around the perimeter of the bud along four microtubular diameters. One of the nine arms is drawn in Fig. 14 as a polygon with four segments that stretch from center to center of five consecutive microtubular precursors. Four segments covered by each arm are essential for the model. If the number is changed to three or five, the resulting structures are markedly different from actual centrioles and basal bodies.

The model assumes that each of the nine arms is rigidly connected to the three specific microtubular precursors in Fig. 14c. Thus it divides the bud into nine identical structures that will be called swivel arms. One of them is shown in Fig. 14d. The free end of each arm is assumed to be able to swivel around a hinge (marked by an asterisk in Fig. 14d) in a fashion shown in Fig. 14e. Note that each hinge coincides with the center of one of the microtubular precursors of the bud. The main idea of the "iris diaphragm" principle of centriole (and basal body)

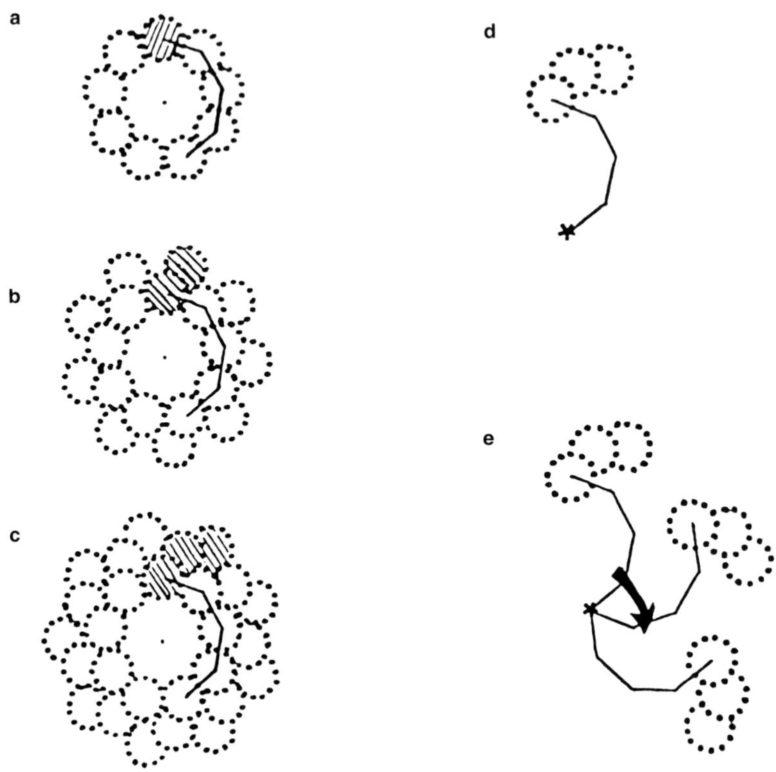

Fig. 14. Earliest stages of the formation of a centriole according to the model (bud formation). (a) Initial closed ring of 9 microtubular precursors indicated by dotted rings of 1 microtubular diameter. One of the future arms is indicated as a polygon. (b) Addition of 9 further precursors in the densest possible packing. (c) Addition of 9 further microtubular precursors in the densest possible packing. As indicated by the cross-hatched structure, the curved shape of the future triplet blades arises automatically. (d) Separate drawing of one of the 9 arms and its hinge (*). (e) Three stages of the postulated swiveling of one of the arms around its hinge.

formation is expressed in the following two postulates. (1) The nine hinges remain fixed in their relative position by an assumed special structure. (2) The nine arms remain rigid while swiveling around their hinges (cf. Fig. 14e). We will show that the known microtubular skeleton results from these postulates in surprisingly accurate detail.

C. Blade Curvature

If we assume that centriolar microtubules and basal body microtubules are composed of 13 protofilaments, as are ordinary microtubules, important constraints on the shape of the future blades result. For example, the B tubule must share 2 of the protofilaments with the A tubule (see Fig. 15a). Since the number of protofilaments is odd, the C tubules cannot be attached along the straight line that connects the centers of A and B (Fig. 15a). It must turn aside in order to share the protofilaments labeled 1 and 1′ in Fig. 15a. Consequently, the resulting triplet is curved, similar in shape to the blades of actual centrioles and basal bodies. The blade curvature can be described by the angle cu = τ (see Fig. 15a), which measures how much the center C of the C tubules is shifted away from the midline connecting the A and B tubules.

A second possibility is to attach the C tubule by sharing the protofilaments labeled 2 and 2′ in Fig. 15a. In this case, the blade curvature cu = 3τ. The two possibilities give rise to different bud structures. If the blade curvature cu = 3τ, the bud structure shown in Fig. 15b arises; if it is cu = τ, then the blades slant out more radially, as shown in Fig. 15c. For subsequent illustrations, the blade curvature is assumed to be cu = 3τ.

In addition to these two possibilities, the blade curvatures could assume values of cu = −τ and cu = −3τ, which correspond to the mirror images of the blade structures just described. Ultimately, these blade curvatures would be required to

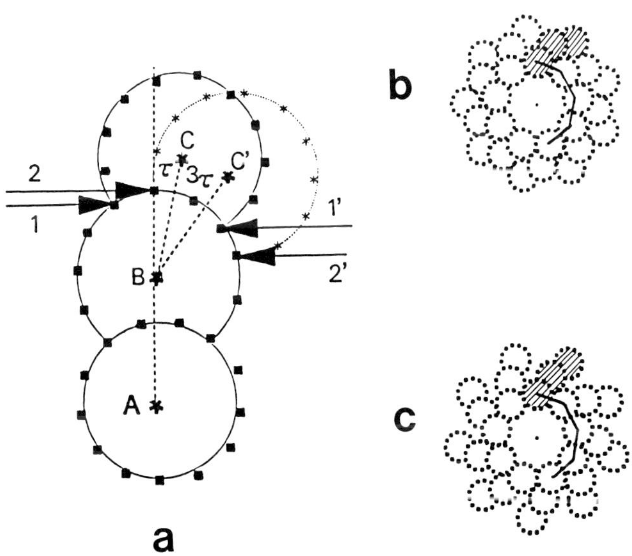

Fig. 15. Derivation of the blade curvature. Each dot represents a microtubular protofilament. (a) Two theoretical possibilities for attaching two C-shaped microtubules (centers marked B, C and C′) to a complete microtubule (center marked A) that are compatible with the postulate of densest packing. The model assumes that protofilaments 1 and 1′ or 2 and 2′ must be shared. As a result, two blade curvatures ($cu = \tau = 360°/26 = 13.85°$ and $cu = 3\tau = 41.5°$ arise. (b) Bud structure if $cu = 3\tau$. (c) Bud structure if $cu = \tau$.

mirror image every other structure of the bud and, in consequence, the entire centriole structure. However, mirror-image centrioles have never been observed. On the contrary, sections through large fields of basal bodies in the surface of ciliated cells and serial sections of centrioles that formed end to end (Berlin and Bowen, 1964; Moser and Kreitner, 1970; Perkins, 1970) show unambiguously that all centrioles and basal bodies have the same handedness. Therefore, we will not consider the possibility that the model is compatible with normal and mirror-image centrioles. Instead, we describe the formation of only one of the two stereoisomers.

D. Unfolding of the Bud

As indicated in Fig. 14e, each rigid arm can rotate in only one direction around its hinge, namely clockwise. Counterclockwise rotation would result in a compression of the microtubular precursors beyond their densest packing, which is not possible. In contrast, the clockwise rotation of the arms opens up the array similar to the opening of the individual segments of an iris diaphragm. Figure 16a–j shows the various stages of the unfolding of the bud. Consistent with experimental data (Anderson and Brenner, 1971; Tamm and Tamm, 1980), the outer diameter and inner lumen of the procentriole increase while the blades slant out in a rather radial direction (Fig. 16a–d). Subsequently, the array reaches a maximum diameter at a swivel angle ϕ_{max} (Fig. 16g). Further rotation turns the blades more tangential, while the radius of the array decreases again (Fig. 16h–j). Eventually the process stops when consecutive blades hit each other front-to-end (Fig. 16j). At this stage the slant of the blades is almost tangential.

E. Necessity of Connectors between Blades

In addition to programming a computer to simulate the unfolding, we built several mechanical models in order to test our hypothesis (Fig. 16k). In particular, we hoped to understand how the bud must be designed in order to prevent entangling of the arms during unfolding. In addition to being rigid and curved, we found that the arms had to fulfill two further conditions.

Fig. 16. Unfolding of the bud by swiveling of each rigid arm around its hinge (maturation). One of the blades is cross-hatched to illustrate its swiveling. Arrow in h points to the reforming ring that has the essential structure of the initial bud. (a–j) Consecutive stages of the unfolding. The blade curvature is $cu = 3\tau = 41.5°$. The fixed blade angle is 70°. The swivel angles ϕ are a: 0°; b: 30°; c: 60°; d: 90°; e: 120°; f: 150°; g: 180°; h: 210°; i: 240°; j: 250°. (k) Mechanical model to illustrate the requirement of connectors (co), which prevent the arms from getting entangled during unfolding. One of the blades is highlighted (arrow) to illustrate the swiveling movement.

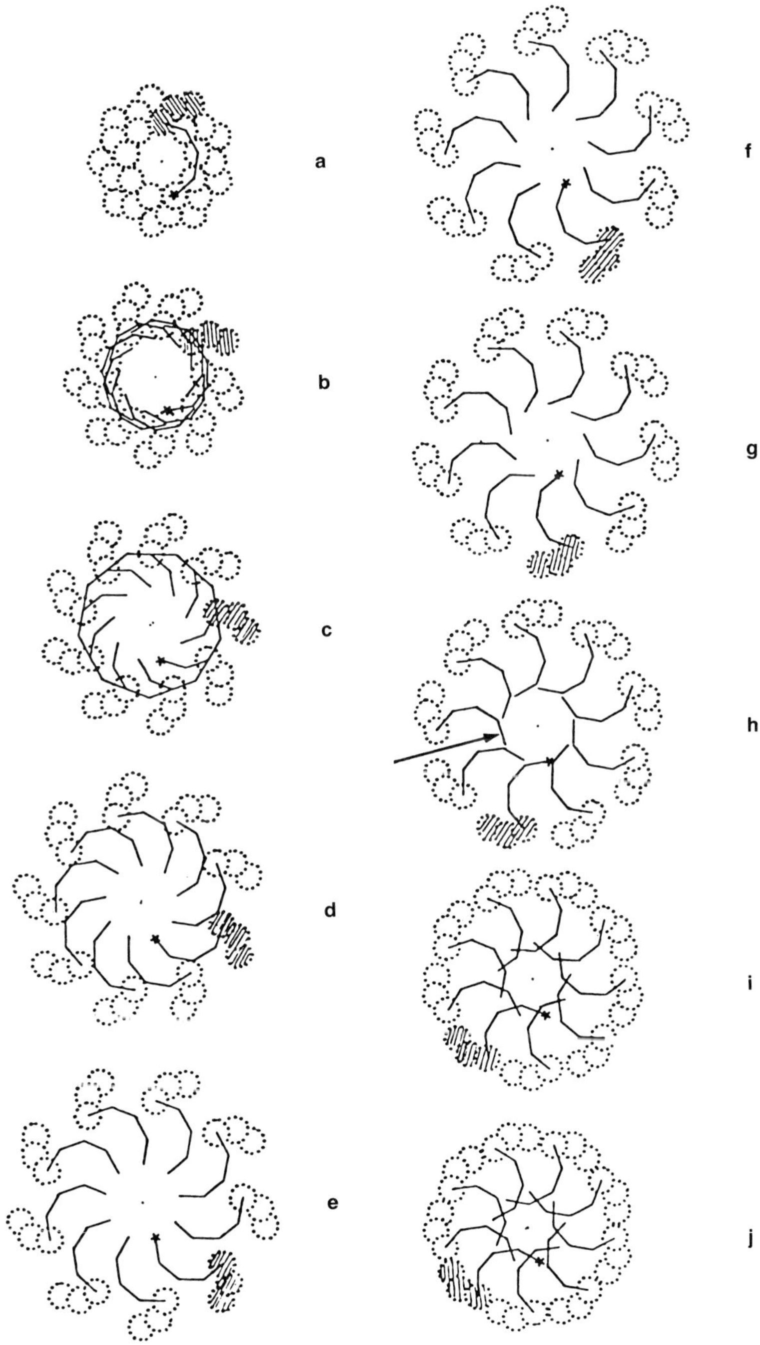

(continued)

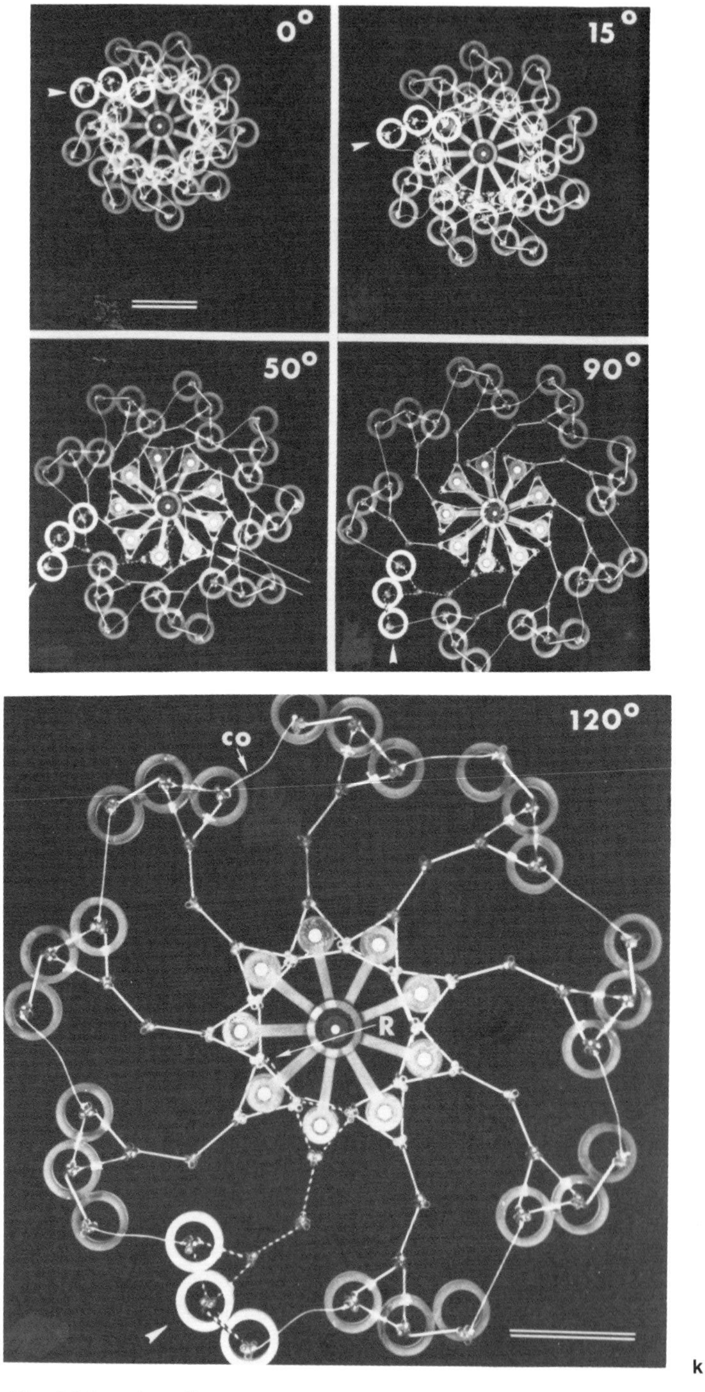

Fig. 16 *(continued)*

1. Consecutive blades had to be connected by flexible fibers (connectors) whose length corresponded to the maximal distance between adjacent blades (i.e., the distance between blades shown in the configuration of Fig. 16g). Figures 16k and 17 indicate the location of such connectors. The mechanical model demonstrated how the connectors prevented the tangle. If a blade approached the preceding one, it had to stretch the connector behind it, thus pulling up the rear blade. This in turn pulled up the blade two places behind it, and so forth all around the circle until the pull had reached the blade in front, which now pulled away from the approaching blade and cleared the space. In this way, entangling of the arms was impossible, and an even opening of the array was guaranteed. Many such connectors must be attached along the entire length of the blades in order to be effective.

Consistent with the model, such connectors are regularly seen between the blades of centrioles and basal bodies (Gibbons and Grimstone, 1960; Dippell, 1968; Allen, 1969; Anderson and Brenner, 1971). The two lower panels of Fig. 2d show such connectors. Obviously, they exist in at least two different planes along the centriolar axis. In longitudinal sections one can detect a large number of fibers that run horizontally between blades.

2. The arms must not lie in a plane but must spiral slightly downward. As shown by the mechanical model, this condition guarantees that the arms cannot touch each other during the swiveling.

F. Axial Growth of Blade Microtubules

As is well known from the literature, the procentrioles and probasal bodies grow in length while their diameter increases. The first microtubule to grow is the A tubule, followed by the B and C tubules (Kalnins and Porter, 1969; Anderson and

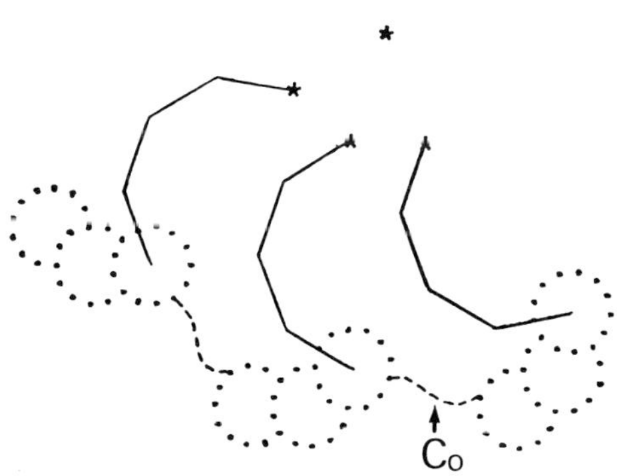

Fig. 17. Schematic drawing of the location of the connectors Co.

Brenner, 1971). Despite large statistical samples and careful observations, the speed of the growth of the blade microtubules is not yet known (Alvey, 1985).

According to our model, the arms are still in the process of swiveling while the blade microtubules grow in length, as shown in Fig. 18. Figure 18 assumes that the blade tubules grow downward, that is, the array of arms will eventually become the distal end of the mature centriole or basal body. This assumption is not vital for the validity of the model but, as pointed out in the next section, it allows a simple explanation for the pitch of the blades, the taper of the entire structure, and perhaps even the swiveling mechanism itself.

G. Pitch, Taper, and Changing Slant of Blades Toward the Distal End

According to the previous section, the model assumes that the swiveling action occurs at the future distal end of the centriole. Consequently, it is reasonable to assume that mechanical resistance of the surrounding cytoplasm will cause the blades at the proximal end to lag behind in their rotation. The lag will not correspond to a large difference in swivel angle because the connectors all along the blades can be expected to couple the movement of the blades reasonably well to one another. Nevertheless, a lag of some time must be expected, and it entails a number of important consequences.

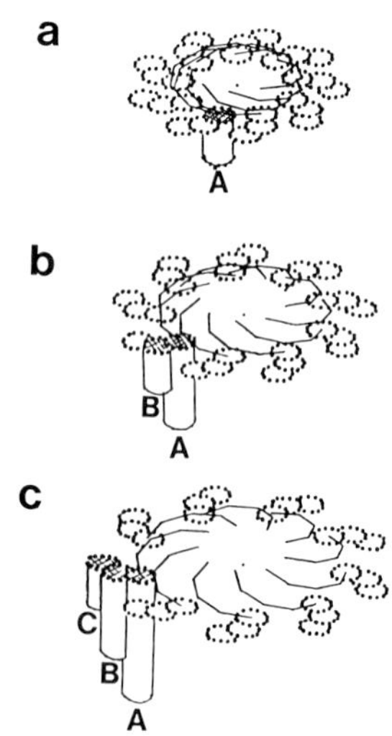

Fig. 18. Schematic elongation of the blade microtubules in the order A, B, and C during the unfolding. (a) Beginning growth of the A tubule. (b) Continued growth of the A tubule and beginning growth of the B tubule. (c) Continued growth of the A and B tubules and beginning growth of the C tubule.

For example, if the swiveling at the distal end has reached the state of Fig. 16i, the rotation at the proximal end may still be in the state of Fig. 16f. As a result (see Fig. 19) (1) the blades are pitched by 10°–15° as in actual centrioles and basal bodies (Anderson and Brenner, 1971); (2) the diameter of the proximal end is larger than the diameter of the distal end (see Figs. 1,19); and (3) at the proximal end, the blades slant out more radially than at the distal end, provided the swiveling angle at the distal end exceeds the angle ϕ_{max} of maximal unfolding. Especially, the slant of the blades at the distal end of basal bodies may become almost completely tangential. This orientation is essential for the function of basal bodies and centrioles as nucleation sites for the axonemes of normal and primary cilia.

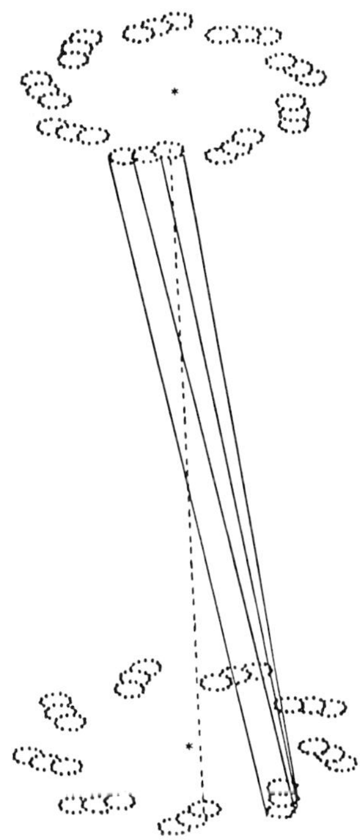

Fig. 19. Origin of pitch, taper, and changing slant of the blades by a lag of swiveling between the distal and proximal end. In the example, the swivel angle at the distal end (*top*) is $\phi_{dist} = 220°$ and at the proximal end (*bottom*) is $\phi_{prox} = 145°$. As a result, each blade is pitched. Since the diameter of the forming centriole decreases after passing through a maximum (cf. Fig. 16g), the diameter at the distal end becomes smaller than at the proximal end, generating the taper of the centriole. In addition, the slant of the blades becomes more tangential after passing the maximal diameter (cf. Fig. 16). Consequently, the slant of the blades at the distal end is more tangential than at the proximal end, as in actual centrioles.

Because the orientation of the outer doublets of ciliary axonemes is tangential and is determined by the orientation of the triplet blades of the basal bodies, the blades at the distal ends of basal bodies must be oriented tangentially. The model accommodates this requirement very easily.

H. Replication

The process of replication is outside the scope of this chapter, which is only concerned with the principle of maturation of centrioles and basal bodies. In particular, the model does not need to suggest any specifics of the location of reproduction because, as mentioned earlier, the location of reproduction is highly variable. Nevertheless, it should at least offer an idea about how the starting structure, namely the bud, could be reproduced by the mature structures. According to present knowledge, procentrioles appear about 4 hr after the onset of DNA replication (Vorobjev and Chentsov, 1982; Alvey, 1985).

At least in principle, the templating of the bud by the mature structure is not very problematic, because the essential geometric feature of the bud, namely the arrangement of the nine microtubular precursors, never changes during the unfolding process. After all, it serves as the fixed set of hinges for the different arms. Particularly during the stages shown in Fig. 16h–j, the configuration of the arms re-establishes the initial ring of nine microtubular precursor sites (see Fig. 16h). Therefore, it seems possible that this reformed ring becomes the template for the new bud structure. Consistent with the assumption that the swiveling arms are located at the distal end of centrioles and basal bodies, we suggest that the new bud is formed from a template at the distal end. Of course, the model cannot predict how the new bud is transported to the various observed sites of maturation outside the parental structure. In particular, basal bodies seem to present a problem for transport, because their distal end is capped by the basal plate and the ciliary axoneme. Therefore, we may have to postulate that the buds for procentrioles travel through the lumen of the basal body from the distal end to the proximal end where they enter the surrounding cytoplasm.

I. Quantitative Aspects of the Model

The qualitative description of the formation of centrioles and basal bodies can be supplemented with a quantitative description. As shown in a previous article (Albrecht-Buehler, 1990), the equations describing the formation and unfolding of the bud quantitatively match the observed data on centrioles and basal bodies in various stages of formation. For example, one can use the equations to fit the computed shape of centrioles with actual electron micrographs of serial sections of centrioles, as illustrated in Fig. 20.

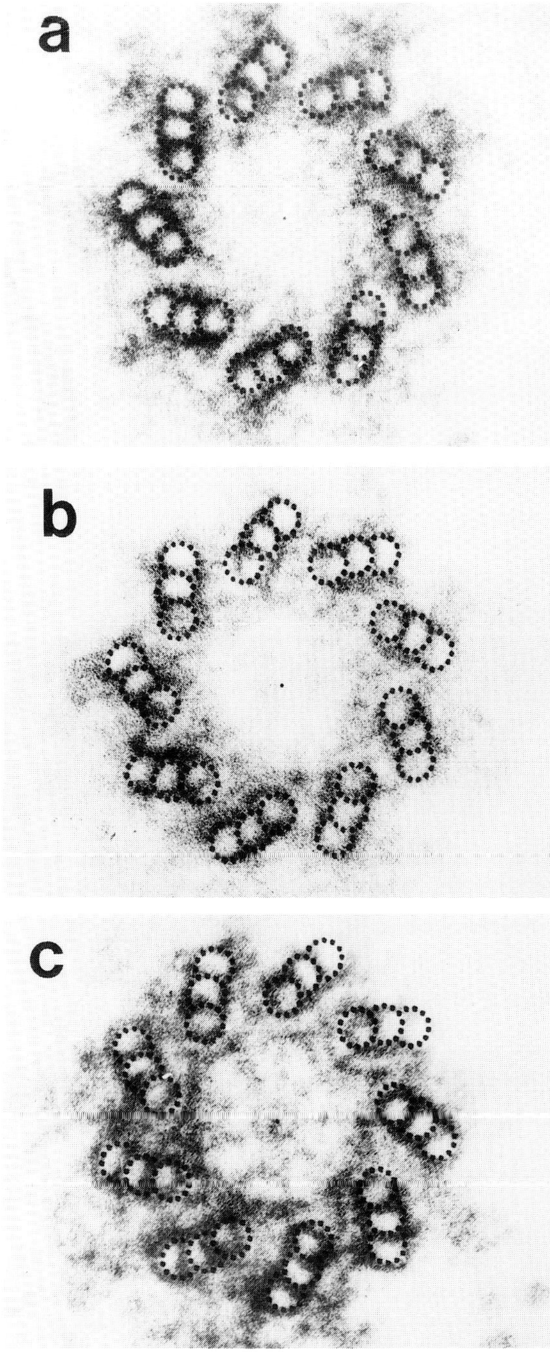

Fig. 20. Fitting actual centriolar cross sections to the model.

IV. DISCUSSION

A. Ninefold Symmetry of Centrioles and Basal Bodies

It is a major shortcoming of the presented arguments that they cannot account for the ubiquitous ninefold symmetry of centrioles and basal bodies. Neither the speculation about the structure of a cellular eye nor the speculation about the formation of centrioles by a movement that resembles the opening of an iris diaphragm requires a specific symmetry of the structures.

One may seek the explanation elsewhere, for example, in the reduction of thermal noise (Albrecht-Buehler, 1981) or in an accident of evolution. However, the final explanation of the structure of centrioles and basal bodies must give more convincing reasons that these for the conservation of the ninefold symmetry. Nevertheless, the speculations were presented in this chapter because they offer explanations for a surprising wealth of details of centriolar geometry.

B. Absorptive Function of Centriolar Blades

The arguments about the function of the pair of centrioles as a cellular eye depend critically on the interpretation of the centriolar blades as blinds that absorb the signals in their thickness. The thickness of the blades is not merely given by one microtubular diameter. As indicated by arrows in Figs. 2c,d, the outsides of the blades of centrioles are usually coated with thick cushions of electron-dense material, effectively increasing the thickness of blades to about 600 Å. Therefore, the requirement of signal absorption in the blade thickness means that the blades must have an extinction coefficient for the postulated infrared signal on the order of k = 1/600 Å = 1.7×10^5 cm^{-1}. This is a very high value for an extinction coefficient in the infrared range. For example, the highest extinction coefficient for water in this range of wavelengths is $k = 8 \times 10^4$ cm^{-1} or about 20 times smaller. Extinctions of such magnitude would presumably require resonance absorption mechanisms or reflectance of the signals by pseudometallic materials. Therefore, the existence of materials with such high extinction coefficients in centrioles and basal bodies will be a critical test for the validity of the presented speculation.

C. Movement as Part of the Assembly of a Cellular Structure

Considering the large number of geometric characteristics of centrioles that can be derived from the iris diaphragm principle, it seems legitimate to take it more seriously than a mere mathematical game. Consistent with conventional ideas about the assembly of other cellular structures, the model assumes that the formation of centrioles and basal bodies begins with the assembly of many macromolecular components into a precursor structure (the bud) using a template (the arrangements of the nine arms after the unfolding; see Fig. 16h). The macromolecular assembly continues with the growth of the blade microtubules. The new feature of the model is the introduction of a movement (the unfolding of the bud) as a major component of the formation of centrioles. Given the current requirement of specimen fixation and dehydration for electron microscopy, this movement cannot be observed directly. However, nearly every intermediary configuration of the blades that is postulated by the model has been observed at one time or another by workers in the field.

D. Reduction of Complex Centriolar Geometry to a Small Number of Mostly Known Constants

A further special feature of the model is its reduction of the quite complex geometry of centrioles and basal bodies to only four basic numbers: (1) the radius of microtubules (120 Å), (2) the rotational symmetry of centrioles (9-fold), (3) the rotational symmetry of microtubules (13-fold), and (4) the number of segments of the initial ring structure covered by the arm (four segments).

For example, the bud structure results from the microtubular radius and the 9-fold symmetry. The blade curvature is a consequence of the radius and the 13-fold symmetry of the microtubules. The arm structure and location of the hinges follows from the microtubular radius, the 9-fold symmetry, and the number of the segments (four) covered by the arm, and so forth.

It may appear that the fixed blade angle is a fifth parameter of the model. However, it is actually a consequence of the packing of microtubular precursors in the bud.

The other features of the centriole structure, for example, pitch of the blades, taper of the centriole, and the automatic termination of the maturation process, are consequences of the swiveling process itself.

E. Speculation about Possible Swiveling Mechanisms

The main assumption of the model of centriole formation is the presence and action of a swiveling mechanism that is expressed in the equations by the swivel angle ϕ. The assumption is compatible with current knowledge.

With respect to the generation of the swiveling motion, there are numerous helical polymers in cells that are capable of producing torque by undergoing conformational changes. One may even think that a eukaryotic analog of the rotary engine of bacterial flagella may turn the swiveling arms. Another way of unfolding the bud could be achieved by the previously mentioned fibers that we called the connectors. In Section E we explained their presence by their ability to prevent the entangling of the swiveling arms during unfolding. It is conceivable that they also act as combined connector–expanders, that is, as rods that elongate by polymerization and thus pry adjacent blades apart while they pull up other blades that are further away. However, this possibility seems less attractive because it would require that the connector–expanders shorten again after the array of arms has achieved its maximal diameter. Another rather intriguing mechanism for generating the swiveling motion may be provided by the simultaneous growth of the blade microtubules themselves. They are pitched and consequently elongate in an oblique direction relative to the axis of the future centriole. Any mechanical resistance in the cytoplasm can be expected to automatically push their base (arms) in the opposite direction. The resulting torque would turn the arms in precisely the required direction of swiveling (cf. Figs. 18,19).

V. SUMMARY

This chapter suggests that centrioles and basal bodies may be considered cellular eyes and that they form by the unfolding of an array of densely packed microtubular precursor structures, resembling the opening of an iris diaphragm (iris diaphragm principle of centriole and basal body formation).

ACKNOWLEDGMENTS

The work presented here has been carried out over several years. During this time I was supported by a number of different grants. In its most recent phase, the work was supported by grants from the Office of Naval Research (N00014-89-J-1700) and the U.S. Army Research Office (ARO 122-89). A major basis of the work was a large number of electron micrographs of centrioles expertly and patiently produced by Anne Bushnell. I am very grateful for her and Amy Alberts' excellent assistance. I should also like to thank Martin Zand for his critical reading of the manuscript.

REFERENCES

Albrecht-Buehler, G. (1981). Does the geometric design of centrioles imply their function? *Cell Motility* **1,** 237–245.

Albrecht-Buehler, G. (1990). The "iris diaphragm"-principle of centriole and basal body formation. *Cell Motil. Cytosk.* **17,** 197–213.

Albrecht-Buehler, G. (1991). Surface extensions of 3T3 cells towards distant infrared sources. *J. Cell Biol.* **114,** 493–502.

Albrecht-Buehler, G., and Bushnell, A. (1979). The orientation of centrioles in migrating 3T3 cells. *Exp. Cell Res.* **120,** 111–118.

Albrecht-Buehler, G., and Bushnell, A. (1980). The ultrastructure of primary cilia in quiescent 3T3 cells. *Exp. Cell Res.* **126,** 427–437.

Allen, R. D. (1969). The morphogenesis of basal bodies and accessory structures of the cortex of the ciliated protozoan *Tetrahymena pyriformis. J. Cell Biol.* **40,** 716–733.

Alvey, P. L. (1985). An investigation of the centriole cycle using 3T3 and CHO cells. *J. Cell Sci.* **78,** 147–162.

Anderson, R. G. W. (1972). The three-dimensional structure of the basal body from the Rhesus monkey oviduct. *J. Cell Biol.* **54,** 246–265.

Anderson, R. G. W., and Brenner, R. M. (1971). The formation of basal bodies (centrioles) in the Rhesus monkey oviduct. *J. Cell Biol.* **50,** 10–34.

Bacetti, B., Dallai, R., and Fratello, B. (1973). The spermatozoa of arthropoda. XXII. The 12+0, 14+0 or aflagellate sperm of *Protura. J. Cell Sci.* **13,** 321–335.

Berlin, J. D., and Bowen, C. C. (1964). Centrioles in the fungus *Albugo candida. Am. J. Bot.* **51,** 650–652.

Brinkley, B. R., and Stubblefield, E. (1970) Ultrastructure and interaction of the kinetochore and centriole in mitosis and meiosis. *In* "Advances in Cell Biology" (D. M. Prescott, L. Goldstein, and E. McConkey, eds.) Vol. 1, pp. 119–185. Academic Press, New York.

Burgos, M. H., and Fawcett, D. W. (1955). Studies on the fine structure of mammalian testis. I. Differentiation of the spermatids in the cat (*Felis domestica*). *J. Biophys. Biochem. Cytol.* **1,** 287–300.

Dippell, R. V. (1968). The development of basal bodies in *Paramecium. Proc. Natl. Acad. Sci. U.S.A.***61,** 461–468.

Gall, J. G. (1961). Centriole replication. A study of spermatogenesis in the snail *Viviparus. J. Biochem. Biophys. Cytol.* **10,** 163–193.

Gibbons, I. R., and Grimstone, A. V. (1960). On flagellar structure in certain flagellates. *J. Biophys. Biochem. Cytol.* **7,** 697–715.

Goldstein, P. (1977). Spermatogenesis and spermiogenesis in *Ascaris lumbricoides* var. *suum. J. Morph.* **154,** 317–338.

Grimes, G. W. (1973). Origin and development of kinetosomes in *Oxytricha fallax. J. Cell Sci.* **13,** 45–53

Kalnins, V. I., and Porter, K. R. (1969). Centriole replication during ciliogenesis in the chick tracheal epithelium. *Z. Zellforsch.* **100,** 1–30.

Komatsu, M., and Fujita, H. (1978). Electron-microscopic studies on the development and aging of the oviduct epithelium of mice. *Anat. Embryol.* **152,** 243–259.

Manton, I., Kowallik, K., and von Stosch, H. A. (1970). Observations on the fine structure and development of the spindle at mitosis and meiosis in a marine centric diatom (*Lithodesmium undulatum*). IV. The second meiotic division and conclusion. *J. Cell Sci.* **7,** 407–443.

Moser, J. W., and Kreitner, G. L. (1970). Centrosome structure in *Anthoceros laevis* and *Marchantia polymorpha. J. Cell. Biol.* **44,** 454–458.

Perkins, F. O. (1970) Formation of centriole and centriole-like structure during meiosis and mitosis in *Labyrinthula (Rhizopodea labyrinthulida) J. Cell Sci.* **6,** 629–653.

Phillips, D. M. (1966). Observations on the spermiogenesis in the fungus gnat *Sciara coprophilia. J. Cell Biol.* **30,** 477–497.

Phillips, D. M. (1974). *In* "Cilia and Flagella" (M. A. Sleigh, ed.) pp. 379–402. Academic Press, London.

Pickett-Heaps, J. (1973). The autonomy of the centriole: Fact or fallacy? *Cytobios.* **3,** 205–214.

Porter, K. R. (1965). The submicroscopic morphology of protoplasm. *The Harvey Lectures* **51,** 175–228.

Prensier, G., Vivier, E., Goldstein, S., and Schrevel, J. (1980). Motile flagellum with a "3+0" ultrastructure. *Science* **207,** 1493–1499.

Shay, J. W. (1974). Ultrastructural observations on spermiogenesis in the fungus gnat *Rhynchosciara* sp. *In* "The Functional Anatomy of the Spermatozoon" (B. A. Afzelius, ed.). Pergamon Press, Oxford.

Sorokin, S. (1962). Centrioles and the formation of rudimentary cilia by fibroblasts and smooth muscle cells. *J. Cell Biol.* **15,** 363–377.

Tamm, S., and Tamm, S. L. (1980). Origin and development of free kinetosomes in the flagellates *Deltatrichonympha* and *Koruga*. *J. Cell Sci.* **42,** 189–205.

Turner, F. R. (1968). An ultrastructural study of plant spermatogenesis. *J. Cell Biol.* **37,** 370–393.

Vorobjev, I. A., and Chentsov, Y. S. (1982). Centrioles in the cell cycle. I. Epithelial cells. *J. Cell Biol.* **93,** 938–949.

Wheatley, D. N. (1966). Cilia in cell-cultured fibroblasts. I. On their occurrence and relative frequencies in primary cultures and established cell lines. *J. Anat.* **105,** 351–362.

Wolfe, W. L., and Zissis, G. J. (1989). "The Infrared Handbook". Environmental Research Institute of Michigan.

4

Microtubule and Intermediate Filament Patterns around the Centrosome in Interphase Cells

I. B. ALIEVA*, E. S. NADEZHDINA*, E. A. VAISBERG†, AND I. A. VOROBJEV*

*A. N. Belozersky Institute of Physico–Chemical Biology
Moscow State University
Moscow 119899
†Institute of Protein Research
Academy of Science USSR
Pushchino 142292

I. INTRODUCTION

The cytoskeletal pattern is quite specific in each cell type. At the same time it changes dramatically under different circumstances, for example, during the in-

English translation copyright © 1992
by Academic Press, Inc.

terphase–mitosis transition, or during cell spreading. Elucidating the mechanisms controlling the formation and maintenance of the different types of cytoskeletal structures is now the goal of numerous investigations. In the present study we will discuss the role of the centrosome in the formation of microtubule (MT) and intermediate filament (IF) networks.

In interphase cells, MTs form a network extending throughout the whole cytoplasm (Brinkley *et al.*, 1980). In general this network is densest in the perinuclear region. In well-spread cells in culture, single MTs forming the network are visible by immunofluorescence. They appear to be many microns in length and to stretch uninterrupted from the central part of the cell to the lamella at the periphery (Osborn *et al.*, 1978; Brinkley *et al.*, 1980,1981).

IFs also are one of the major components of the cytoskeleton. Unlike MTs, which are assembled from tubulin, IFs in different cell types are formed from different IF proteins (Traub, 1985). Two types of IF, those assembled from cytokeratin and those assembled from vimentin, have been more frequently studied in cultured cells. In the interphase cells, they form networks extending throughout the major part of the cytoplasm. In some cell types, the IF network often has an appearance similar to the MT network. For example, in cells of many cultured fibroblast lines the MT and vimentin-type IF are tightly associated, and in the lamelloplasm, where they are easily observed, the MT and IF patterns may be almost identical (Geiger and Singer, 1980). However, high concentrations of IF may occur not only near the nucleus but in the peripheral parts of the cytoplasm.

It is well known that the centrosome plays a key role in the formation of MTs (see Vorobjev and Nadezhdina, 1987, for review). Some data indicate that the centrosome may also be involved in the formation of the IF network (Goldman *et al.*, 1980; Kalnins *et al.*, 1985). We will first describe our work on MTs in the region around the centrosome and then discuss our findings on the role of the centrosome in the formation of the IF network.

II. CENTROSOME AND MICROTUBULES

The centrosome in mammalian interphase cells consists of two centrioles: a parent centriole around which most MTs assemble and a daughter centriole around which relatively few MTs assemble. The older parent centriole carries pericentriolar satellites from which MTs radiate, whereas the more recently formed daughter centriole generally lacks them. In addition to the two centrioles, the centrosome may contain other foci further from the centriole from which MTs also radiate and other structures, such as striated rootlets (reviewed by Vorobjev and Nadezhdina, 1987). These foci will be referred to as free foci. All of the sites from which the MTs assemble constitute the MT organizing center (MTOC) of the centrosome.

To study the formation of the MT network, the pre-existing one can be depolymerized with Colcemid and similar drugs. It was shown that, after removal of these drugs, MTs began to regrow from the MTOC in the centrosomal region toward the plasma membrane in an aster-like fashion (Brinkley *et al.*, 1976; Osborn and Weber, 1976a; Spiegelman *et al.*, 1979). There is evidence that MTs also grow from the centrosome in untreated cells (Brinkley *et al.*, 1981; Kirschner and Mitchison, 1986).

Recently it has been shown that MTs *in vivo* are in a dynamic equilibrium with unpolymerized tubulin. This finding has led to the suggestion that any change in the MT network is the result of disassembly or assembly of individual MTs showing inherent instability (reviewed in Kirschner and Mitchison, 1986; Alberts *et al.*, 1989). However, the problem of how the MT pattern is established, and whether the behavior of the MT network can be explained solely on the basis of dynamic instability of MTs attached to the MTOC in the centrosome, still needs to be examined.

A variety of techniques has been applied to the study of MT dynamics in living cells. Results from experiments using microinjection of labeled tubulin, fixation after various periods of time, and subsequent examination of the microinjected cells by electron microscopy or immunofluorescence (Kirschner and Mitchison, 1986; Geuens *et al.*, 1989) have been particularly useful in studying MT dynamics.

The results obtained by this approach showed that the injected exogenous tubulin is rapidly incorporated into MTs at their distal ends, since at short time intervals numerous small segments of MTs were labeled throughout the cytoplasm. Some short labeled MTs were also found in the centrosomal region. These probably represent the formation of new MTs directly from the centrosome (Soltys and Borisy, 1985; Kirschner and Mitchison, 1986; Geuens *et al.*, 1989). The rate of MT elongation measured with this technique in most cultured cells is rather high—up to 4 μm/min—except for cultured cells from the nervous system, in which the process is ten times slower—0.3 μm/min (Okabe and Hirokawa, 1988).

By microinjecting fluorescein-labeled tubulin (Kirschner and Mitchison, 1986; Geuens *et al.*, 1989) and examining the cells by video-enhanced light microscopy (Sammak and Borisy, 1988) it was possible to observe MT dynamics directly in the lamelloplasm at the periphery of well-spread cells. Such direct observation of MT behavior in living cells showed that single MTs in the lamella undergo length excursions independently, and exhibit what was termed "tempered instability" (Sammak and Borisy, 1988). However, this approach cannot be used to analyze the distribution and dynamics of MTs around the centrosome because in this region the MTs are too close to each other (Soltys and Borisy, 1985).

It has been suggested that all MTs grow directly from the centrosome. A cell, however, may contain up to 2000 MTs (Schliwa *et al.*, 1979; Schulze and Kirschner, 1986) and almost all of them are constantly growing at their tips (Schulze and Kirschner, 1986; Geuens *et al.*, 1989). *In vitro* experiments show that relatively few MTs (about 20–30) polymerize on centrosomes in the presence of tubulin (Kuriyama and Borisy, 1981; Mitchison and Kirschner, 1984). From these data it is difficult to imagine that, in a living cell, all the MTs are attached to the centrosome. It would be important to determine, therefore, how many MTs are attached to the centrosome at any one time and what the origin of the remaining ones is.

First direct evidence for the number of MTs associated with the centrosome *in vivo* was provided by Schliwa and co-workers (1982) from studies of whole-mount preparations of human neutrophils by electron microscopy. They found that, in untreated neutrophils, about 20 MTs radiate from the centrosome in an aster-like fashion. The number of MTs increases in cells treated with the synthetic chemoattractant peptide *N*-formylmethionylleucylphenylalanine (FMLP) or the tumor-promoting drug 12-*O*-tetradecanoylphorbol 13-acetate (TPA) (Schliwa *et al.*, 1983). Using immunofluorescent techniques, other investigators similarly found that the mean number of MTs in neutrophils was 32.4 ± 4.5 per cell and that all of them seemed to be connected to the centrosome (Anderson *et al.*, 1982).

Using an alternative approach, Gudima and co-workers (1983a,b, 1986) counted the number of MTs in each ultrathin section containing the mature parent centriole, which is more active in initiating the assembly of MTs. Multiplying this parameter with the average number of centriole-containing sections (5–8), one can obtain an estimate of the total number of centrosome-attached MTs. For both granulocytes and macrophages, the mean number of MTs radiating from the centrosome, as obtained by this technique, was about 30 (Gudima *et al.*, 1988) and therefore similar to that obtained using other approaches (Anderson *et al.*, 1982; Schliwa *et al.*, 1982; Cassimeris *et al.*, 1986). Macrophages and granulocytes thus have a very sparse MT network in comparison with that found in some types of cultured cells (Frankel, 1976; Cassimeris *et al.*, 1986), in which it is difficult, if not impossible, to estimate the number of MTs in the vicinity of the centrosome by immunofluorescence because of their high density. Whole mounts of lysed cells likewise cannot be used to study MTs around the centrosome by electron microscopy in cultured cells such as fibroblasts, because the centrosome is too close to the nucleus (Kuriyama and Borisy, 1981).

To overcome this problem, Gudima and co-workers (1983a,b,1986,1988) analyzed serial ultrathin sections to estimate the number of MTs radiating from the centrosome in different types of cultured cells. For analysis of MTs radiating from the centrosome, Gudima and co-workers (1983a,b,1986,1988) selected three groups of MTs that were counted separately: (1) short MTs, which did not extend beyond a 0.8-μm radius from the centriole; (2) medium length MTs, which extended beyond 0.8 μm but not beyond 2.0 μm from the centriole; and (3) long MTs, which extended beyond the 2.0-μm radius. It was found that the mean number and length of MTs changed during cell attachment, spreading, and migration on the glass surface. An inverse correlation was observed between the number of MTs around the centrosome and the number of MTs in the rest of the cytoplasm during the development of the MT network while the cells were spreading.

In fully spread cultured mouse embryo fibroblasts (MEF), short MTs around the centrosome were rare (0.1 ± 0.1 MT per section) and the average number of long MTs in this region was 4.2 ± 1.6 (per section). In suspension, the number of long MTs radiating from the centrosome decreased to 2.1 ± 1.6 MT per section, but the short MTs increased to 6.2 ± 4.1 MT per section around the older and more active parent centriole. The total number of MT profiles in the centrosomal region also increased. The long MTs apparently start to grow after the cells begin to spread on the substratum and their number reaches a maximum in polarized fibroblasts (8.9 ± 4.5 MT per section). At the intermediate radial spreading stage, the total number of MT profiles increases with distance from the centrosome (Gudima *et al.*, 1983b). Similar assembly of numerous short MTs around centrioles was observed in cultured renal epithelial cells after suspension (Gudima *et al.*, 1986). Thus in both cell types MT assembly was initiated in the centrosomal region after detachment of the cells from the substratum, during which the pre-existing extensive cytoplasmic MT network is reduced in size (Osborn and Weber, 1976b).

An examination of the dynamics of MT assembly around the centrosome after their depolymerization by cooling (Vorobjev and Chentsov, 1983) showed that the number of MTs radiating from the centrosome reaches a peak at a time when only a few MTs are present in the lamelloplasm and then slowly decreases to control levels. This result suggested to us that many of the MTs in cultured cells are not attached to the centrosome and have both of their ends free (Vorobjev and Chentsov, 1983). Summarizing the data, we concluded that in different types of cells MTs between 10 and 100 are attached to the centrosome.

A study comparing the MT volume densities in different regions of the cytoplasm (centriolar and noncentriolar, perinuclear and peripheral) indicated a greater volume of MTs in the peripheral regions than expected if all of the MTs attached to the MTOC in the centrosome extended radially to the plasma membrane (Zimmer *et al.*, 1981). To explain these results, the authors suggested that the MTs are very long, bend at the periphery of the cell, and return to the nucleus (Zimmer *et al.*, 1981). These data, however, are also consistent with our idea that not all of the MTs are connected to the centrosome.

Other more direct evidence indicates that this is the case in some cell types. For example, free MTs are found in segments of axons, as shown by reconstructions from serial sectioning (Tsukita and Ishikawa, 1981). Such MTs were also detected by immunofluorescence microscopy in cytoplasts lacking centrosomes (Karsenti *et al.*, 1984) and in aging Chinese hamster fibroblasts (Raes *et al.*, 1984). The simplest explanation for the origin of free MTs would be that free MTs spontaneously polymerize in the cytoplasm throughout the cell (Karsenti *et al.*, 1984). However, some data (discussed subsequently) do not support this idea. To try to understand the origin of these free MTs, we have recently analyzed the distribution of MTs in the centrosomal region in a number of different cell types in culture, before and after treatment with various agents. The results of these studies are presented in the following sections.

A. Microtubule Distribution around the Centrosome in Cultured Cells

1. Distribution of Microtubules in Normal Pig Kidney Embryo (PK) Cells before and after Treatment with Various Agents: Immunofluorescence Studies

Examination of PK cells by immunofluorescence with antibodies to tubulin shows that these cells have a MT network that is typically found in cultured cells (Fig. 1a). The microtubules are more numerous around the nucleus and relatively sparse closer to the cell periphery, where many MTs run parallel to the cell border. The position of the centrosome is generally difficult to detect. After a 10 min treatment with taxol (10 μ*M*), the MT network became much more prominent and MT numbers increased, especially in the lamelloplasm, so that single MTs were difficult to resolve (Fig. 1c). Some changes in the MT pattern were also observed after treatment with various agents, including metabolic inhibitors. The most striking changes were observed after treatment with *p*-trifluoromethoxy-phenylhydrazone (FCCP). Incubation of PK cells in 20 μ*M* FCCP for 30 min resulted in a partial depolymerization of MTs in the lamelloplasm, especially those MTs that ran parallel to cell borders. Only MTs radiating from the juxtanuclear region remained (Fig. 1d). In many cells, a single MTOC became apparent. After ouabain treatment (1 m*M* for 30 min), only minor changes in the distribution of MTs were found. Single MTOCs could now be resolved in some cells, however (Fig. 1e).

2. Stereoscopic Analysis of Microtubule Distribution around the Centrosome

Immunofluorescence studies of many types of cultured cells show that MTs radiate from the centrosome. They do so from different types of foci that one may observe in the centrosome by electron microscopy (see review by Vorobjev and Nadezhdina, 1987). In the cell lines we have studied, these foci included the heads of pericentriolar satellites, lateral surfaces of both centriolar cylinders, and free foci. Although located in the centriolar area, the free foci are not connected directly to the centrioles.

Fig. 1. Double immunofluorescence of PK cells with antibody to tubulin (a) showing staining of MT and antibody to vimentin (b) showing staining of IF. The distribution of MT seen by immunofluorescence with antibody to tubulin in PK cells treated with (c) taxol (10 μ*M*, 10 min), (d) FCCP (20 μ*M*, 30 min) and (e) ouabain (1 m*M*, 30 min). Arrows indicate MT radiating from the centrosome. Bar: 10 μm.

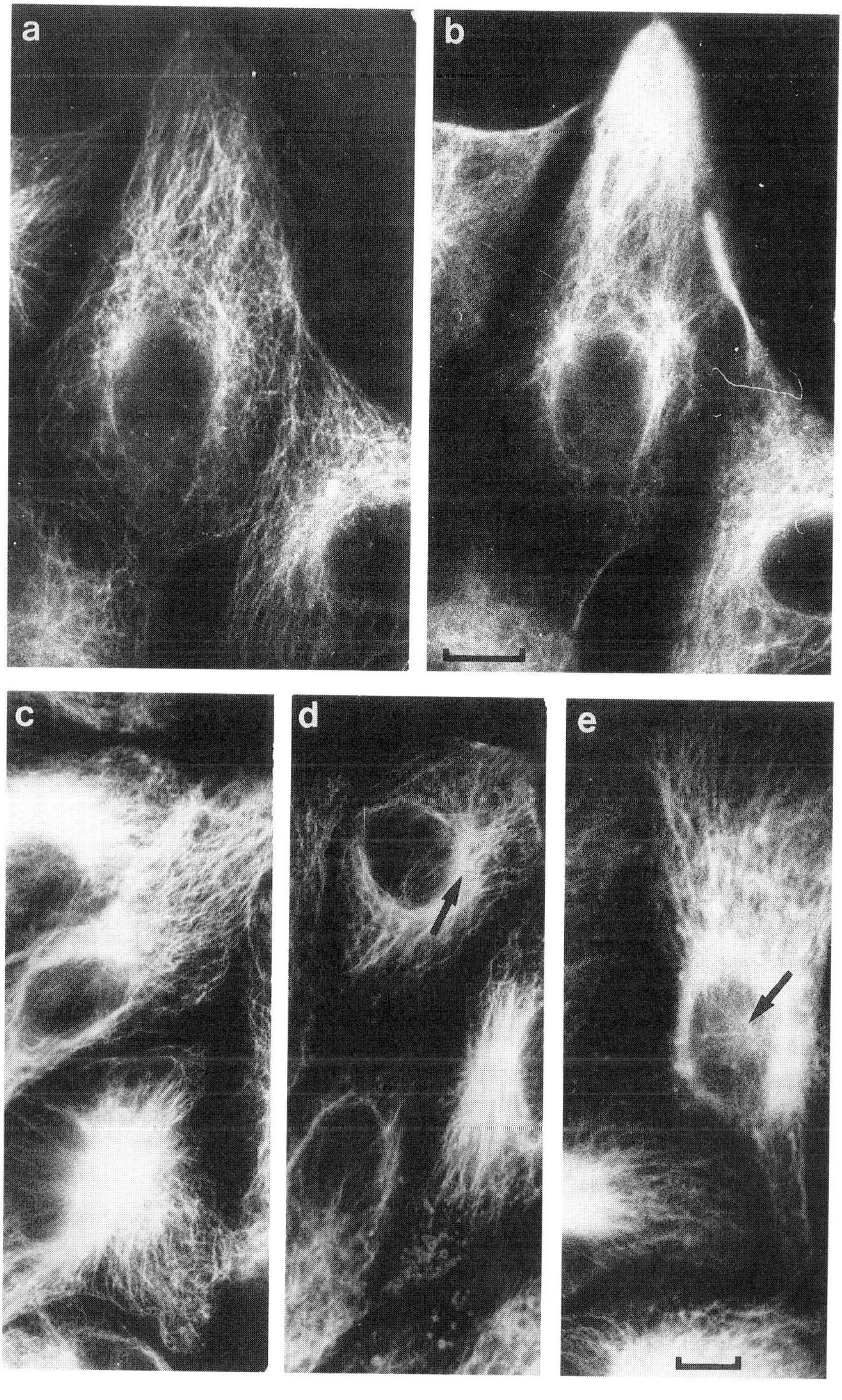
a
b
c
d
e

For a detailed study by electron microscopy of the distribution of MTs in the centrosomal region, cells were extracted for 15 min at 37°C in a MT-stabilizing solution (Gudima *et al.*, 1988) prior to fixation. Fixation procedure, embedding, and sectioning and electron microscopy were essentially standard (Weakley, 1972). The stereopairs of serial semithick sections (0.25 μm) obtained were photographed at a tilt angle of 10°. Each series included all the sections containing both centrioles and two additional sections: one above the upper centriole and another below the lower centriole. The region studied was 4.6 × 3.6 μm in area and 0.8–1.5 μm in depth, depending on the orientation of centrioles. With the help of stereopair viewing, we succeeded in identifying all MTs in the volume studied. For the reconstructions, MT profiles were traced from the prints, section by section, to translucent film sheets. This procedure generated a two-dimensional projection of the MTs around the centrosome. On the reconstructions (Fig. 2–6), only MTs radiating from the centrosome are shown. MTs oriented in any other direction in relation to the centrosome were omitted. MTs in the centrosomal region of 5 control cells from each cell line and in 10 cells after each treatment were traced. The results are given as the mean number of MTs ± SD.

Observations made using stereomicroscopy show that only some of the MTs radiating from the centrosome are attached to MT nucleating sites such as centriolar cylinders, associated pericentriolar satellites, and free foci. Numerous other MTs in fact, had their proximal ends free. To help us understand the orientation of each MT, we imagined them lengthened from their proximal ends.

According to their length, position, and the orientation of their proximal ends, we divided the MT into five classes: Class 1, short MTs less than 2 μm in length whose proximal ends are less than 0.1 μm from one of the MT nucleating sites from which MTs radiate; Class 2, long MTs more than 2 μm in length, similarly associated with MT nucleating sites, differing from the first class only in their length; Class 3, short free MTs less than 2 μm in length with both ends at a distance more than 0.1 μm from any of the MT nucleating sites (the proximal ends of these MTs, however, are oriented toward one of these sites); Class 4, long free MTs in classes 1–4 will be referred to as MTs radiating from the centrosome; and Class 5, free MTs that pass through the space studied but are not oriented toward any of the nucleating sites. The number of MTs in this last class was rather small in all cell types studied.

a. Microtubule Pattern around the Centrosome in Various Cell Types. We have analyzed the MT pattern around the centrosome in MEFs and cells of three different cell lines: L fibroblasts, bovine tracheal epithelial cells (TR), and PK cells.

In all four cell types examined, almost all MTs in the vicinity of the centrosome either are attached to one of the MT nucleating foci (Class 1 and 2 MT) or are free but oriented with their proximal ends toward the foci (Class 3 and 4 MT). Only a few MTs (Class 5 MT) that pass by the centrosome were found in the 2-μm radius around it.

In MEFs, L fibroblasts, and TR cells, MTs are attached to the three different types of foci, that is, pericentriolar satellites, walls of the centrioles, and free foci. In TR cells, however, the free foci nucleate 3 times fewer MTs than in fibroblasts; in PK cells, such free foci are absent.

The results obtained also showed that the total number of MTs associated with the centrosome (Class 1–4 MT) in MEFs was 96.8 ± 16.9; in L fibroblasts, 62.6 ± 5.4; in TR cells, 61.6 ± 18.7; and in PK cells, only 31.0 ± 3.6. The number of free MTs (Class 3 and 4 MT) in these cells was 37.4 ± 6.3; 14.6 ± 3.5; 20.4 ± 10.2, and 14.0 ± 0.2, respectively. Examples are shown of the distribution of MTs around the centrosome of TR cells (Fig. 2a,b) and PK cells (Fig. 3a,b).

The majority of the MTs attached to one of the MT nucleating foci in the centrosome is rather short (<2 μm, Class 1 MT). Most long MTs have their proximal ends free and belong to Class 4 rather than to Class 2. The number of long centrosome-attached MTs (Class 2) was, in fact, less than 5 in all the cell types studied, varying from 3.4 ± 1.2 in MEFs to 0.4 ± 0.2 in PK cells.

To insure that the large number of free MTs observed in the centrosomal region did not result from their disassembly or fragmentation during lysis, taxol (10 μ*M*), a drug known to inhibit MT depolymerization completely (Schiff and Horwitz, 1980), was added to the standard lysis solution. The total number and distribution of the different classes of MT in the centrosomal region of PK cells lysed in taxol-containing medium was similar to that observed without taxol.

b. Microtubule Pattern around the Centrosome in Taxol-Treated Cells. It has been shown previously that prolonged incubation in taxol leads to the inactivation of the MT nucleating capacity of the centrosome and results in the formation of bundles of free MTs in the cytoplasm of taxol-treated cells (Brenner and Brinkley, 1982). We expected that a shorter treatment (10 min, 10 μ*M*) would result in minor changes in the MT network that would reflect MT dynamics and enable us to estimate the rate of MT turnover in the centrosome region. The distribution of MTs after a short incubation of cells in taxol is shown in Fig. 4a,b.

After 10 min incubation of PK cells in taxol, the total number of MTs around the centrosome doubled from about 30 to 70. Despite the fact that the total number of MTs doubled, the number of short unattached Class 3 MTs actually decreased fivefold, whereas the number of short MTs attached to the centrosome (Class 1 MT) doubled. The number of long MTs, both free (Class 4 MT) and attached to the various foci (Class 2 MT), also increased. The number of free Class 4 MTs increased slightly more than twofold, whereas the number of Class 2 MTs attached to the centrosome increased by a factor of 5 over control values obtained from cells that were not incubated in taxol. The number of Class 5 MTs, that is, those not oriented toward any of the MT initiation foci, increased by a factor of three after the taxol treatment.

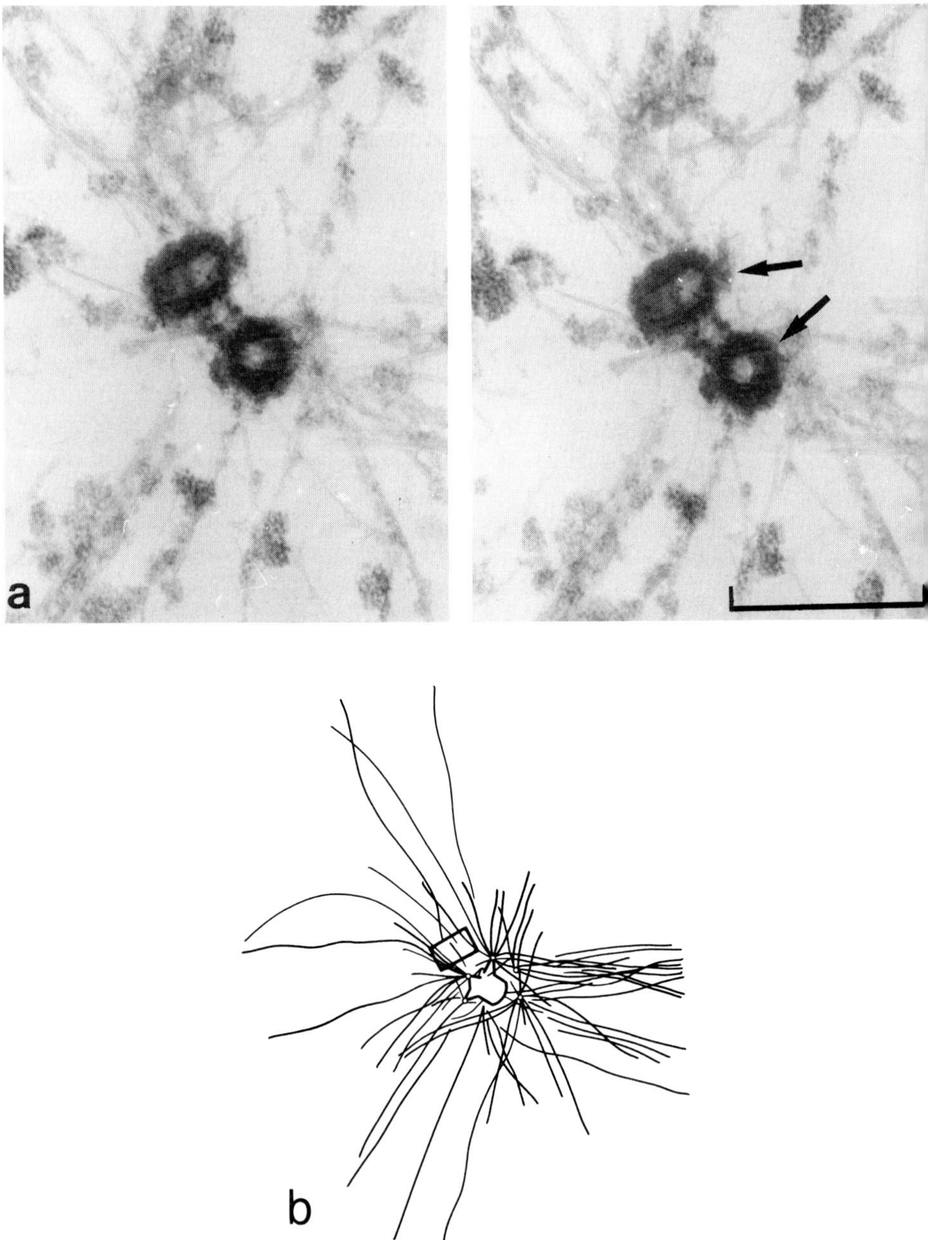

Fig. 2. (a) Stereopair of micrographs of the centrosomal region in TR cell. Tilt angle, 10°. Arrows indicate the centrioles. Bar: 0.5 μm. (b) Profiles of MT oriented toward foci in the centrosome, reconstructed from five serial sections.

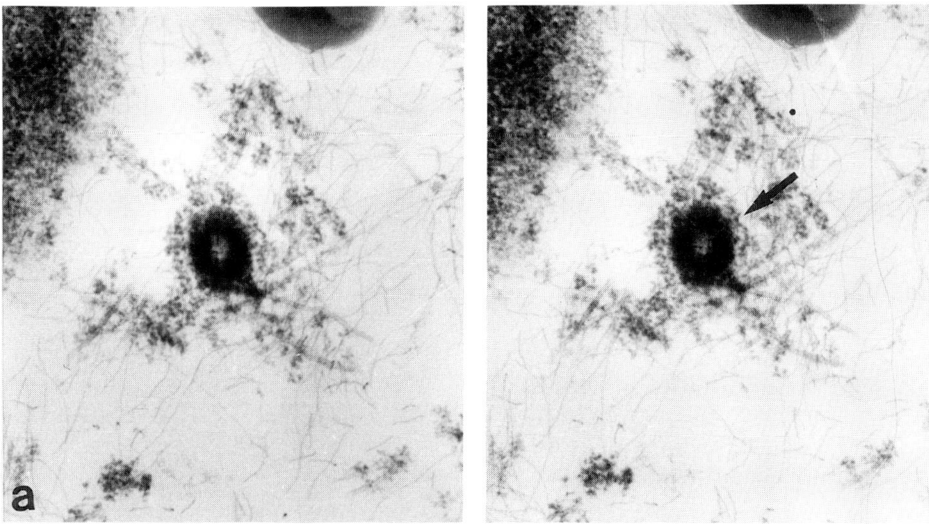

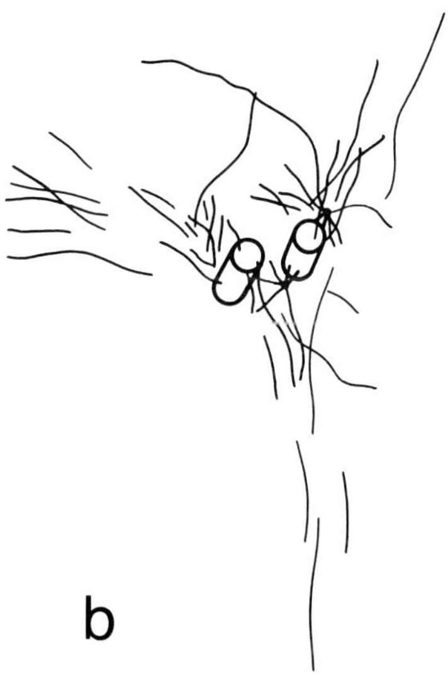

Fig. 3. (a) Stereopair of electron micrographs through the centrosomal region in PK cell. Tilt angle, 10°. Arrow indicates one of the two centrioles. Bar: 0.5 μm. (b) Profiles of MT oriented toward foci in the centrosome, reconstructed from five serial sections.

c. Microtubule Pattern around the Centrosome of PK Cells after Treatment with Various Agents. In these experiments, PK cells were exposed to 30 min to the various agents to observe the effect of this exposure on the distribution of MTs around the centrosome. The results obtained were compared with those obtained from untreated cells. The agents examined included dinitrophenol (DNP)

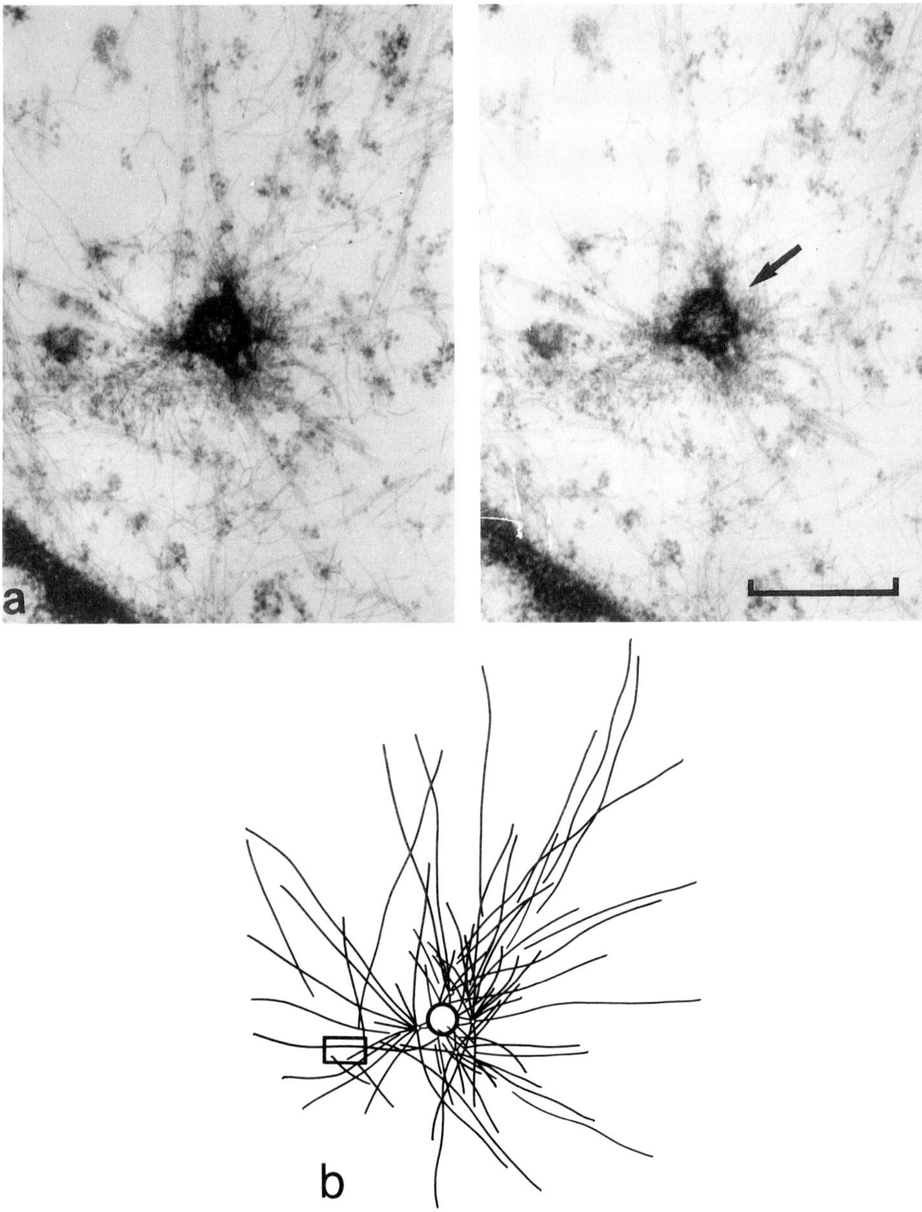

Fig. 4. (a) Stereopair of electron micrographs of the centrosomal region of PK cell after treatment with taxol (10 μ*M*, 10 min). Tilt angle, 10°. Arrow indicates centriole. Bar: 0.5 μm. (b) Profiles of MT oriented toward foci in the centrosome, reconstructed from five serial sections.

alone (800 μ*M*) and DNP (800 μ*M*) in combination with deoxyglucose (DOG) (1 μmM), FCCP (20 μ*M*), the calcium ionophore A23187 (0.1 μ*M*), sodium azide (20 m*M*), and ouabain (1 m*M*). Typical MT patterns in the centrosomal region after treatment with FCCP and sodium azide are shown in Fig. 5a,b.

The data obtained show that the total number of MTs radiating from the centrosome increased under the action of all the drugs studied. The maximal increase, slightly more than twofold, was observed after 30 min treatment with FCCP. Other agents had a smaller effect. In their ability to induce an increase in the total number of MTs, the agents may be arranged as follows: FCCP > taxol (10 min) ≈ ouabain ≈ A23187 ≈ sodium azide > DNP > DNP + DOG > control.

Most of this increase was due to an increase in the number of MTs radiating from pericentriolar satellites. In contrast to the results of taxol treatment, the number of MTs in Classes 1–4 increased almost proportionally with the increase in the total number of MTs after the treatment of cells with any of these agents. For the free MTs, the increase in the number of the longer Class 4 MTs was greater than the increase in the number of the shorter Class 3 MTs. The number of long Class 2 MTs attached to the centrosome increased proportionally to the general increase in the MT numbers, and increases as much as after the taxol treatment.

The number of MTs running out of the area studied also increased severalfold after all these treatments. Their numbers after FCCP, A23187, and ouabain treatments were similar to that seen after taxol treatment. The number of Class 5 MTs,

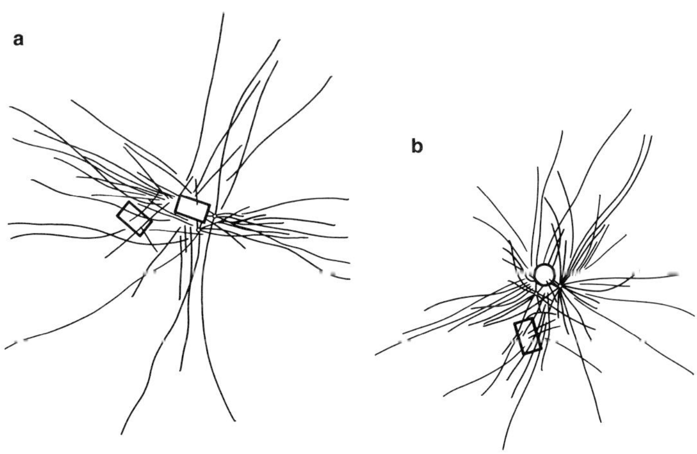

Fig. 5. Profiles of MT oriented toward foci in the centrosomal region of PK cell after treatment with (a) FCCP (20 μ*M*, 30 min) and (b) sodium azide (20 μ*M*, 30 min), reconstructed from six and five serial sections, respectively.

however, did not increase after these treatments, as after taxol treatment, but instead remained the same as in controls.

B. Microtubule Pattern around the Centrosome Fits the Conveyor Hypothesis of Microtubule Assembly

The general consensus in the literature (see Brinkley *et al.*, 1981; Kirschner and Mitchison, 1986; Alberts *et al.*, 1989 for review) is that a centrosome produces MTs in random directions and that their minus ends are continuously anchored in the centrosome and thus protected from depolymerization. Data obtained from four types of cultured cells, both fibroblast and epithelial cells, however, show that most of the MTs attached to MTOCs in the centrosome are comparatively short, not more than 2 μm in length, and do not extend to the peripheral lamelloplasm. In contrast, most of the long MTs that do extend to the lamelloplasm have their proximal ends free and are not connected to the MTOCs in the centrosome. Dynamic instability of MTs (Mitchison and Kirschner, 1984) does not explain the MTs with both ends free that have been described here and by others (Karsenti *et al.*, 1984; Raes *et al.*, 1984). Nevertheless, in the vicinity of the centrosome almost all MTs, both long and short, are either attached to, or have one of their ends pointing toward a MTOC in the centrosome.

The MT system in living cells is in a quasistationary state (Soltys and Borisy, 1985; Schulze and Kirschner, 1986) with individual MTs rapidly growing and shrinking at their tips. Direct dark-field microscopy shows that MT shrinkage may be very rapid and catastrophic (Kristofferson *et al.*, 1986) or moderate (Horio and Hotani, 1986). MTs with both ends free in the cytoplasm rapidly disappear (Sammak *et al.*, 1987; Tao *et al.*, 1988). Those MTs with MT-associated proteins depolymerize much more slowly *in vitro* than MTs polymerized from tubulin alone (Horio and Hotani, 1986). Direct experiments have shown that MTs elongate at their distal ends (Soltys and Borisy, 1985; Schulze and Kirschner, 1986). Labeling after a short pulse indicated that only a few newly formed MTs were connected to the centrosome (Soltys and Borisy, 1985; Schulze and Kirschner, 1986). The dynamics of MT growth in the vicinity of the centrosome could not be determined. The authors concluded that after a few minutes almost all MTs in the cell became labeled with the microinjected tubulin (Soltys and Borisy, 1985; Schulze and Kirschner, 1986; Geuens *et al.*, 1989). The mean rate of MT turnover was estimated to be between 4 and 20 min (Saxton *et al.*, 1984; Schulze and Kirschner, 1986; Sammak *et al.*, 1987; Tao *et al.*, 1988).

Analysis of the reconstitution of the MT network after cooling (Vorobjev and Chentsov, 1983) shows that, in the first few minutes of recovery, the number of MTs attached to various MTOCs in the centrosome is twice as great as the number normally present. This led us to suggest that two systems of MTs exist in an interphase cell (Vorobjev and Chentsov, 1983). One system of MTs forms a network in the cytoplasm in which the MTs do not radiate from definite foci. The other consists of MTs radiating from the centrosome.

Later, both of these MT systems were observed directly (Karsenti *et al.*, 1984; Racs *et al.*, 1984). To relate these two populations of MTs to each other, we suggested that the centrosome works as a conveyor (Vorobjev and Chentsov, 1983; Vorobjev and Nadezhdina, 1987). The central idea was that MTs growing from MTOCs in the centrosome are not continually anchored there by their minus ends. Instead, they have a certain potential to detach from the MTOC. As a result, each MT has a mean time during which it is attached to the centrosome. A new MT may start growing from each of the MT nucleating sites in the centrosome immediately after the previous MT leaves it. MTs that detach and move away from the centrosome either depolymerize rapidly or exist for various periods of time in the cytoplasm, forming the cytoplasmic MT network. The conveyor assembly hypothesis thus allows an explanation of the inverse relationship between the density of MTs in the centrosome, which reflects the activity of the MTOCs and the density of MTs in the rest of the cytoplasm (Frankel, 1976; Vorobjev and Nadezhdina, 1987).

At steady state, MTs with both ends free should disassemble (Kirschner, 1980); this mean time of existence depends on their mean depolymerization rate. MTs connected to the centrosome are more resistant to Colcemid or nocodazole treatments than free MTs (Weber and Osborn, 1981; Karsenti *et al.*, 1984; Vorobjev and Chentsov, 1985). A well-developed MT network will, therefore, develop only in those cells in which the growth rate of MTs from the centrosome is significantly higher than the disassembly rate of the free MTs.

The free MTs nevertheless can persist for some time in the absence of centrosomes; such MTs, however, have a low capacity for maintaining and regenerating themselves. For example, in centrosome-free fibroblast fragments, the number of MTs decreased considerably by 3 hr after microsurgery but disappeared completely only after 6 hr, whereas the fragments themselves remained viable for 8–10 hr (Gelfand *et al.*, 1985). After nocodazole treatment, only a few MTs polymerized in cytoplasts from L929 fibroblasts lacking centrioles (Karsenti *et al.*, 1984). In detached fibroblast fragments and in parts of axons, MTs were unable to reform after depolymerization (Gelfand *et al.*, 1985; Baas and Heidemann, 1986). In the axons, however, MT reassembly could occur if the depolymerization was incomplete and MT fragments remained (Gaas and Heidemann, 1986). These data support the idea that most of the cytoplasmic MTs with both ends free do not arise by spontaneous assembly but by detaching from the MTOC in the centrosome. The fate of such MTs may be different. They can be stabilized by MAPs or by capping their ends to prevent depolymerization.

As discussed earlier, the concept of dynamic instability alone does not completely explain the observed behavior of MTs. However, after being supplemented with the concept of conveyor assembly of MTs, it does give a satisfactory explanation of the data presented in this chapter.

According to this view, the MT pattern results from three processes going on simultaneously: initiation of MT assembly in the centrosome; detachment of MTs from the centrosome; and the dynamic instability of their ends modulated by MAPs, capping, and other factors. Working as a conveyor, the centrosome thus provides a mechanism for the renewal of the MT network. Centrosome activity, determined by counting the number of MTs growing simultaneously from the centrosomal region, reflects the rate of rebuilding of the MT network. This activity can be increased by exposing cells to different agents, for example, the metabolic inhibitors, as described previously. This activity is also altered during cell spreading on the glass surface (Gudima *et al.*, 1983b,1986,1988) and by concanavalin A stimulation of lymphocytes (Brown *et al.*, 1985).

C. Effect of Metabolic Inhibitors and Other Agents on the Microtubule Initiating Activity of the Centrosome

MT disassembly is regulated by ATP both *in vivo* and *in vitro* (Bershadsky and Gelfand, 1981; Spurck *et al.*, 1986), being inhibited by the depletion of ATP levels (Bershadsky and Gelfand, 1981; De Brabander *et al.*, 1982b; Spurck *et al.*, 1986). Thus, metabolic inhibitors depleting the ATP pool dramatically reduce the rate of MT depolymerization in the presence of Colcemid or vinblastine (Bershadsky and Gelfand, 1981). Also, in the presence of sodium azide during nocodazole treatment, numerous free MTs persist whereas they disappear much earlier without sodium azide (De Brabander *et al.*, 1982b).

Maro and Bornens studied the changes in the HeLa cell cytoskeleton induced by the oxidative phosphorylation uncouplers FCCP and DNP (Maro and Bornens, 1982). They described a complete disruption of the cytoplasmic MT network within 1 hr in cells treated with FCCP and a similar but weaker effect with DNP. After both treatments, MTOCs with short MTs radiating from them became visible. In contrast, sodium azide had no effect on the MT network (Maro and Bornens, 1982).

We found that treatments with DNP and FCCP did not result in the depolymerization of the existing cytoplasmic MT complex in PK cells, and that treatments with sodium azide gave similar results. The differences observed in the two studies may be due to the shorter treatment time used in our experiments. They may also be due to differences in the stability of cytoplasmic MTs among various normal and transformed cell lines, as described elsewhere (Brinkley *et al.*, 1981; Zimmer *et al.*, 1981). Treatment of interphase PtK_1 cells with DNP/DOG (5–10 min) also does not induce significant changes in the MT pattern. Only a slight decrease in the number of MTs and fragmentation of MTs was observed in some cells (Spurck *et al.*, 1986).

Our explanation of the results is that all the treatments we used (except ouabain) decreased ATP levels and thus stabilized cytoplasmic MTs (Bershadsky and Gelfand, 1981). From the results obtained we conclude that, among the four classes of MTs, the most unstable MTs are the long Class 4 MTs with both ends free, since their numbers increased by a larger percentage than the numbers of classes after all treatments. The Class 3 MTs (short MT with both ends free) seem to be transitional. Their number continuously increases with MT detachment from MT nucleating foci in the centrosome and at the same time is diminished as the MTs elongate and become members of Class 4. Our experiments with taxol support this hypothesis.

If one accepts the hypothesis of MT assembly by a conveyor mechanism, one can conclude that these agents promote, either directly or indirectly, the assembly of MTs in the centrosome but do not affect the frequency by which they detach from the MT nucleating foci. In cells treated with uncouplers of oxidative phosphorylation, the number of long attached MTs and short MTs increased proportionally. At the same time, the number of MTs that were not oriented toward any of the nucleating sites in the centrosome remained the same. These agents also induced an increase in the total number of MTs but not in the number of MTs running beyond the area studied, which remained the same as in the controls and was less than in taxol-treated cells. This indicates that the uncouplers did not promote MT elongation.

Concerning the dynamics of MT detachment from MT nucleating sites our results indicate the following. After treatment with taxol, the total number of MTs around the centrosome increased, but the number of short Class 3 MTs with both ends free decreased. At the same time there was a dramatic increase in attached long Class 3 MTs, which would explain the decrease in the number of free short Class 2 MTs after taxol treatment relative to controls. We conclude, therefore, that the stabilization of MTs by taxol also inhibits MT dissociation from the nucleating structures.

Numerous free MTs packed in bundles described after prolonged taxol treatment (De Brabander *et al.*, 1981,1982a,b) result from spontaneous MT assembly in the cytoplasm. Also, the absence of the centrosome attached MTs after such treatment indicates that the pre-existing MTs in such cells had become dissociated from the centrosome.

Incubation of activated lymphocytes with taxol (1–4 hr) also induced a dramatic reorganization of MT and IF (Brown *et al.*, 1985; see Chapter 10). The total number of MTs radiating from the centrosome was calculated and shown to increase about fivefold (from 40–45 to 160–200) after the taxol treatment. In our experiments, the total number of MTs did not increase as much, but only doubled.

This may be due to differences in the incubation time, in the cell type used, or in the method used to count the MTs.

Treatment with the other agents used in our experiments resulted in an increase in the number of MT nucleating foci known as pericentriolar satellites and also in the number of MTs themselves. A preferential increase in the number of long Class 4 MTs with both ends free took place. Thus, unlike taxol, these agents did not seem to stabilize the attachment of MTs to the centrosome, since there was an increase in the number of free MTs simultaneously with an increase in the attached ones. Although the agents used stabilized MTs indirectly as a result of depletion of ATP levels, the relatively large increase in long Class 4 MTs with both ends free indicates that the metabolic inhibitors may actually promote the separation of MTs from the nucleating sites.

Our data concerning the dynamics of MTs around the centrosome appear to be in contrast to those of Sammak and Borisy (1988), who showed a rapid turnover of MTs at the cell's periphery. To reconcile these observations, we suggest that the MTs near the centrosome are in a different environment than those in the lamelloplasm. Such a proposal was made by De Brabander and co-workers (1982a,b), who considered another model for the initiation of MT assembly, namely, that the threshold for MT assembly in the vicinity of centrosomes is lower than that in other regions of the cytoplasm, and suggested that some of the MTs formed in the vicinity of the centrosome have been captured and remain attached to the centrosome while growing from their distal end. To understand how MT assembly is initiated by the centrosome, additional experiments are necessary.

III. CENTROSOME AND INTERMEDIATE FILAMENTS

IF proteins, contrary to tubulin and microtubule associated proteins (MAPs) are insoluble in solutions at physiological ionic strength. Almost nothing is known about their polymerization *in vivo* nor how the IF network is rebuilt after disassembly. IFs may form a continuous network extending throughout the cytoplasm or they may be packed in several bundles or aggregates. Under certain conditions, the IF network collapses into an aggregate in the perinuclear region of the cell (Traub, 1985).

Collapse of IFs around the nucleus occurs after treatment of cells with antitubulin drugs or vanadate (Goldman, 1971; Goldman and Knipe, 1972; Franke *et al.*, 1978; Wang and Choppin, 1981). A similar collapse takes place after the microinjection of antibodies to keratin or vimentin (Eckert and Daley, 1981; Gawlitta *et al.*, 1981; Klymkowsky *et al.*, 1983). In some cases, the collapse of the IF network is reversible and, after removal of the drug, the IFs spread back through the cytoplasm and reacquire the normal pattern of distribution (Herman and Albertini, 1982; Geuens *et al.*, 1983).

It was thought that the formation of the perinuclear IF aggregates and the subsequent dispersal of the IFs to their normal pattern on removal of the drug indicated that an IF organizing center exists inside the aggregate (Eckert *et al.*, 1982). However, the aggregates were very large and rather homogeneous, and no specific IF organizing structures could be found inside these aggregates by electron microscopy.

The perinuclear aggregates of IFs just described often surround the centrosomes or are situated near these organelles (Starger *et al.*, 1978; Blose, 1980; Goldman *et al.*, 1980). In cultured astrocyte precursor cells, the presence of glial-specific IFs were first detected in the perinuclear region before they appeared in the rest of the cytoplasm. Double labeling indicated that the perinuclear site of assembly of this type of IF corresponds to the centrosomal region (Kalnins *et al.*, 1985). It is, therefore, possible that the centrosome is an IF organizing center (IFOC) (Goldman *et al.*, 1980), as well as a MTOC.

On the other hand, the centrosome may only appear to be such a center because the IFs are distributed along it and form crosslinks with the MTs radiating from the centrosome. To prove that the centrosome is also an IFOC, it would be necessary to show that the growth of IFs can occur from the centrosomal region in the absence of the IF aggregate and without disruption of the pre-existing MT network. We have tried to determine whether the centrosome is an IFOC.

A. Intermediate Filament Foci in the Centrosome

Examination of the L fibroblasts by electron microscopy shows that in interphase cells the IF network is condensed around the centrosome. In the centrosome, numerous IFs are found around both centrioles, around free foci from which MTs radiated, and along the MTs. In regions more than 0.5 μm away from the centrioles, the IFs run parallel to the MTs. Although the IFs are often in close proximity to the centrioles and MTs, direct contacts between the IFs and these structures are seldom seen. Using stereo analysis, however, we have succeeded in finding definite foci from which IFs radiate in the centrosomes. They are moderately dense granules 0.020–0.025 μm in diameter (Fig. 6) with 3–7 IFs radiating from each granule. These foci are situated within 0.3 μm of the centrioles. In other parts of the cytoplasm, the IFs are generally packed into curved bundles in a semiparallel arrangement, as described elsewhere (Goldman, 1971; Goldman *et al.*, 1980). We never found a focus from which the IFs radiate that simultaneously served as a focus for MTs. The organizing centers therefore appear to be two distinct structures.

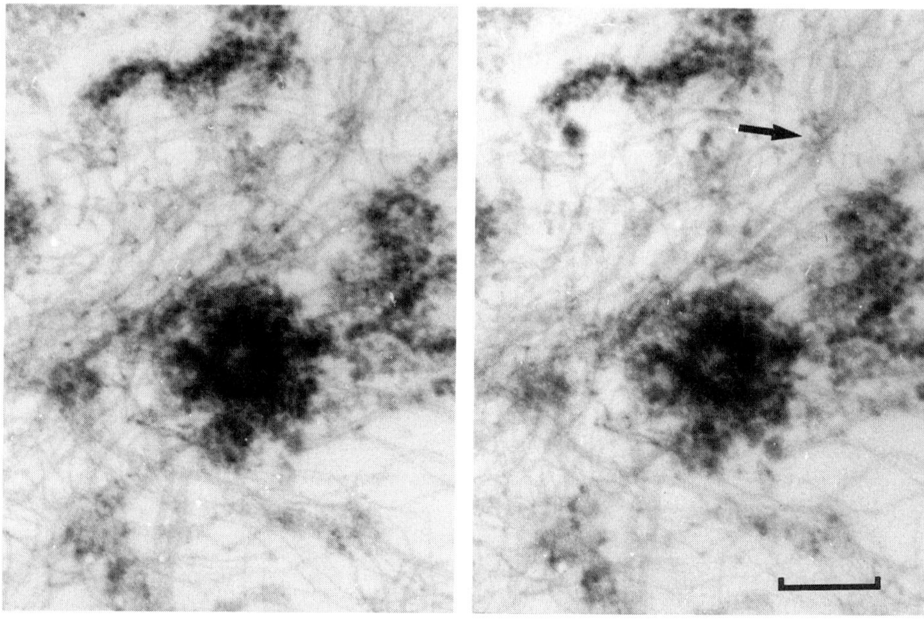

Fig. 6. Stereopair of electron micrographs of a semithin section through the centrosomal region in L fibroblasts showing focus from which IF radiate (arrow). Bar: 0.2 μm.

B. Effect of Ultracentrifugation of Living Cells on Their Intermediate Filament System: Identification of Intermediate Filament Organizing Centers

Centrifugation of PK cells on coverslips at 20,000g for 2 hr led to a centrifugal shift of nuclei. The regions with the highest concentration of MTs moved together with nuclei and always occupied a region at their centripetal poles. At the same time, the IF system went through dramatic changes (Fig. 7a,b). After centrifugation, most of the IFs occupied the centripetal part of the cytoplasm, forming large aggregates in which MTs were absent (Fig. 7b). Some bundles of IFs connected the aggregates with the nuclei.

Electron microscopy of centrifuged cells confirmed that the vimentin aggregates in the centripetal parts of the cells consist of densely packed IFs (data not shown). It also showed that this region contained lipid granules, commonly found in cultured PK cells, closely bound to the aggregated IFs. During centrifugation the granules float and probably take the IFs with them. Close links between lipid granules and IFs have been described elsewhere (Mayerson and Brumgaugh, 1981; Franke *et al.*, 1978). IFs may also move to the centripetal pole in response to the shift in nuclear position or to the deformation of the cell by the centrifugation. After return of the centrifuged cells to normal culture conditions, a rearrangement of IFs takes place; in 2–3 hr the IF pattern becomes normal again. We studied this return to the normal pattern by examining cells fixed at different time intervals after the end of centrifugation.

When examined 40–60 min after centrifugation, foci in many cells were found near the centripetal pole of the nucleus. These foci appeared as dots or small patches from which IFs radiated (Fig. 8a,b), and form what we shall refer to as the IFOC.

In some cells it was possible to observe simultaneously the aggregates in the lamelloplasm in the centripetal part of the cell and the IFOC near the nucleus. Generally, however, only one or the other of these structures was present in each cell. When both were present, bundles of IFs connecting the IFOC near the nucleus with the IF aggregate at the periphery of the cell were always present. The size of the IFOC and the centripetal aggregate had an inverse relationship: the larger was the one, the smaller was the other.

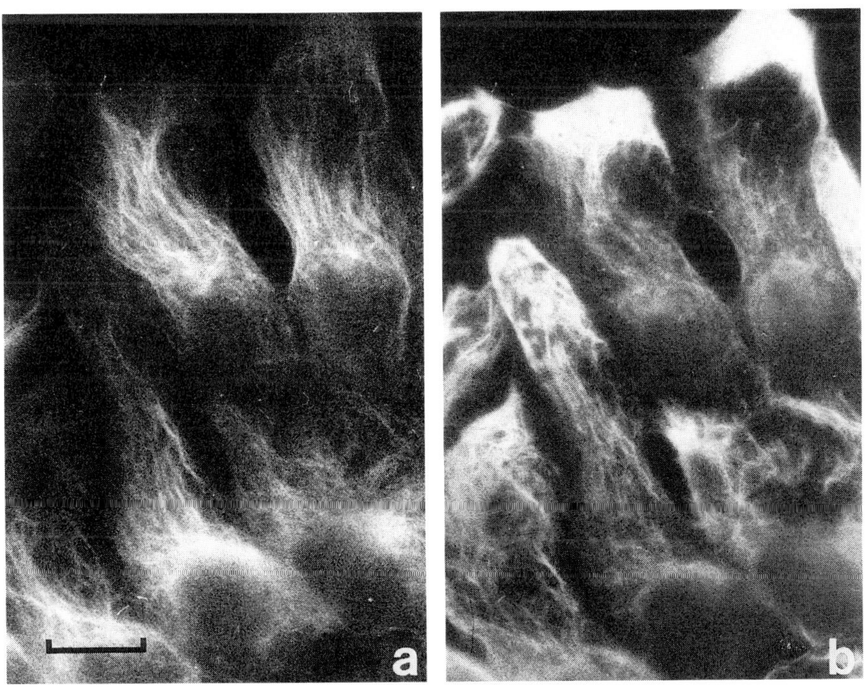

Fig. 7. Double immunofluorescence staining with antibodies to (a) tubulin and (b) vimentin, showing the distribution of MT (a) and IF (b) in PK cells after centrifugation (20,000 g, 2 hr). IF aggregates are visible in the centripetal part of the cells toward the top in b. Bar: 10 μm.

As mentioned earlier, the centrosomal region may be identified by immunofluorescence with antibodies to tubulin because of the high density of MTs radiating from it. Double labeling with antibodies to tubulin and vimentin indicated that, as far as can be determined by light microscopy in PK cells, the IFOC and MTOC are located in the same position (Fig. 8). In some cells, two MTOCs per cell were present. These cells also had two IFOCs located in the same position. The diameter of the vimentin IFOC was always much greater than that of the centrosome. Electron microscopy of the IFOC-containing regions showed that they contain the centrosome surrounded by numerous IFs. On ultrathin sections, the concentration of IFs in these regions seemed to be similar to that in the centrosomal regions of noncentrifuged control cells. Double labeling with antibodies to tubulin and vimentin indicated that some of the MTs and IFs radiating from the centrosomal region ran parallel to each other.

IFOCs were also found in the centrosomal region of PK cells fixed 40 min after 3–4 hr centrifugation (Fig. 8c,d). In some cells, the IFOC appeared even after a 6 hr centrifugation. In the others, only the centripetal aggregate of IFs was seen.

C. Relationship of Centrosome to Intermediate Filaments

The data on the relationship between the IFOC and the centrosome are controversial. In some cell lines, the centrosome and IF aggregates appearing as a result of IF collapse are observed in the same region, but in others they seem to be separated, though not far from each other (Eckert *et al.*, 1982,1984a).

Observations made on IF aggregates in the perinuclear region led to the suggestion that the IF aggregates contain the IFOCs, which play a key role in the organization and distribution of IFs (Goldman *et al.*, 1980; Eckert *et al.*, 1982,1984a,b). Unfortunately, the IFOCs are difficult to observe after the formation of perinuclear aggregates of IF. The situation is quite different in the centrifuged cells, in which we have identified the IFOCs in the centrosomal region by light and electron microscopy. There the great majority of IFs have aggregated at a site far from the nucleus and the suggested sites for the IFOC. This enabled us to observe distinct IFOCs during the reformation of the IF network. Visualization of the IFOC, especially during the longer centrifugations, may also be possible because cells are adapting to an increased gravitational force. Contrary to nocodazole or taxol treatment (Maro *et al.*, 1984), which also greatly alter the MT pattern, centrifugation of living cells led to the reorganization of the IF without

---

Fig. 8. Double immunofluorescence staining with antibodies to (a,c) tubulin and (b,d) vimentin, showing the distribution of MT (a,c) and IF (b,d) in PK cells 40 min after centrifugation at 20,000 g for 2 hr (a,b) and 3 hr (c,d). Arrows indicate the IF organizing centers. Bar: 10 μm.

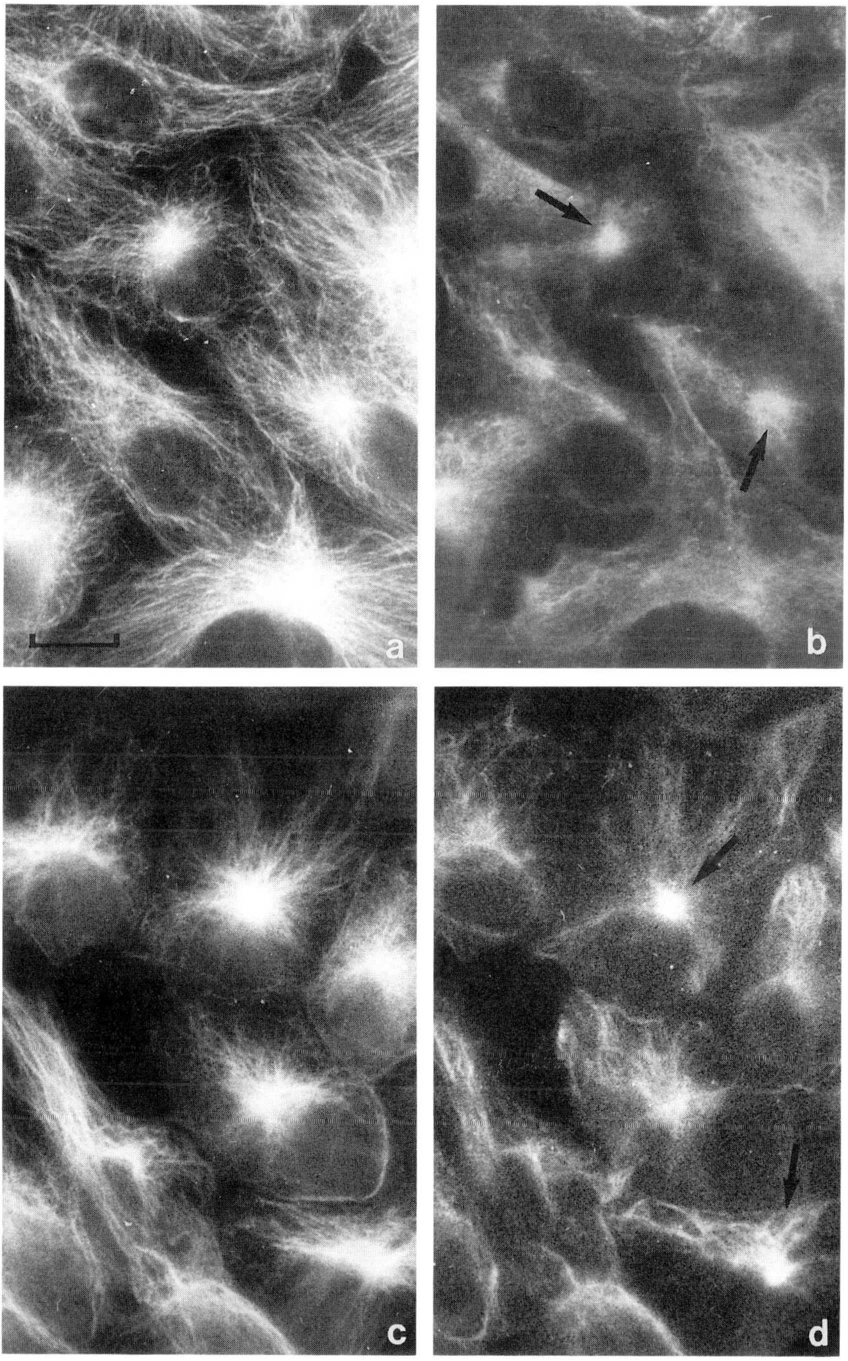
a
b
c
d

dramatic changes in the MT pattern. This difference in MT organization may influence the dynamics of IF network reformation and the relationship of the IF to the centrosome.

As mentioned previously, in L fibroblasts the IF network is condensed around the centrosome. In these cells, we succeeded in finding foci from which IFs radiate without any manipulation of the living cells. These foci were distinct from the MT nucleating centers of the centrosome. The observations made indicate that IFs may assemble in the centrosomal region and, with the help of MTs, disperse throughout the cytoplasm.

Data are still insufficient to draw definite conclusions about the position and functions of the foci from which the IFs radiate. These foci may not exist continuously. Rather, they may be transient structures that function only during rapid changes in the IF network. The IFOCs may appear only when there is a reorganization of the pre-existing IFs or polymerization of new ones. They may not be present in the IF aggregates.

In general, the centrosome seems to be a specific organelle capable of initiating the assembly of both MTs and IFs. It does not, however, determine the pattern of distribution of MTs and IFs in the cytoplasm. The distribution of these components of the cytoskeleton is regulated by other cytoplasmic factors and depends very little on the centrosome. Under the action of different agents the centrosome can be induced to produce extra MTs and possibly IFs.

REFERENCES

Alberts, B., Bray, D., Lewis, J., Raff, M., Roberts, K., and Watson, J. D. (1989). Molecular Biology of the Cell, 2d Ed. Garland, New York.

Anderson, D. C., Wible, L. J., Hughes, B. J., Smith, C. W., and Brinkley, B. R. (1982). Cytoplasmic microtubules in polymorphonuclear leucocytes: Effects of chemotactic stimulation and colchicine. *Cell* **31,** 719–729.

Baas, P. W., and Heidemann, S. R. (1986). Microtubule reassembly from nucleating fragments during regrowth of amputated neurites. *J. Cell Biol.* **103,** 917–927.

Bershadsky, A. D., and Gelfand, V. I. (1981). ATP-dependent regulation of cytoplasmic microtubule disassembly. *Proc. Natl. Acad. Sci. U.S.A.* **78,** 3610–3613.

Blose, S. H. (1980). The microtubule organizing centers, MTOCs, are surrounded by ten nanometer filament ring in endothelial cells in interphase and mitosis. *Eur. J. Cell Biol.* **22,** 373–379.

Brenner, S. L., and Brinkley, B. R. (1982). Tubulin assembly sites and the organization of microtubule arrays in mammalian cells. *Cold Spring Harbor Symp. Quant. Biol.* **46,** 241–254.

Brinkley, B. R., Fuller, G. M., and Highfield, D. P. (1976). Tubulin antibodies as probes for microtubules in dividing and non-dividing mammalian cells. *In* "Cell Motility" (R. Goldman, T. D. Pollard, and J. Rosenbaum, eds.) pp. 435–456. Cold Spring Harbor Laboratory Press, Cold Spring Harbor, New York.

Brinkley, B. R., Fistel, S. H., Marcum, J. M., and Pardue, R. L. (1980). Microtubules in cultured cells: Indirect immunofluorescent staining with tubulin antibody. *Int. Rev. Cytol.* **63,** 59–95.

Brinkley, B. R., Cox, S. M., Pepper, D. A., Wible, L., Brenner, S. L., and Pardue, R. L. (1981). Tubulin assembly sites and organization of cytoplasmic microtubules in cultured mammalian cells. *J. Cell Biol.* **90,** 554–562.

Brown, D. L., Little, J. E., Chaly, N., Schweitzer, I., and Paulin-Levasseur, M. (1985). Effects of taxol on microtubule organization in mouse splenic lymphocytes and on response to mitogenic stimulation. *Eur. J. Cell Biol.* **37,** 130–139.

Cassimeris, L. U., Wadsworth, P., and Salmon, E. D. (1986). Dynamics of microtubule depolymerization in monocytes. *J. Cell Biol.* **102,** 2023–2032.

De Brabander, M., Geuens, G., Nuydens, R., Willebrords, R., and De Mey, J. (1982a). Taxol induces the assembly of free microtubules in living cells and blocks the organizing capacity of the centrosomes and kinetochores. *Proc. Natl. Acad. Sci. U.S.A.* **78,** 5608–5612.

De Brabander, M., Geuens, G., Nuydens, R., Willebrords, R., and De Mey, J. (1982b). Microtubule stability and assembly in living cells: The influence of metabolic inhibitors, taxol, and pH. *Cold Spring Harbor Symp. Quant. Biol.* **46**, 227–240.

Eckert, B. S., and Daley, R. D. (1981). Response of cultured epithelioid cells (PtK_1) to microinjection of antibody specific to keratin. *J. Cell Biol.* **41,** 227–230.

Eckert, B. S., Daley, R. D., and Parysek, L. M. (1982). *In vivo* disruption of the cytokeratin cytoskeleton in cultured epithelial cells by microinjection of antikeratin: Evidence for the presence of an intermediate filament organizing center. *Cold Spring Harbor Symp. Quant. Biol.* **46**, 403–412.

Eckert, B. S., Caputi, J. E., and Warren, R. H. (1984a). Dynamics of keratin filaments and the intermediate filament distribution in PtK_1 cells by double immunofluorescence. *Cell Motil.* **4,** 169–181.

Eckert, B. S., Caputi, J. E., and Brinkley, B. R. (1984b). Localization of the centriole and keratin intermediate filaments in PtK_1 cells by double immunofluorescence. *Cell Motil.* **4,** 241–247.

Franke, W. W., Schmid, E., Osborn, M., and Weber, K. (1978). Different intermediate-sized filaments distinguished by immunofluorescence microscopy. *Proc. Natl. Acad. Sci. U.S.A.* **75,** 5034–5038.

Frankel, F. R. (1976). Organization and energy-dependent growth of microtubules in cells. *Proc. Natl. Acad. Sci. U.S.A.* **73,** 2798–2802.

Gawlitta, W., Osborn, M., and Weber, K. (1981). Coiling of intermediate filaments induced by microinjection of a vimentin-specific antibody does not interfere with locomotion and mitosis. *Eur. J. Cell Biol.* **26,** 83–90.

Geiger, B., and Singer, S. J. (1980). Association of microtubules and intermediate filaments in chicken gizzard cells as detected by double immunofluorescence. *Proc. Natl. Acad. Sci. U.S.A.* **77,** 4769–4773.

Gelfand, V. I., Glushankova, N. A., Ivanova, O. Y., Mittelman, L. A., Pletyushkina, O. Y., Vasiliev, J. M., and Gelfand, I. M. (1985). Polarization of cytoplasmic fragments microsurgically detached from mouse fibroblasts. *Cell Biol. Int. Rep.* **9,** 883–892.

Geuens, G., De Brabander, M., Nyudens, R., and De Mey, J. (1983). The interaction between microtubules and intermediate filaments in cultured cells treated with taxol and nocodazole. *Cell Biol. Int. Rep.* **7,** 35–47.

Geuens, G., Hill, A. M., Levilliers, N., Adoutte, A., and De Brabander, M. (1989). Microtubule dynamics investigated by microinjection of *Paramecium* axonemal tubulin: Lack of nucleation but proximal assembly of microtubules at the kinetochore during prometaphase. *J. Cell Biol.* **108,** 939–953.

Goldman, R. D. (1971). The role of three cytoplasmic fiber in BHK21 cell motility. I. Microtubules and the effect of colchicine. *J. Cell Biol.* **51,** 752–762.

Goldman, R. D., and Knipe, D. M. (1972). Functions of cytoplasmic fibers in muscle cell motility. *Cold Spring Harbor Symp. Quant. Biol.* **46,** 523–534.

Goldman, R. D., Hill, B. F., Steinert, P., Whitman, M. A., and Zackaroff, R. V. (1980). Intermediate filament–microtubule interactions: Evidence in support of a common organization center. *In*

"Microtubules and Microtubule Inhibitors" (M. De Brabander and J. De Mey, eds.) pp. 91–102. Elsevier, New York.

Gudima, G. O., Vorobjev, I. A., and Chentsov, Y. S. (1983a). The behavior of the cell center during fibroblast spreading. *Biol. Nauki (Russian)* **N7,** 45–50.

Gudima, G. O., Vorobjev, I. A., and Chentsov, Y. S. (1983b). Change in the structure of the cell centre upon polarization and migration of fibroblasts. *Tsytologia (Russian)* **25,** 883–888.

Gudima, G. O., Vorobjev, I. A., and Chentsov, Y. S. (1986). The behavior of the cell center during epithelial cell spreading. *Tsytologia (Russian)* **28,** 576–581.

Gudima, G. O., Vorobjev, I. A., and Chentsov, Y. S. (1988). Centriolar location during blood cell spreading and motion *in vitro*: An ultrastructural analysis. *J. Cell Sci.* **89,** 225–241.

Herman, B., and Albertini, D. F. (1982). The intracellular movement of endocytic vesicles in cultured granulosa cells. *Cell Motil.* **2,** 583–597.

Horio, T., and Hotani, H. (1986). Visualisation of the dynamic instability of individual microtubules by dark-field microscopy. *Nature (London)* **321,** 605–607.

Kalnins, V. I., Subrahmanyan, L., and Fedoroff, S. (1985). Assembly of glial intermediate filament protein is initiated in the centriolar region. *Brain Res.* **345,** 322–326.

Karsenti, E., Kobayashi, S., Mitchison, T., and Kirschner, M. (1984). The role of centrosome in organizing of microtubule array: Properties of cytoplasts containing or lacking centrosomes. *J. Cell Biol.* **98,** 1763–1776.

Kirschner, M. W. (1980). Implications of treadmilling for the stability and polarity of actin and tubulin polymers *in vivo*. *J. Cell Biol.* **86,** 330–334.

Kirschner, M. W., and Mitchison, T. (1986). Beyond self-assembly: From microtubules to morphogenesis. *Cell* **45,** 329–342.

Klymkowsky, M. W., Miller, R. H., and Lane, E. B. (1983). Morphology, behavior, and interaction of cultured cells after the antibody-induced disruption of keratin filament organization. *J. Cell Biol.***96,** 494–500.

Kristofferson, D., Mitchison, T., and Kirschner, M. W. (1986). Direct observation of steady-state microtubule dynamics. *J. Cell Biol.* **102,** 1007–1019.

Kuriyama, R., and Borisy, G. G. (1981). Microtubule-nucleating activity of centrosomes in Chinese hamster ovary cells is independent of the centriole cycle but coupled to the mitotic cycle. *J. Cell Biol.* **91,** 822–826.

Maro, B., and Bornens, M. (1982). Reorganization of HeLa cell cytoskeleton induced by an uncoupler of oxidative phosphorylation. *Nature (London)* **295,** 334–336.

Maro, B., Paintrand, M., Sauron, M. -E., Paulin, D., and Bornens, M. (1984). Vimentin filaments and centrosomes. Are they associated? *Exp. Cell Res.* **150,** 452–458.

Mayerson, P. L., and Brumbaugh, J. A. (1981). Landever, a chick melanocyte mutant with defective melanosome translocation: A possible role for 10-nm filaments and microfilaments, but not microtubules. *J. Cell Sci.* **51,** 25–51.

Mitchison, T., and Kirschner, M. (1984). Microtubule assembly by isolated centrosomes. *Nature (London)* **312,** 232–237.

Okabe, S., and Hirokawa, N. (1988). Microtubule dynamics in nerve cells: Analysis using microinjection of biotinylated tubulin into PC-12 cells. *J. Cell Biol.* **107,** 651–664.

Osborn, M., and Weber, K. (1976a). Cytoplasmic microtubules appear to grow from an organizing structure towards the plasma membrane. *Proc. Natl. Acad. Sci. U.S.A.* **73,** 867–871.

Osborn, M., and Weber, K. (1976b). Tubulin-specific antibody and the expression of microtubules in 3T3 cells after attachment to a substratum. Further evidence for the polar growth of cytoplasmic microtubules *in vivo*. *Exp. Cell Res.* **103,** 331–340.

Osborn, M., Born, T., Koitsch, H-J., and Weber, K. (1978). Stereoimmunofluorescence microscopy. I. Three-dimensional arrangement of microfilaments, microtubules, and tonofilaments. *Cell* **14,** 477–488.

Raes, M., De Brabander, M., and Remacle, J. (1984). Polyploid cells in ageing hamster fibroblasts *in vitro*: Possible implication of the centrosome. *Eur. J. Cell Biol.* **35,** 70–80.

Sammak, P. J., Gorbsky, G. J., and Borisy, G. G. (1987). Microtubule dynamics *in vivo*: A test of mechanisms of turnover. *J. Cell Biol.* **104,** 395–405.

Sammak, P. J., and Borisy, G. G. (1988). Direct observation of microtubule dynamics in living cells. *Nature (London)* **332,** 724–726.

Saxton, M. D., Stemple, D. L., Leslie, R. J., Salmon, E. D., Zavortink, M., and McIntosh, J. R. (1984). Tubulin dynamics in cultured mammalian cells. *J. Cell Biol.* **99,** 2175–2186.

Schiff, P. B., and Horwitz, S. B. (1980). Taxol stabilizes microtubules in mouse fibroblast cells. *Proc. Natl. Acad. Sci. U.S.A.* **77,** 1561–1565.

Schliwa, M., Euteneuer, U., Herzog, W., and Weber, K. (1979). Evidence for rapid structural and functional changes of the melanophore microtubule-organizing center upon pigment movements. *J. Cell Biol.* **83,** 623–632.

Schliwa, M., Pryzvanski, K. B., and Euteneuer, U. (1982). Centrosome splitting in neutrophils: An unusual phenomenon related to cell activation and motility. *Cell* **31,** 705–717.

Schliwa, M., Pryzvanski, K. B., and Borisy, G. G. (1983). Tumor promoter-induced centrosome splitting in human polymorphonuclear leukocytes. *Eur. J. Cell Biol.* **32,** 75–85.

Schulze, E., and Kirschner, M. (1986). Microtubule dynamics in interphase cells. *J. Cell Biol.* **102,** 1020–1031.

Soltys, B. J., and Borisy, G. G. (1985). Polymerization of tubulin *in vivo*: Direct evidence for assembly onto microtubule ends and from centrosomes. *J. Cell Biol.* **100,** 1682–1689.

Spiegelman, B. M., Lopata, M. A., and Kirschner, M. W. (1979). Multiple sites for the initiation of microtubule assembly in mammalian cells. *Cell* **16,** 239–252.

Spurck, T. P., Pickett-Heaps, J. D., and Klymkowsky, M. W. (1986). Metabolic inhibitors and mitosis. II. Effects of dinitrophenol/deoxyglucose and nocodazole on the microtubule cytoskeleton. *Protoplasma* **131,** 60–74.

Starger, J. M., Brown, W. E., and Goldman, R. D. (1978). Biochemical and immunological analysis of rapidly purified 10-nm filaments from baby hamster kidney (BHK-21) cells. *J. Cell Biol.* **78,** 93–108.

Tao, W., Walter, R. J., and Berns, M. W. (1988). Laser-transacted microtubules exhibit individuality of regrowth, however most free new ends of the microtubules are stable. *J. Cell Biol.* **107,** 1025–1036.

Traub, P. (1985). Intermediate Filaments, Springer-Verlag, Berlin.

Tsukita, S., and Ishikawa, H. (1981). The cytoskeleton in myelinated axons: Serial section study. *Biomed. Res.* **2,** 424–437.

Vorobjev, I. A., and Chentsov, Y. S. (1983). The dynamics of reconstitution of microtubules around the cell center after cooling. *Eur. J. Cell Biol.* **30,** 149–153.

Vorobjev, I. A., and Chentsov, Y. S. (1985). Centrioles and microtubules in interphase cells under Colcemid action: Dose- and time-dependent effects. *Tsytologia (Russian)* **27,** 1101–1105.

Vorobjev, I. A., and Nadezhdina, E. S. (1987). The centrosome and its role in the organization of microtubules. *Int. Rev. Cytol.* **106,** 227–293.

Weber, K., and Osborn, M. (1981). Microtubule and intermediate filament networks in cells viewed by immunofluorescence microscopy. *In* "Cytoskeletal Elements and Plasma Membrane Organization" (G. Poste and G. L. Nicolson, eds.) pp. 1–53. North-Holland, Amsterdam.

Wang, E., and Choppin, P. W. (1981). Effect of vanadate on intracellular distribution and function of 10-nm filaments. *Proc. Natl. Acad. Sci. U.S.A.* **78,** 2363–2367.

Weakley, B. S. (1972). A Beginner's Handbook in Biological Microscopy. Churchhill Livingstone, Edinburg.

Zimmer, D. B., Turner, D. S., Goldstein, M. A., and Brinkley, B. R. (1981). A quantitative ultrastructural study of microtubules in transformed and nontransformed cells *in vitro*. *Cell Biol. Int. Rep.* **5,** 1115–1125.

5

Monoclonal Antibodies to Microtubule-Organizing Center Antigens

RYOKO KURIYAMA

Department of Cell Biology and Neuroanatomy
University of Minnesota
Minneapolis, Minnesota 55455

I. INTRODUCTION

Many of the microtubule-mediated functions in eukaryotic cells are regulated through the formation and breakdown of highly ordered microtubule-containing structures. The key element in the organization of such complex and dynamic arrays is the microtubule-organizing center (MTOC) (Pickett-Heaps, 1969), which consists of electron-dense cytoplasmic foci from which microtubules originate. In animal cells, interphase and mitotic microtubules are formed from a centrosome composed of a pair of centrioles and a surrounding amorphous cloud of pericentriolar material. Plants and lower eukaryotes, on the other hand, contain MTOCs very different in structure from animal centrosomes; therefore, these structures have been given different names. This structural diversity, as revealed by ex-

Copyright © 1992 Academic Press, Inc.
All rights of reproduction in any form reserved.

tensive observations of mitosis/meiosis structures in a variety of organisms (Heath, 1981), is in fact one of the most prominent features of MTOCs. Despite such structural differences, MTOCs organize microtubules in a similar manner in a wide variety of cells. It is thus reasonable to assume that these morphologically diverse structures would be closely related to each other in their composition. Analysis of their molecular components would therefore provide insight into the mechanism of MTOC function and its regulation.

Several approaches are possible for such studies. Centrosomes have been isolated and their microtubule-nucleating activity has been demonstrated *in vitro* (Gould and Borisy, 1977). The centrosomal capacity to initiate microtubule assembly is removed by protease treatment, suggesting that the activity resides in protein components (Ring *et al.,* 1980; Roobol *et al.,* 1982; Kuriyama, 1984). Although this approach is simple and straightforward, it does not reveal much about the specific MTOC components involved. A more direct approach would be purification and characterization of MTOCs by biochemical fractionation. However, until recently it was not possible to purify a sufficient number of centrosomes/MTOCs free from other subcellular components for reliable biochemical analysis (Mitchison and Kirschner, 1984; Bornens *et al.,* 1987, Komesli *et al.,* 1989; Rout and Kilmartin, 1990; see Chapter 1), because very few are present in each cell and because their activity is relatively unstable. Thus, although mammalian centrosomes are biochemically the best characterized MTOCs, our knowledge about their molecular composition is still rather poor. An alternative strategy that has been quite powerful in identifying molecular components is characterization of specific antibodies to MTOC antigens. As a matter of fact, an extraordinarily large number of papers describing antibodies specific to MTOCs, mainly in animal centrosomes, have accumulated during the last few years. In this review on antibodies to MTOC antigens, I will first attempt to survey the existing antibody probes, classify them into several categories, and review briefly some of the results obtained. Available centrosomal antibodies are summarized in tabular form and our results on characterization of monoclonal antibodies directed against several morphologically distinct MTOCs are presented. Finally, I will discuss what we have learned about MTOCs using antibodies as probes and how we can use these probes for future studies.

II. ANTIBODY PROBES SPECIFIC FOR CENTROSOMES

Numerous sera or antibodies have been reported that label MTOCs when examined by indirect immunofluorescence microscopy. These sera have been screened mainly on centrosomes in cultured mammalian cells and their cross-reactivity with other types of MTOCs has not, with a few exceptions, been examined.

A. Rabbit Nonimmune and Human Autoimmune Sera

In 1978, Connolly and Kalnins first reported that about 10% of the sera collected from rabbits before immunization with materials of interest already included antibodies reactive to centrioles/basal bodies (Connolly and Kalnins, 1978). Centrosomal antibodies are also frequently found in sera from patients with connective diseases, especially the CREST (calcinosis, Raynaud's phenomenon, esophageal dysmotily, sclerodactyly, and telangiectasias) syndrome. By screening over 100 serum samples, it was concluded that specific antibodies to centrosomal antigens are present in 1–4% of patient sera (Tuffanelli *et al.*, 1983; Senécal *et al.*, 1985). These findings provided us with a very simple and efficient way to obtain specific reagents useful for studying the structure and function of this cellular organelle. A list of the rabbit nonimmune and human autoimmune sera used thus far to examine the subcellular distribution of centrosomes by indirect immunofluorescent microscopy is presented in Table I. Some nonimmune sera seem to stain both centrioles and basal bodies. For instance, Z-3 and Z-7, rabbit sera screened by Connolly and Kalnins, showed fluorescent granules, 0.6–0.9 μm long, corresponding in size to centrioles (1978). The vast majority of the sera, however, labels the whole centrosomal structure. For example, serum 5051 from a scleroderma patient was the first autoantibody characterized extensively (Calarco-Gillam *et al.*, 1983). The specificity of 5051 for the pericentriolar material in centrosomes was elegantly demonstrated using meiotic spindles in mouse eggs, which do not contain centrioles (Calarco-Gillam *et al.*, 1983). The antibody cross-reacts with a number of mammalian cell lines (Calarco-Gillam *et al.*, 1983; Sellitto & Kuriyama, 1988b), *Xenopus* (Calarco-Gillam, *et al.*, 1983), chicken osteoclasts (Moudjou *et al.*, 1989a), sea urchin eggs (Schatten *et al.*, 1985), crane fly (Bastmeyer *et al.*, 1986), and yeast (Snyder and David, 1988; Snyder, 1989), and has greatly advanced our knowledge of the distribution and behavior of the pericentriolar material in a wide variety of organisms. 5051 also has been shown to immunostain spindle poles and regions of the nuclear envelope in higher plant cells. The lack of apparent focal structures for microtubule assembly in plant cells has often led to suggestions that, unlike animal cells, plant cells might not need MTOCs for organization of microtubules. However, the presence of antigens that cross-react with serum 5051 (Clayton *et al.*, 1985; Wick, 1985) strongly indicates that at least some conserved components of MTOCs are involved in microtubule organization in higher plants.

Although many anticentrosomal antibodies obtained by screening of nonimmune/autoimmune sera recognize centrosomes in both interphase and mitotic cells, several examples of naturally occurring autoantibodies react exclusively with components in mitotic centrosomes (Table I). Of particular interest is the one called POPA (prophase-originating polar antigen) characterized by Sager *et al.*, (1986). Detailed analyses have provided evidence that the antibody reacts with a

TABLE I

Nonimmune/Autoimmune Sera Specific for Centrosomes

Source	Name, molecular weight of components identified, immunostaining characteristics[a]	References
Rabbit	Centriole-specific, 50K (*1, 2*)	Connolly & Kalnins (1978)
	0013: 180K–250K, 130K, 50–65K (*3–9*);	Maunoury (1978)
	14K, 17K (*10, 11*)	Lin *et al.,* (1981)
		Kronebusch & Singer (1987)
	NRS-01 antiserum	Hiraoka *et al.,* (1989)
Human		Brenner & Brinkley (1982)
		Tuffanelli *et al.,* (1983)
	5051: PCM-specific (12–22)	Calarco-Gillam *et al.,* (1983)
		Dales & Wallace (1985)
	112K, pI 5.8	Bastmeyer & Russell (1986)
		Senécal & Rauch (1988)
Mitosis-specific human autoimmune sera	NuMA specific	Price *et al.,* (1984)
		McCarty *et al.,* (1984)
		Senécal *et al.,* (1985)
	POPA: 110K/115K	Sager *et al.,* (1986)
	Sea urchin egg-specific	Leslie & Wilson (1987)
	200K–230K	Maekawa *et al.,* (1991)

[a]Related references are given by number in parentheses and refer to the following: (1) Wang *et al.,* 1979; (2) Turksen *et al.*, 1982; (3) Maunoury, 1979; (4) Maro *et al.*, 1984; (5) Tassin *et al.*, 1985; (6) Courvalin *et al.,* 1986; (7) Sauron *et al.,* 1986; (8) Gosti *et al.,* 1987; (9) Komesli *et al.,* 1989; (10) Segarini *et al.,* 1983; (11) Fung and Kasamatsu, 1985; (12) Maro *et al.,* 1985; (13) Clayton *et al.,* 1985; (14) Schatten *et al.,* 1985; (15) Wick, 1985; (16) Bastmeyer *et al.,* 1986; (17) Houliston *et al.,* 1987; (18) Sellitto and Kuriyama, 1988b; (19) Snyder and Davis, 1988; (20) Snyder, 1989; (21) Hiraoka *et al.,* 1989; (22) Moudjou *et al.,* 1989a.

115/110 K component located at the spindle poles of metaphase and anaphase cells. The antigen appears in discrete cytoplasmic foci in early prophase, but moves to the spindle poles during later stages of mitosis. Since those foci appear to be closely related to the site of microtubule formation, the antigen may represent one of the unique components necessary for the organization of spindle microtubules during mitosis.

B. Polyclonal and Monoclonal Antibodies to Preselected Proteins

The second group of centrosomal probes includes polyclonal and monoclonal antibodies generated against preselected proteins. As summarized in Table II, a variety of proteins have been identified in the centrosomal area, including microtubule-associated proteins (MAPs) in brain tissues (MAP1, Sherline and Mascardo, 1982a; Sato *et al.,* 1983; Bonifacino *et al.,* 1985; Luca *et al.,* 1986; De Mey *et al.,* 1987), Chinese hamster ovary (CHO) mitotic cells (Hoffman and Mullins, 1990), and *Drosophila* embryos (Kellogg *et al.,* 1989). Various cytoskeletal elements such as tubulin (Pepper and Brinkley, 1977; Oakley *et al.,* 1990), *Dictyostelium* dynein (Koonce and McIntosh, 1990), and intermediate filament-like fibers (Schatten *et al.,* 1987; Steffen and Linck, 1989) have also been identified in the centrosomes. Other proteins include different types of enzymes (PNP, purine nucleoside phosphorylase, Oliver *et al.,* 1981; cAMP-dependent protein kinases, Nigg *et al.,* 1985; De Camilli *et al.,* 1986; Joachim and Schwoch, 1988; p34^{cdc2} kinases, Bailly *et al.,* 1989; Riabowol *et al.,* 1989; Rattner *et al.,* 1990), transformation-associated proteins (Kasamatsu and Nehorayan, 1979; Blair Zajdel and Blair, 1988), cell cycle control proteins (cyclin, Pines and Hunter, 1989; p13^{suc1}, Bailly *et al.,* 1989), coritcal proteins prepared form protozoa (centrin, Salisbury *et al.,* 1986; see Chapter 6; Klotz *et al.,* 1986; Moudjou *et al.,* 1989b), a nucleocytoskeletal-like protein from yeast (Hurt, 1988), S-100b (Uchida and Endo, 1988), and the transferin receptor (Willingham and Pastan, 1985) (see also Chapter 2). A high molecular weight nuclear protein, NuMA (nucleus mitotic apparatus), characterized by Lydersen and Pettijohn (1980) is restricted to the nuclear matrix in interphase and becomes associated with spindle poles during mitosis. Using immunoblot analyses, antibodies specific to the NuMA protein were frequently found in sera from patients with connective tissue diseases (Price *et al.,* 1984), suggesting that the NuMA protein is a human autoantigen.

Kinesin is a microtubule-binding motor protein essential for intracellular particle transport in eukaryotes (Vale *et al.,* 1985). Neighbors *et al.,* (1988) reported that the polyclonal antikinesin antibody specifically reacts with centrosomes in interphase and mitotic mammalian cells in culture. *bimC4* and *cut7* are mitotic mutants defective in separation of duplicated spindle pole bodies (SPBs) during mitosis in *Aspergillus* (Enos and Morris, 1990) and the fission yeast *Schizosaccharomyces pombe* (Hagan and Yanagida, 1990), respectively. These two genes have recently been cloned and sequenced, and it was concluded that the genes encode a homolog of the heavy chain of kinesin. These results indicate that the microtubule motor kinesin may be localized at the MTOC region and involved in the mechanism of MTOC separation or spindle morphogenesis.

It should be noted that most of the antibodies listed in Table II were not originally intended as specific markers for centrosomal structures. Their cross-reactivity with centrosomes was noticed later during the process of screening on mitotic and interphase cells. The one exception is the 51K protein characterized by Toriyama *et al.,* (1988). This component was identified in granular material of the mitotic poles of sea urchin spindles and purified from the high salt extract of

TABLE II

Polyclonal and Monoclonal Antibodies to Preselected Proteins

Immunogens	Molecular weight of components identified Immunostaining characteristics[a]	References
Brain MAP1		Sherline & Mascardo (1982a)
	>300K, nucleus specific	Sato *et al.*, (1983)
	280K	Bonifacino *et al.*, (1985)
		Luca *et al.*, (1986)
Other MAPs	*Drosophila* embryos	Kellogg *et al.*, (1989)
	Mitotic CHO cells	Hoffman & Mullins (1990)
Steroids	Autoimmune?	Nenci & Marchetti (1978)
S-100b		Uchida & Endo (1988)
Transferrin receptor	92K	Willingham & Pastan (1985)
Protozoa Cortex protein	Rootlet proteins, 40K, 175K	Klotz *et al.*, (1986)
	Centrin: 115K–120K (1–5)	Salisbury *et al.*, (1986)
	200K Calcium-binding protein	Moudjou *et al.*, (1989b)
SV40V3P	Autoimmune? 14K, 17K (6, 7)	Kasamatsu & Nehorayan (1979)
p53	Transformation-associated protein	Blair Zajdel & Blair (1988)
NuMA	Nuclear matrix, 300K (8, 9)	Lydersen & Pettijohn (1980)
Nuclear protein	Yeast NSP1, 86K	Hurt (1988)
Tubulin		Pepper & Brinkley (1977)
	α-tubulin	Oakley *et al.*, (1990)
Dynein	*Dictyostelium* specific	Koonce & McIntosh (1990)
Kinesin		Neighbors *et al.*, (1988)
IF-like protein	*Drosophila* vimentin	Schatten *et al.*, (1987)
	Tektin filaments	Steffen & Linck (1989)
PNP	Purine nucleoside phosphorylase	Oliver *et al.*, (1981)
Protein kinase	52K, 56K, Golgi-specific	Nigg *et al.*, (1985)
		De Camilli *et al.*, (1986)
	Rat parotid glands	Joachim & Schwoch (1988)
$p34^{cdc2}$		Bailly *et al.*, (1989)
		Riabowol *et al.*, (1989)
$p13^{suc1}$		Bailly *et al.*, (1989)
Cyclin		Pines & Hunter (1989)
51K protein	Sea urchin eggs, EF-1α? (10–12)	Toriyama *et al.*, (1988)

[a]Related references are given by number in parentheses and refer to the following: (1) Salisbury *et al.*, 1987; (2) Baron and Salisbury, 1988; (3) Salisbury *et al.*, 1988; (4) Lechtrek *et al.*, 1989; (5) Wright *et al.*, 1989; (6) Kasamatsu *et al.*, 1980; (7) Shyamala *et al.*, 1982; (8) Price *et al.*, 1984; (9) van Ness and Pettijohn, 1984; (10) Ohta *et al.*, 1988b; (11) Ohta *et al.*, 1988a; (12) Ohta *et al.*, 1990.

isolated sea urchin mitotic spindles of which it is a major component. Recent detailed immunofluorescence observations have concluded that the 51K protein is located not only in the centrosomal region but also in all the other microtubule-containing regions in sea urchin eggs (Ohta *et al.*, 1988a).

The fact that centrosomal structures are recognized by such a wide range of antibody probes raises a simple but fundamental question. Are all these antigens true centrosome-specific components? Immunoblot analyses have demonstrated that many antibodies recognize proteins that differ in molecular weight from the original antigens. For example, antibodies raised against centrin, a calcium-binding phosphoprotein with an apparent molecular mass of 20 kDa isolated from the striated flagellar root of the unicellular green algae *Tetraselmis* (Salisbury *et al.*, 1986), recognize a 165K centrosome-associated protein in cultured mammalian cells (Baron and Salisbury, 1988; see Chapter 6). Centrin and the high molecular weight centrosomal component, although not identical, are closely related and appear to share common epitopes. In the case of monoclonal antibodies, however, the antigens in centrosomes and the original antigens may be entirely unrelated to each other, since a single determinant recognized by a monoclonal antibody is frequently shared among essentially unrelated proteins.

C. Monoclonal Antibodies Prepared from Heterogeneous Subcellular Fractions

Centrosomes are fragile single-copy structures and routine biochemical fractionation may not be good enough to completely purify them for the purpose of polyclonal antibody production. To overcome this problem, hybridoma technology, combined with immunocytochemistry, has been employed using partially purified heterogeneous MTOC fractions as immunogens. Because an antibody secreted by a hybridoma clone reacts with a single epitope, contaminants will be removed during the process of antibody screening (Kohler and Milstein, 1976). Another great advantage of this approach is that it is possible to raise specific antibodies against hitherto unknown minor molecular components of centrosomes, the function of which may be difficult to assess without the availability of antibodies. This shotgun approach was first employed with success by Ring *et al.*, in 1980 (Table III). They prepared the interphase centrosomal fraction from mouse neuroblastoma cells (N115) by taking advantage of the large number of centrosomes (10–20 per cell) in a relatively small volume of cytoplasm in these cells. Supernatants from 45 of the hybridoma cultures screened specifically stained the centrosomes in N115 cells. Among them, two hybridoma clones, C25.9 and C29.2, stained centrosomes as bright dots, often with a dark point at the center. They also cross-reacted with a number of other MTOCs, including mouse

TABLE III

Monoclonal Antibodies Prepared from Heterogenous Subcellular Fractions

Immunogens	Antibodies, molecular weight of components identified, immunostaining characteristics[a]	References
Neuroblastoma MTOCs		Ring *et al.*, (1980)
Interphase centrosomes	CTR 435: PCM-specific	Bailly *et al.*, (1989)
	CTR532: 170K protozoa micronuclei and spindles	Keryer *et al.*, (1989)
	PtK centrosomes, 32K, IF-like (1)	Joswig & Petzelt (1990)
Sea urchin spindles	SU4: 25K, 180K, 190K SU5: 50K, similar to EF-1α; (2)	Kuriyama & Borisy (1985)
CHO spindles	CHO1 to CHO7 (3–5)	Sellitto & Kuriyama (1987)
Dictyostelium NABs	AX1 to AX7	Sellitto & Kuriyama (1987)
Nuclear matrix	P11: 170K	Chaly *et al.*, (1984)
Mitotic HeLa extracts	MPM-1, MPM-2: family of phosphoproteins (6–20)	Davis *et al.*, (1983)
	MPM-13: 43K	Rao *et al.*, (1989)
Drosophila nuclei	Bx63 and Rb 188: *Drosophila*-specific (21–25)	Frash *et al.*, (1986)
Yeast SPBs	110K, 90K, 80K	Rout & Kilmartin (1990)

[a]Related references are given by number in parentheses and refer to the following: (1) Joswig *et al.*, 1990; (2) Kuriyama *et al.*, 1990b; (3) Sellitto and Kuriyama, 1988a; (4) Kuriyama, 1989; (5) Nislow *et al.*, 1990; (6) Vandré *et al.*, 1984; (7) Vandré *et al.*, 1986; (8) Wordeman and Cande, 1986; (9) Hecht *et al.*, 1987; (10) Keryer *et al.*, 1987; (11) Millar *et al.*, 1987; (12) Engle *et al.*, 1988; (13) Davis *et al.*, 1989; (14) Hiraoka *et al.*, 1989; (15) Kuang *et al.*, 1989; (16) Le Guen and Crozet, 1989; (17) Vandré and Borisy, 1989; (18) Masuda *et al.*, 1990; (19) Kuriyama *et al.*, 1990a; (20) Centonze and Borisy, 1990; (21) Freeman *et al.*, 1986; (22) Whitfield *et al.*, 1988; (23) Raff and Glover, 1988; (24) Raff and Glover, 1989; (25) Gonzalez *et al.*, 1990.

fibroblast and CHO cell centrioles, *Tetrahymena* basal bodies, and neuroblastoma kinetochores.

As listed in Table III, monoclonal anticentrosomal probes have also been generated against interphase centrosomes from lymphocytes (Bailly *et al.*, 1989; Keryer *et al.*, 1989) and PtK cells (Joswig and Petzelt, 1990), mitotic spindles isolated from sea urchin eggs (Kuriyama and Borisy, 1985) and CHO cells (Sellitto and Kuriyama, 1987), and MTOC-containing fractions of the cellular slime mold, *Dictyostelium discoideum* (Sellitto and Kuriyama, 1987). Most recently, Rout and Kilmartin (1990) achieved a several-thousand-fold enrichment for SPBs with attached spindles using yeast cell extracts. Monoclonal antibodies raised against this material have identified three components (110K, 90K, and 80K) whose detailed localization in the SPB structures was examined by immunoelectron microscopy (Rout and Kilmartin, 1990).

Monoclonal antibodies that bind to centrosomes also have been identified during the process of screening for antibodies directed against subcellular fractions other than centrosomes/MTOCs. They include nuclear matrices of mouse splenic lumphocytes (Chaly *et al.*, 1984), *Drosophila* nuclei (Frash *et al.*, 1986), and mitotic HeLa cell extracts (Davis *et al.*, 1983). Since interphase centrosomes attach to the nucleus very firmly (Bornens, 1977; Nadezhina *et al.*, 1979; Kuriyama and Borisy, 1981a), isolated nuclei used as immunogens could be associated or contaminated with centrosomal structures. Alternatively, centrosomes and nuclei may share immunologically related or common elements.

Davis *et al.*, (1983) prepared monoclonal antibodies (MPM-1 and MPM-2) against a mitotic HeLa cell extract. The antibodies recognized a class of phosphoproteins abundant in mitotic cells and were found to stain several discrete mitotic structures, such as centrosomes, kinetochores, and midbodies, leading to the conclusion that components of centrosomes are among the cellular proteins specifically phosphorylated at the transition from interphase to mitosis (Vandré *et al.*, 1984). M-phase-associated phosphoproteins seem to be widely distributed since MTOC-components reacting with MPM-1/MPM-2 have so far been identified in nematodes (Hecht *et al.*, 1987), *Drosophila* (Millar *et al.*, 1987), *Aspergillus* (Engle *et al.*, 1988), slime molds (Vandré *et al.*, 1986), *Paramecium* (Keryer *et al.*, 1987), sea urchin eggs (Kuriyama *et al.*, 1990a), mammalian embryos (Hiraoka *et al.*, 1989; Le Guen and Crozet, 1989), higher plants (Vandré *et al.*, 1986), diatoms (Wordeman and Cande, 1986), the fission yeast *S. pombe* (Masuda *et al.*, 1990), in addition to cultured mammalian cells (Vandré *et al.*, 1986). The antibodies also recognize the brain MAP1 proteins (De Mey *et al.*, 1987). As shown in Table II, several different antibodies generated against MAP1 show positive staining of centrosomes. Since many such antibodies react with polypeptides of molecular weight different form that of the original MAP1 molecules, an immunological cross-reactive domain, most likely a unique subset of phosphorylated epitopes, appears to be shared among MAP1 and the centrosomal components. Despite the fact that cells contain numerous phosphoproteins, only certain phosphorylated components are recognized by the MPM-1/MPM-2 and the MAP1 antibodies, which appear to be highly selective. The conserved phosphorylated epitope recognized must therefore display strong antigenicity.

Rao and colleagues (1989) have further extended their experiments to obtain another centrosomal antibody (MPM-13) using mitotic HeLa cell extracts as immunogens. MPM-13 recognizes proteins in a major 43K band and a variable number of minor bands at 56K that are associated with the Colcemid-sensitive pericentriolar material of mammalian centrosomes.

III. MONOCLONAL ANTIBODIES TO MICROTUBULE-ORGANIZING CENTER-CONTAINING STRUCTURES

A. Mitotic Sea Urchin Spindles

Since the monoclonal antibodies have proven to be effective in identifying specific organellar components, even when crude heterogeneous fractions are used as immunogens, we attempted to obtain specific reagents to several different types of MTOC-containing structures by raising monoclonal antibodies against these organelles. We first used sea urchin eggs to prepare immunogens since they provide a naturally synchronizable homogeneous population of dividing cells and are available in substantial quantities (Kuriyama and Borisy, 1985). We isolated mitotic spindles and taxol-induced asters from eggs of the sea urchin *Strongylocentrotus purpuratus.* After immunization of BALB/c mice, spleen cells were taken and fused with NS-1 myeloma cells. Hybridoma supernatants secreting antibodies were screened in two stages by indirect immunofluorescence staining: first on isolated sea urchin mitotic spindles in 96-well plates to rapidly identify potential positive hybridomas, the second on whole mitotic eggs on coverslips to distinguish between spindle-specific staining and adventitious contamination. Of the 94 positive wells in the first screening, 5 hybridomas were deemed to exhibit specific staining of the mitotic apparatus in whole eggs.

We selected two antibodies, SU4 and SU5, for further analysis. Figure 1 illustrates localization of tubulin (A–C), and the antigens corresponding to SU4 (D–F) and SU5 (G–I) in whole mitotic cells (first and second rows), and isolated spindles (third row). Mitotic spindles have prominent microtubule arrays radiating from each aster. SU4 and SU5 stained structures primarily in the central region of the aster, resulting in a dumbbell-shaped staining pattern. The patterns of SU4 and SU5 staining were, however, slightly different from each other because SU5 also showed positive staining in the spindle region between the asters, whereas the staining by SU4 was confined to the astral region. Both the antibodies seem to react primarily with the centrospheric region of the isolated spindle (F, I). Immunoblot analysis showed that hybridoma SU4 recognized a major 190K polypeptide as well as bands of 180K and 20K, whereas hybridoma SU5 stained a species with an apparent molecular weight of 50K.

In order to gain more information about the 50K component identified by the SU5 antibody, we used the antibody to isolate cDNA sequences corresponding to the centrosphere protein by screening a λgt11 cDNA expression library, prepared by using RNA from sea urchin blastulas (Kuriyama *et al.,* 1990b). Three positive clones were identified from 10^6 recombinant phages screened, and one of the clones contained a 1.8-kb insert fragment. The β-galactosidase fusion protein produced by this clone had an apparent molecular weight of 150K indicating that

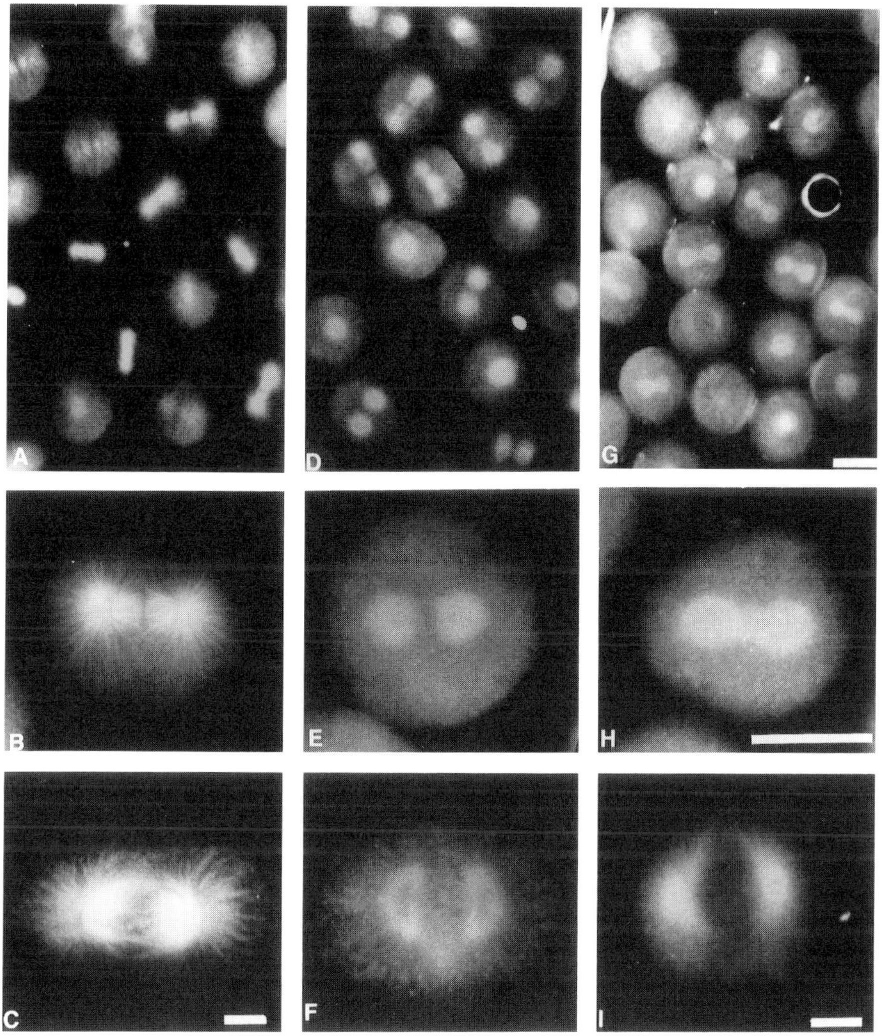

Fig. 1. Immunofluorescence staining of mitotic sea urchin eggs and isolated spindles with monoclonal antibody to yeast α-tubulin (Kilmartin *et al.,* 1982) (A–C), and antibodies SU4 (D–F) and SU5 (G–I). Whole mitotic eggs are presented in the first and second rows, at low magnification (A, D, G) to show the uniformity of the staining reaction and at high magnification (B, E, H) to show structural details. Mitotic spindles shown in the third row were isolated in a medium containing 1 *M* glycerol, 5 m*M* MES, 1 m*M* EGTA, 2 m*M* $MgSO_4$, and 0.05% Triton X-100 at pH 6.15 (Sakai *et al.,* 1977). Both SU4 and SU5 react primarily with the centrospheric region of the isolated structures. Bar: A, D, G, 50 μm; B, E, H, 50 μm; C, F, I, 10 μm. [Reproduced from Kuriyama and Borisy, *Journal of Cell Biology* (1985), **101,** 526, by copyright permission of the Rockefeller University Press.]

an estimated 34K of the SU5-reactive polypeptide was coded by the inserted cDNA. We sequenced this 1.8-kb cDNA fragment and found that the predicted amino acid sequence of the SU5 antigen showed striking similarity to the peptide chain elongation factor (EF-1α) found in yeast, fungus, shrimp, insect, mouse, and human cells (Kuriyama *et al.,* 1990b).

This homology between the SU5 antigen in centrospheres and EF-1α is unexpected; there are three possible explanations for such similarity. First, the 50K protein immunoreactive to SU5 could indeed be EF-1α, and the centrosphere in sea urchin eggs happens to be its natural location. In fact, numerous ribosomal particles are found in association with spindle microtubules (Zimmerman, 1960; Harris, 1975); a recent report by Suprenant *et al.,* (1989) has shown the direct association of ribosomes with sea urchin microtubules both *in vivo* and *in vitro.* Sea urchin eggs are filled with yolk granules and the mitotic spindle is a region free of the large yolk granules. Major soluble cytoplasmic components, such as EF-1α (Slobin, 1980; Lenstra and Bloemendal, 1983), would therefore tend to localize around yolk-free regions, such as mitotic spindles. Such passive accumulation of microinjected fluorescently tagged bovine serum albumin has been demonstrated around mitotic spindles in cleaving sea urchin eggs (Wang and Taylor, 1979). However, recent phylogenetic studies by le Maire and collaborators have suggested that the SU5 antigen is not the true EF-1α of the sea urchin, since the sequence of the SU5 centrosphere protein is not at the expected position in the phylogenetic tree of EF-1α (M. le Maire, personal communication).

The second possibility is that the 50K protein is an intrinsic centrosphere protein whose similarity to EF-1α is not coincidental. That is, the EF-1α gene may have acquired and maintained another function as a centrosphere component without loss of its original role as the peptide chain elongation factor. This concept of "gene sharing" (Piatigorsky and Wistow, 1989) was suggested by the observation that amino acid sequences share homologies among various lens crystalline proteins and common cellular enzymes such as lactate dehydrogenase and α-enolase (Wistow and Piatigorsky, 1988). Such shared amino acid homologies among apparently unrelated proteins have been seen in many other instances (Chaput *et al.,* 1988; Geetha-Habib *et al.,* 1988; Tong *et al.,* 1988, Numata *et al.,* 1991).

Another possible explanation is that EF-1α may be a member of a multigene family. Mao and Wang (1988) have shown that statin, a protein found only in nonproliferating cells, is 80% homologous to the human EF-1α at the nucleotide sequence level. A gene product of a temperature-sensitive (ts) mutant *gstl* in *Saccharomyces,* that is defective for the G_1–S transition, also has been shown to contain consensus sequences for a target site of cAMP-dependent protein kinase(s) and for GTPase that has an extensive homology to polypeptide chain elongation factor EF-1α (Kikuchi *et al.,* 1988). The major protein of *Xenopus* previtellogenic oocytes, thesaurin a, is also homologous to EF-1α (Viel *et al.,* 1987). It was recently suggested that the 51K protein, a major nontubulin protein component in mitotic spindles of sea urchin eggs (Toriyama *et al.*, 1988. Table II) is also a GTP-binding protein, structurally and functionally related to EF-1α (Ohta *et al.*, 1990). It is not, however, known at this time how the 51K protein is related to the SU5 centrosphere protein.

B. Nucleus-Associated Bodies from *Dictyostelium* and Mammalian Centrosomes

The structure of MTOCs is known to be highly diverse. For example, the MTOC in the cellular slime mold *Dictyostelium discoideum* is called a nucleus-associated body (NAB) and is composed of a disk-shaped multilayered core embedded in electron-dense granular material (Roos, 1975). The NAB is located next to the nucleus, and the nucleus–NAB complex can be isolated as a coherent unit (Kuriyama *et al.*, 1982). The isolated NABs retain their microtubule-nucleating activity *in vitro* (Fig. 2); we have demonstrated that, although structurally quite distinct from mammalian centrosomes, NABs and centrosomes share common functional properties such as the ability to initiate microtuble assembly *in vitro* (Kuriyama *et al.*, 1982; Kuriyama, 1984). This would suggest the presence of conserved functional elements between centrosomes and NABs. We have generated a library of monoclonal antibodies against these two morphologically distinct MTOCs in order to compare their macromolecular composition and to evaluate the functional properties of each corresponding antigen (Sellitto and Kuriyama, 1987).

Mitotic spindles of CHO cells were chosen to generate antibodies to the mammalian centrosome for the following reasons: (1) extensive morphological as well as functional analyses of centrosomes are available on this particular cell type; (2) the cells have a rapid doubling time, which makes it easier to prepare large numbers of mitotic cells from CHO cells than from other lines of cultured mammalian cells, and (3) we have already established a one-step procedure for the isolation and purification of mitotic spindles from CHO cells in a medium containing the microtubule-stablizing drug taxol (Kuriyama *et al.*, 1984). Electron microscopic observations have shown that these isolated spindles include well-preserved mitotic centrosomes at each pole. Both the isolated CHO spindle and nucleus–NAB complex fractions were absorbed on nitrocellulose strips and surgically implanted under the skin of mice. By screening hybridoma supernatants with indirect immunofluorescence staining on whole CHO cells or *Dictyostelium* amebas, we obtained a total of 13 different hybridoma clones that labeled centrosomes or NABs (Table IV). A set of antibodies CHO1–CHO7, was raised against CHO spindles and a second set of antibodies, AX1–AX7, was generated against *Dictyostelium* nucleus–NAB complexes.

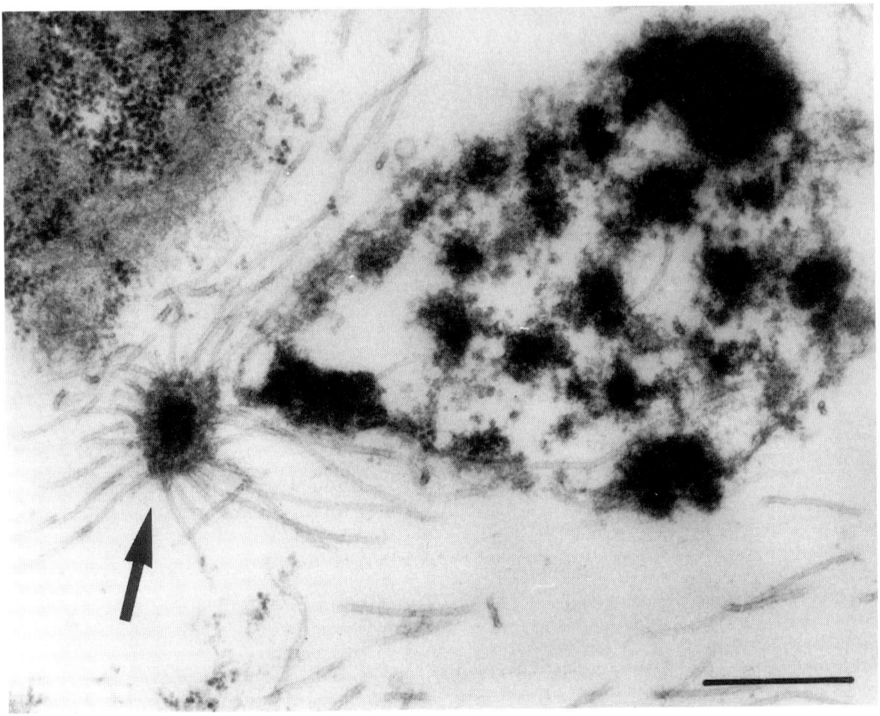

Fig. 2. Electron micrograph of a thin section of the nucleus–NAB complex isolated from *Dictyostelium*. Ameboid cells were lysed with chilled extraction medium (0.3 *M* sucrose, 0.1 *M* PIPES, 2m*M* $MgSO_4$, 1 mg/ml leupeptin, 1 mg/ml PMSF, and 0.25% Triton X-100 at pH 6.8). The isolated nucleus–NAB complexes were incubated with brain microtubule proteins *in vitro*. Note the number of microtubules polymerized onto the NAB (arrow). Bar 5 μm. [Reproduced from Kuriyama *et al.*, *Cell Motility* (1982), **2**, 264, by copyright permission of Wiley–Liss.]

1. Cross-Reactivity of Antibodies to MTOC Antigens between Species

Two of the monoclonal antibodies raised, CHO3 and AX3, stain MTOCs in both CHO cells and *Dictyostelium* amebas (Figs. 3, 4). The CHO3 antibody labels interphase centrosomes and nuclei, as shown in Fig. 3A′. At the onset of mitosis, the corresponding antigen localizes at the spindle poles and persists in that position throughout the mitotic process (Fig. 3B′). The intensity of fluorescence at the mitotic poles remains invariable through the different stages of mitosis. *Dictyostelium* nuclei are also labeled with the antibody (Fig. 3C′) along with MTOCs. AX3 was originally identified as an antibody probe specific for microtubules in *Dictyostelium* (Fig. 4A, B). The corresponding antigen is also localized at the mammalian centrosomes during interphase (Fig. 4D′). In mitotic spindles fixed

TABLE IV

Monoclonal Antibodies Prepared against CHO Spindles (CHO1–CHO7) and Nucleus–NAB Complexes from *Dictyostelium* (AX1–AX7)

Antibody	MTOC Interphase	MTOC Mitotic	Cross-reactivity with other structures
CHO1	yes	no	Midbodies, nuclei, spindle microtubules
CHO2	yes	yes	None
CHO3	yes	yes	Nuclei, midbodies, *Dictyostelium* MTOCs, MAP1
CHO4	yes	yes	Intermediate filaments
CHO5	yes	yes	Golgi apparatus
CHO6	yes	yes	Stress fibers
CHO7	yes	yes	None
AX1	yes	no	None
AX3	yes	yes	Microtubules, centrosomes
AX4	yes	yes	Interphase nuclei
AX5	no	yes	Interphase nuclei
AX6	yes	yes	Interphase nuclei
AX7	no	yes	Cytoplasmic granules

with glutaraldehyde, however, the AX3 antigen distributes along the entire length of spindle fibers (Fig. 4E′), indistinguishable from tubulin staining (Fig. 4E″). Digestion of fixed cells with alkaline phosphatase completely eliminates CHO3 staining (Fig. 3D′, E′, F′) and severely reduces the immunofluorescence signal from the AX3 antibody (Fig. 4C′). Thus, both antibodies are specific for phosphorylated moieties that are conserved between the two species. The CHO3 antibody recognizes a family of high molecular weight polypeptides both in CHO spindles and ameba lysates. We also have identified a 225K centrosomal phosphoprotein, located in numerous granular aggregates at the periphery of the centrosphere in dividing sea urchin eggs, with this antibody (Kuriyama, 1989). On the other hand, the AX3 antibody labels a 280K heat-stable MAP in *Dictyostelium* (Kimble and Kuriyama, unpublished observations). High molecular weight components immunoreactive to AX3 are also identified in HeLa cells, *Drosophila* and sea urchin eggs (Kimble and Kuriyama, unpublished observations).

2. *Cross-Reactivity of Antibodies to MTOC Antigens with Other Subcellular Structures*

One of the most striking results seen in cells stained with monoclonal antibodies to MTOC antigens is their cross-reaction with other organelles (Table IV). That is, many antibodies label not only MTOCs but also one of the following sub-

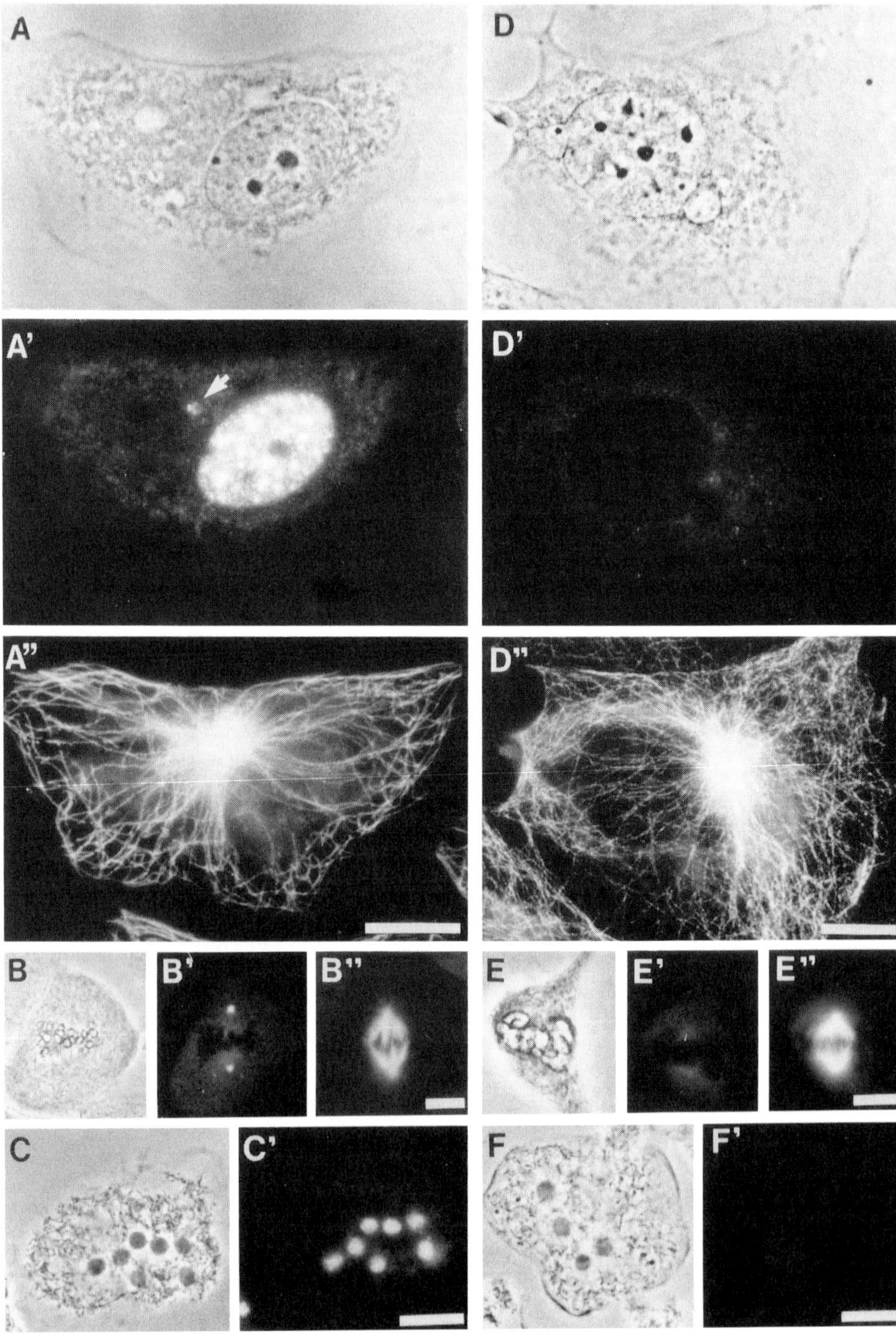
A
A'
A"
B
B'
B"
C
C'
D
D'
D"
E
E'
E"
F
F'

cellular structures: cytoskeletal elements such as microtubules (AX3; Fig. 4A, E′); intermediate filaments (CHO4); stress fibers (CHO6); and the midbody matrix (CHO1; Sellitto and Kuriyama, 1988a); Golgi apparatus (CHO5); nuclei (CHO1, CHO3; Fig. 3A′, C′; AX4–AX6), and unidentified cytoplasmic granules (AX7). As shown in Fig. 5, CHO1 stains centrosomes (Fig. 5A′) and nuclei (Sellitto and Kuriyama, 1988a). During mitosis, however, the CHO1 antigen undergoes a drastic change in distribution (Fig. 5A′). It is diffusely distributed in the spindle region in metaphase cells, with a slightly higher concentration around the metaphase plate (Fig. 6A′). Then it starts to be reorganized into discrete short fibers aligned along the spindle axis during anaphase (Fig. 6C′). With the progression into cytokinesis, the fibers labeled with the CHO1 antibody become brighter in the furrowing region, and then gradually shorten and coalesce (Fig. 6D′–F′) into a bright fluorescent dot at the center of the intercellular bridge between the two daughter cells (Fig. 6G′). Our recent results from *in vitro* co-purification experiments further indicate that the CHO1 antigen appears to be a novel mitotic MAP and that this particular distribution of the MAP in the spindle is required for mitosis (Nislow *et al.*, 1990). The presence of a MAP, or a closely related protein, in a MTOC is quite plausible. Since mitotic as well as interphase microtubules originate from MTOCs, some MAPs with affinity for the minus end of microtubules could be embedded in the MTOCs to provide conditions favorable for initiation of microtubule formation. Indeed one of the first identified antibodies specific for centrosomes was a brain MAP1 antibody (Sherline and Mascardo, 1982a). A MAP from mitotic CHO cells (Hoffman and Mullins, 1990) and several MAPs from *Drosophila* embryos isolated by microtubule affinity chromatography also recently have been shown to localize at centrosomes (Kellogg *et al.*, 1989).

In Fig. 7, CHO cells stained with antibodies CHO4 (Fig. 7 A,B) and CHO5 (Fig. 7 C,D) are shown. In addition to centrosomes, which can be clearly identified as dots, CHO4 antibody also stains filamentous structures (Fig. 7A′), and CHO5 antibody stains structures in the perinuclear region in close proximity to the centrosomes (Fig. 7C′). In mitotic CHO cells treated with the CHO4 antibody, cage-like structures around mitotic spindles, composed of intermediate filaments as reported by Zieve *et al.*, (1980) and Blose and Bushness (1982), are stained. Treatment of CHO cells with nocodazole results in a bundling of the CHO4-positive filaments (Fig. 7B′), suggesting that the filaments revealed by the CHO4 antibody indeed represent intermediate filaments. The perinuclear aggregates that

Fig. 3. Localization of CHO3 antigen(s) in CHO cells (A, B, D, E) and *Dictyostelium* amebas (C, F). (*Left*) control cells; (*right*) cells treated with alkaline phosphatase. The same cells at interphase (A–A″, D–D″) and at mitosis (B–B″, E–E″), and ameboid cells (C–C″, F–F″) are seen by phase-contrast (A–F), immunofluorescence staining with CHO3 antibodies (A′–F′), and staining with antibodies to tubulin (A″–F″). Arrow indicates the position of interphase centrosome. Alkaline phosphatase digestion completely eliminated CHO3 staining (D′, E′, F′). Bars: 10 μm.

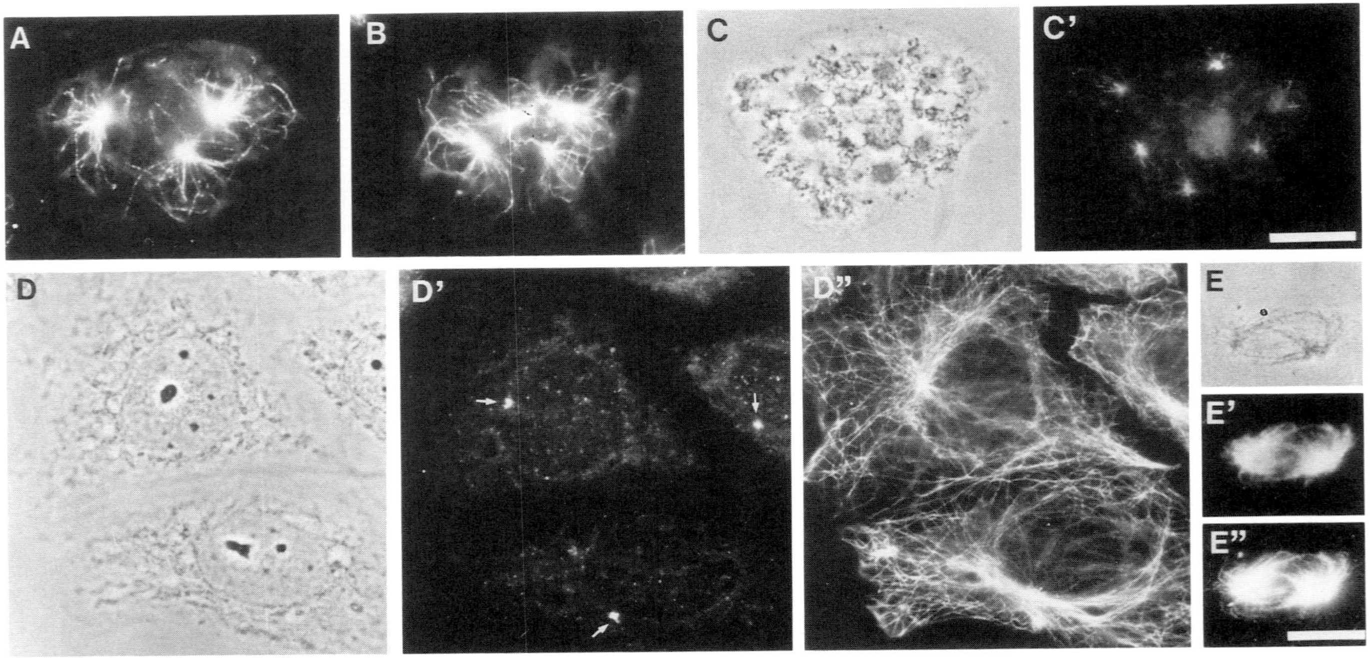

Fig. 4. Localization of AX3 antigen(s) in *Dictyostelium* amebas (A–C) and CHO cells (D–E). Ameboid cells fixed with paraformaldehyde are stained with AX3 (A, C′) and tubulin (B) antibodies. The cell in C was digested with alkaline phosphatase prior to staining with the AX3 antibody. CHO cells fixed with cold methanol (D) and the isolated CHO spindles fixed with glutaraldehyde (E) are seen by phase-contrast (D, E) and immunofluorescence staining with AX3 antibodies (D′, E′) and antibodies to tubulin (D″, E″). Arrows indicate the position of interphase centrosomes. Bars: 10 μm.

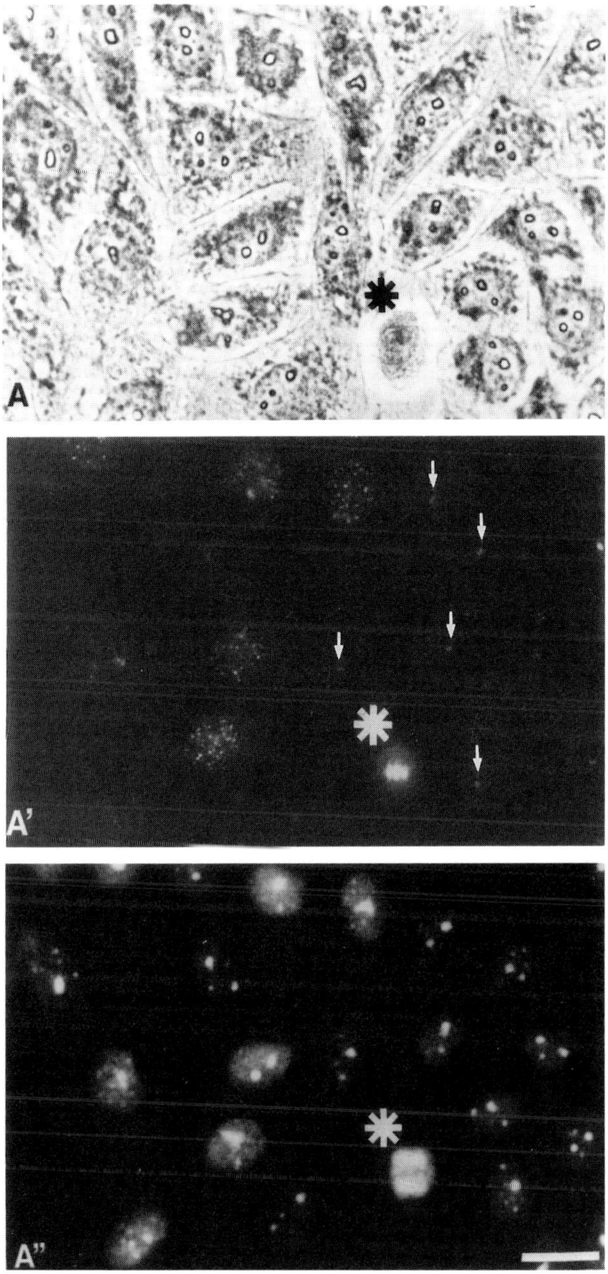

Fig. 5. Localization of CHO1 antigen in whole CHO cells. The same cells are seen by phase-contrast (A) and fluorescence microscopy after double staining with CHO1 antibody (A′) and ethidium bromide (A″). The asterisk indicates the position of a mitotic cell. Some of the interphase nuclei are stained by CHO1 in a speckled pattern (A′). CHO1 antibody also labels interphase centrosomes (arrows) and midbodies. Bars: 10 μm.

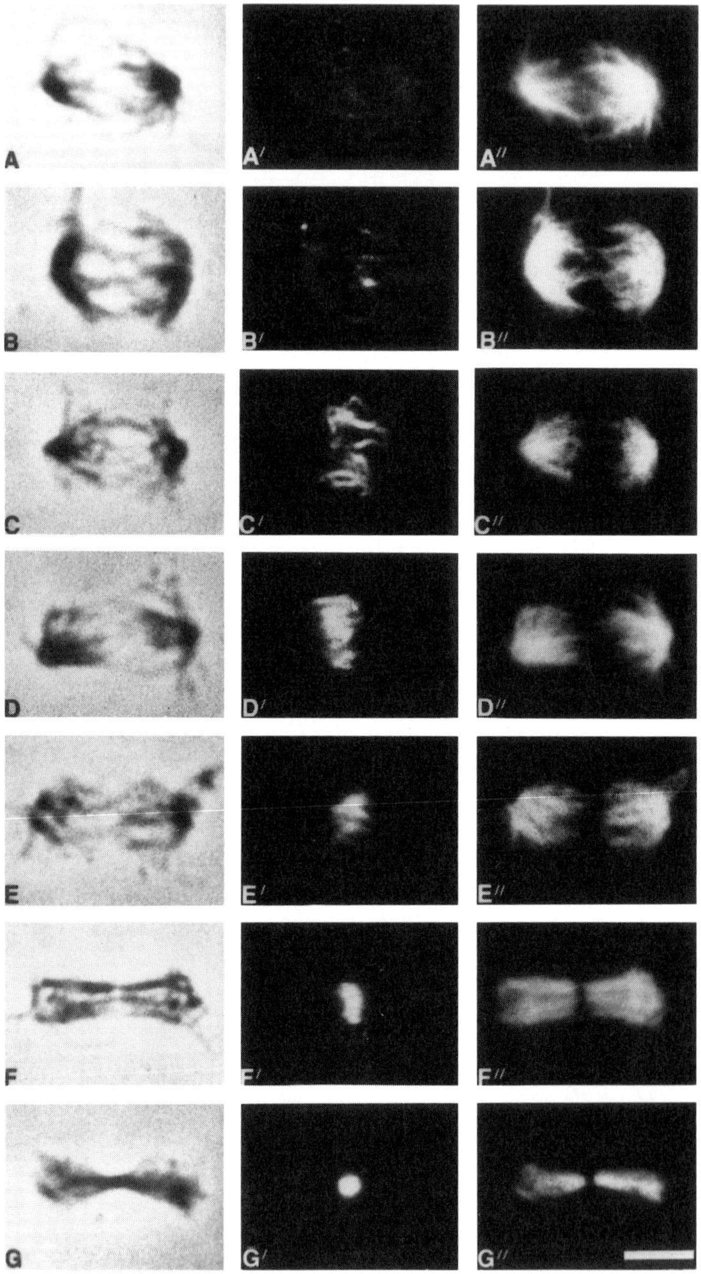
A
A′
A″
B
B′
B″
C
C′
C″
D
D′
D″
E
E′
E″
F
F′
F″
G
G′
G″

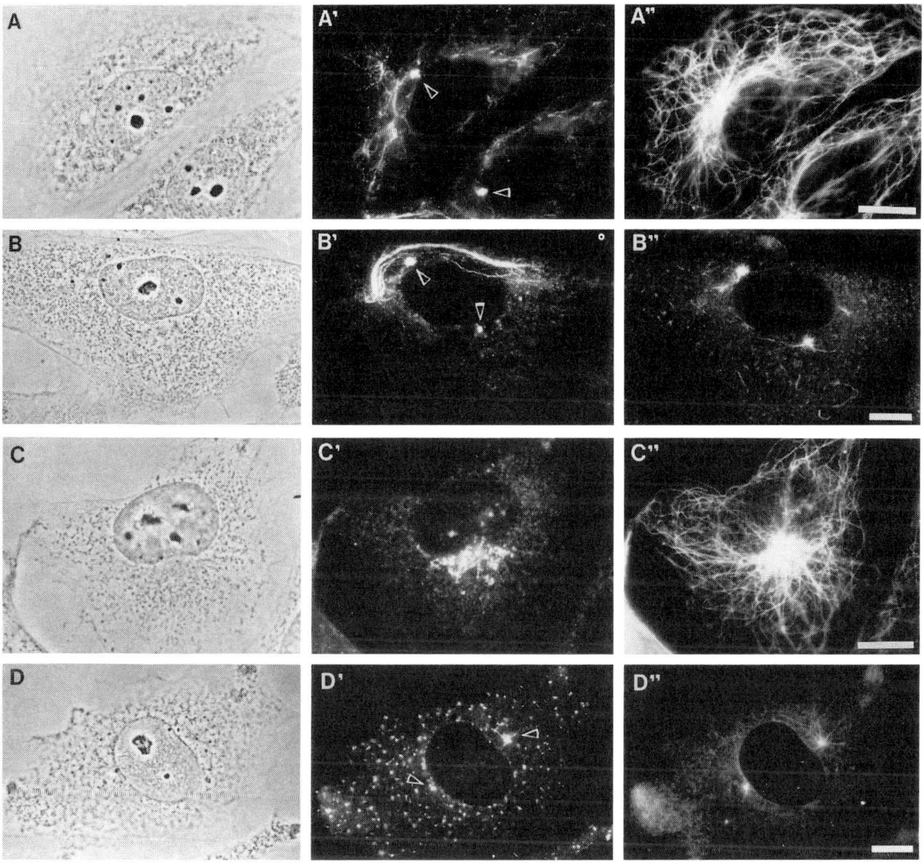

Fig. 7. Immunofluorescence staining of CHO cells with CHO4 (A, B) and CHO5 (C, D) antibodies which cross-react with intermediate filaments (CHO4) and Golgi apparatus (CHO5). The same cells are seen by phase-contrast (A–D) and immunofluorescence microscopy after double staining with either CHO4 (A′–B′) or CHO5 (C′–D′) and tubulin (A″–D″) antibodies. Staining of the centrosomes is clearly seen in interphase cells (arrowheads). Cells in B–B″, D–D″ were treated with 0.5 μg/ml nocodazole for 2 hr before fixation. Note that nocodazole treatment resulted in the bundling of intermediate filaments (B′) and in the dispersion of the Golgi elements from the perinuclear region (D′). Bars: 10 μm.

Fig. 6. Localization of the CHO1 antigen in isolated mitotic spindles and midbodies of CHO cells. The same isolated structures are seen by phase-contrast (A–G) and immunofluorescence microscopy after double staining with CHO1 antibody (A′–G′) or antibody to tubulin (A″–G″). Note the labeling of the fibers in the central part of the spindle and of the midbody (G′) by the CHO1 antibody. Bar: 5 μm. [Reproduced from Sellitto and Kuriyama, *Journal of Cell Biology* (1988a), **106,** 434, by copyright permission of the Rockefeller University Press.]

are immunoreactive with the CHO5 antibody are scattered through the cytoplasm in nocodazole-treated CHO cells (Fig. 7D′), indicating that these aggregates correspond to the Golgi complex. After the dispersion of the Golgi elements in nocodazole-treated cells, the CHO5-positive centrosomes are more clearly seen (Fig. 7D′).

The observed cross-reactions of antibodies to MTOC antigens with other cellular organelles should not be surprising since monoclonal antibodies recognize single antigenic determinants that might be shared among unrelated proteins that are components of different subcellular structures (Table IV). However, some of the cross-reactions might be more than coincidental. Several of the antibodies raised against centrosomes isolated from N115 neurobastoma cells (Ring *et al.*, 1980) and human lymphocytes (Buendia *et al.*, 1990) label both centrosomes and intermediate filaments. A monoclonal antibody generated against a *Drosophila* intermediate filament protein specifically labels centrosomes in sea urchin eggs (Schatten *et al.*, 1987). Steffen and Linck (1989) have shown recently that a novel set of proteins named tektins shares common properties with intermediate filament proteins. Antitektin antibodies also are able to react with components present in centrosomes of animal cells. Immunofluorescence staining of mammalian centrosomes was demonstrated by Hurt using antiyeast nucleoskeletal-like protein (NSP1) antibodies (Table II). The flanking amino-terminal region of the protein encoded by the *NSP1* gene is known to have similarity to intermediate filament proteins (Hurt, 1988). Recent results from the nucleotide sequencing of DNA encoding the 110K protein of yeast SPBs indicate that the protein has homology to other proteins containing a coiled-coil motif (Kilmartin and Rout, 1990). Since intermediate filament proteins also include the same motif (Steinert and Roop, 1988), intermediate filament-like proteins may serve as structural elements of MTOCs.

Structural association between centrosomes and the Golgi apparatus has long been recognized in various types of cells (Sandoval *et al.*, 1984; Kreis, 1990). Colocalization of the CHO5 antigen in the two structures could indicate that MTOCs and the Golgi apparatus not only maintain a structural association but also share common epitopes or proteins. In fact, a cAMP-dependent protein kinase is associated with both structures (Nigg *et al.*, 1985; De Camilli *et al.*, 1986), as is the CHO5 antigen. Structural integrity of the Golgi apparatus at the perinuclear position is thought to be maintained by cytoplasmic microtubules (Sandoval *et al.*, 1984; Kreis, 1990). Detailed microscopic analysis has recently provided evidence that scattered Golgi elements move back to the centrosomal region along repolymerized microtubules upon removal of the drug from nocodazole-treated cells (Ho *et al.*, 1989). Some motor protein(s), responsible for the translocation of the Golgi elements along the microtubules, may attach to individual vesicles. The presence of such Golgi-associated microtubule proteins has recently been demonstrated in rat liver (Bloom and Brashear, 1989) as well as in cultured mammalian cells (Allan and Kreis, 1986; De Camilli *et al.,* 1986; Sterner and Olmsted, 1989). Such molecules could be stored in the centrosomal region in close proximity to the Golgi apparatus, or they may directly bind to centrosomes through their affinity for the minus end of microtubules.

Nuclear components are among the most common cross-reacting antigens seen with MTOC antibodies. Of the 13 MTOC antibodies discussed earlier, 5 (CHO1, CHO3, AX4–AX6) cross-react with nuclei. CHO3 antibody labels nuclei only in proliferating cells. Quiescent cells obtained by exposing cells to a culture medium without fetal calf serum showed no immunofluorescence staining of nuclei with the CHO3 antibody (Sellitto and Kuriyama, 1988c). It is also worthwhile to point out that the nuclear staining seen with CHO3 antibodies in interphase *Dictyostelium* amebas disappears as the cells enter mitosis. Similar cross-reactions between centrosomes and nuclei also have been seen with autoimmune sera (Price *et al.,* 1984; Dales and Wallace, 1985; Senécal *et al.,* 1985; Courvalin *et al.,* 1986; Maekawa *et al.,* 1990a) and antibodies raised against nuclear components (Lydersen and Pettijohn, 1980; Chaly *et al.,* 1984; Frash *et al.,* 1986; Hurt, 1988), transformation-associated proteins (Kasamatsu and Nehorayan, 1979; Blair Zajdel and Blair, 1988), p34^{cdc2} (Bailly *et al.,* 1989; Riabowol *et al.,* 1989), p13^{suc1} (Bailly *et al.,* 1989), cyclin (Pines and Hunter, 1989), purified interphase centrosomes (Keryer *et al.,* 1989), a mitotic HeLa cell extract (Davis *et al.,* 1983; Vandré *et al.,* 1984), and brain MAP1 (Sato *et al.,* 1983; Bonifacino *et al.,* 1985; De Mey *et al.,* 1987). These occurrences seem to be too frequent for the explanation that common epitopes, conserved among essentially unrelated proteins, are present by coincidence in the two organelles. Alternatively, similar molecules may indeed be present in both nuclei and centrosomes. In fact, using a human autoimmune serum as a probe, we have recently identified polypeptides with the same apparent molecular weights in both organelles (Maekawa *et al.,* 1990a). Close association between centrosomes and nuclei was also suggested by a different line of evidence. The centrosome generally occupies a juxtanuclear position, which makes it easy for the two organelles to interact (Bornens, 1977; Nadezhdina *et al.,* 1979; Kuriyama and Borisy, 1981a). Centriolar duplication is facilitated by this physical association between the centrosome and the nucleus (Kuriyama and Borisy, 1981a). Likewise, nuclear events (DNA replication) appear to be associated with centriolar events such as the separation of the two centrioles (Sherline and Mascardo, 1982a,b) and ciliation (Tucker *et al.,* 1979). These results indicate that centrosomal components may participate in nuclear activities and vice versa.

3. Cell Cycle Dependency of Antigen Localization at MTOCs

Although all the monoclonal antibodies recognize either centrosomes or NABs, they are highly heterogeneous in their ability to bind MTOC components. As summarized in Table IV, only three *Dictyostelium* antibodies (AX3, AX4, AX6) and six CHO antibodies (CHO2–CHO7) stain these structures in both interphase and mitotic cells. The remaining four antibodies discriminate between mitotic (AX5, AX7) and interphase (CHO1, AX1) MTOCs. The appearance of specific MTOC components at distinct stages of the cell cycle is not unlikely since the structure and activity of centrosomes has been shown to change during the cell cycle (Rattner and Phillips, 1973; Snyder and McIntosh, 1975; Kuriyama and Borisy, 1981b; Rieder and Borisy, 1982; Vorobjev and Chentsov, 1982). Of particular interest to us are MTOC components specific for mitotic stages. MTOCs in mitotic but not interphase cells of *Dictyostelium* can be stained with AX5 and AX7 antibodies. Centrosomes in mitotic animal cells are labeled by the antibodies raised against mouse nuclear matrix fractions (Chaly *et al.,* 1984) and several kinds of human autoantibodies (for example, POPA, Sager *et al.,* 1986; NuMA, Lydersen and Pettijohn, 1980; Price *et al.,* 1984; see Tables I, II). We have also screened sera from CREST patients and found one autoantibody (SP-H) that labels mitotic centrosomal components with an apparent molecular weight of 210K to 230K (Maekawa *et al.,* 1991). During interphase, the SP-H antigen is seen in a number of discrete dots in cells arrested at metaphase by nocodazole. These dots start to aggregate on removal of the microtubule-depolymerizing drug, and are eventually formed at spindle poles. Staining of the same cells with other centrosomal antibodies, CHO3 and 5051, showed that the three different centrosomal antigens did not entirely colocalize (Maekawa *et al.,* 1990).

Centrosomal proteins are known to be preferentially phosphorylated at the transition from interphase to mitosis (Vandré *et al.,* 1984). We have isolated two antibodies, CHO3 and AX3, that react with centrosomal phosphoproteins in CHO cells during mitosis. These two antibodies provide distinct immunostaining patterns in cells and recognize different sets of immunoreactive polypeptides on Western blots. These results could imply that the CHO3 and AX3 antibodies react with different antigenic sites, for example, different posttranslational modifications in components of mitotic centrosomes. This would not be surprising since several different enzymes and modifying factors, such as protein kinases (Nigg *et al.,* 1985; De Camilli *et al.,* 1986; Joachim and Schwoch, 1988; Bailly *et al.,* 1989; Riabowol *et al.,* 1989) and cell cycle controlling factors (Bailly *et al.,* 1989; Pines and Hunter, 1989), are specifically associated with mitotic centrosomes. For example, the 225K centrosomal phosphoprotein in sea urchin eggs identified by the CHO3 antibody is quickly phosphorylated and dephosphorylated during mitosis in a rapid succession by the action of protein kinase(s) and phosphatase(s), respectively (Kuriyama, 1989). Molecular identification of the individual domains that react with CHO3 and AX3 antibodies could be important in evaluating the role of protein phosphorylation in the mechanism of cell cycle regulation.

4. *Effect of Antibodies to MTOC Antigens on in Vitro Microtubule Formation*

Several reports discuss the effect of antibodies on the function of centrosomes. Pepper and Brinkley (1977) demonstrated that a nontubular form of tubulin could be an intrinsic component of mammalian centrosomes. This suggestion has received additional support from a recent report showing localization of gamma tubulin in the MTOC of *Aspergillus* (Oakley *et al.,* 1990). Exposure of centrosomes to tubulin antibodies inhibits their ability to nucleate microtubules *in vitro* (Pepper and Brinkley, 1979). A calcium-binding 62–64K protein (Moudjou *et al.,* 1989b) and MPM-2 reactive phosphoprotein (Centonze and Borisy, 1990) also are required for microtubule formation *in vitro.* Toriyama *et al.,* (1988) prepared antibodies to the 51K protein of sea urchin mitotic spindles. When the antibody was added to the homogenized mitotic apparatus fraction, the number of asters reconstructed from the organizing centers was observed to decrease to one-third of the control. One of the monoclonal antibodies raised against the 51K protein also inhibits mitotic spindle formation in microinjected sea urchin eggs (Ohta *et al.,* 1988b). We also carried out *in vitro* inhibition tests of microtubule formation in order to examine whether the components identified by our centrosomal antibodies are involved in microtubule nucleation from centrosomes. We have obtained the preliminary results that the CHO2 antibody appears to be effective in inhibiting the ability of centrosomes to nucleate microtubules *in vitro:* pretreated cells with CHO2 tend to nucleate shorter and much smaller numbers of microtubules from the centrosome. Although antibodies other than CHO2 examined thus far are negative in these *in vitro* inhibition tests, the corresponding antigens do not necessarily fail to participate in microtubule formation. It may be that the epitopes recognized by the antibodies are not involved in the process of microtubule assembly.

IV. CONCLUDING REMARKS

A fairly large number of anti-MTOC antibodies is now available and the subcellular localization of corresponding antigens has been examined by indirect immunofluorescence microscopy. We must now summarize what has been learned about MTOCs by use of these probes. First, use of the antibodies confirmed that MTOCs are highly complex multicomponent structures. Components identified by each antibody vary greatly. Components show a high degree of heterogeneity in terms of their subcellular distribution. Some antigens show cell cycle dependency in their distribution; others do not. Some cross-react with other cellular organelles; others specifically stain MTOCs. Some components appear to be involved in microtubule nucleation; others are obviously not. Differences in the localization of antigens corresponding to each antibody are most obvious at the transition from interphase to mitosis (Maekawa *et al.,* 1990b). Therefore, it is always important to keep in mind that the behavior and distribution of a particular MTOC component detected by one antibody, for example, by 5051 or MPM-2 two of the most widely used MTOC probes reveals only some aspects of centrosomes and MTOCs and should not be regarded as a comprehensive characterization of the centrosome.

Many components are present in the different MTOCs. Are they all specific constituents involved in MTOC function? Because of the possibility of adventitious cross-reactions via common epitopes shared among essentially nonrelated proteins, some of the polypeptides reacting with the monoclonal antibodies to MTOC antigens may indeed be nonspecific. The number of antibodies reacting with MTOC antigens is far greater than the number of components suggested as necessary for microtubule growth as determined by *in vitro* inhibition tests. Clearly, a large number of components is not directly involved in microtubule assembly but may be associated only passively with the centrosomes. For example, polystyrene (Hamaguchi *et al.,* 1986) or fluorescent carboxylated beads (Wadsworth, 1987) microinjected into echinoderm eggs move toward the minus end of astral microtubules, and thereby accumulate around sperm astral centers or spindle poles. Mitotic poles of the spindle function as the loci for equipartition, and many subcellular components or structures that are to be segregated in the progeny in an equal manner, including chromosomes, could have a tendency to associate with the centrosomes and spindle poles. One example of such a nonspecific association could be the virus-like particles found in the pericentriolar material in certain lines of mammalian cells, including CHO cells (Gould and Borisy, 1977). It is also possible that the extra components might be essential for MTOC functions other than microtubule nucleation that have not yet been defined. For example, the product of yeast gene *SPA2,* isolated using the human anti-PCM antibody (5051)(Calarco-Gillam *et al.,* 1983), is involved in control of cell division and does not function in microtubule nucleation (Snyder, 1989). Microinjection of CHO3 antibodies into living cells also indicates that the epitope recognized by the CHO3 antibody might have an essential role in mitotic progression although it does not appear to be directly involved in microtubule assembly (Sellitto and Kuriyama, unpublished observations).

Much progress has been made in our understanding of MTOCs by the identification of a significant number of molecular components using specific antibodies as probes. Future research should focus on establishing the function of each constituent. As we have seen, such functional evaluation has already started by the use of *in vitro* assays for inhibition of microtubule formation by antibodies and by microinjection of antibodies into living cells to determine their effect on various cell functions. In order to analyze protein functions in more detail, and identify roles of MTOCs other than microtubule organization, however, it will be necessary to use antibody probes in conjunction with molecular genetic techniques. This approach is being taken to study DNA fragments encoding proteins immunoreactive with human autoimmune sera (Snyder and Davis, 1988; Snyder, 1989; Balczon and Zimmer, 1990; Doxsey *et al.*, 1990), a monoclonal anti-32K centrosomal component antibody (Joswig *et al.,* 1990), Bx63 antibodies (Whitfield *et al.,* 1988), and SU5 (Kuriyama *et al.,* 1990b). The proteins centrin (Huang *et al.,* 1988a,b), yeast NSP1 (Hurt, 1988), and yeast 110K components of SPBs (Kilmartin and Rout, 1990) have been isolated and characterized. Centrosomes may well be the central bodies of the cell, as their name indicates, but they are still covered with a thick veil. Antibody probes are obviously one of the most powerful tools to date for studying this mysterious organelle.

ACKNOWLEDGMENTS

I wish to thank C. Sellitto for her invaluable help in the preparation and characterization of monoclonal antibodies. Thanks are also due to M. Kimble and T. Maekawa for their thoughtful discussions and criticisms. This work was supported by NSF DCB8510151 and NIH GM41350.

REFERENCES

Allan, V., and Kreis, T. E. (1986). A microtubule-binding protein associated with membranes of the Golgi apparatus. *J. Cell Biol.* **103,** 2229–2239.

Bailly, E., Dorée, M., Nurse, P., and Bornens, M. (1989). p34^{cdc2} is located in both nucleus and cytoplasm; part is centrosomally associated at G_2/M and enters vesicles at anaphase. *EMBO J.* **8,** 3985–3995.

Balczon, R., and Zimmer, W. E. (1990). The molecular cloning of cDNA encoding a mammalian centrosome antoantigen. *J. Cell Biol.* **111,** 180a (*Abstract*).

Baron, A. T., and Salisbury, J. L. (1988). Identification and localization of a novel, cytoskeletal, centrosome-associated protein in PtK_2 cells. *J. Cell Biol.* **107,** 2669–2678.

Bastmeyer, M., and Russell, D. G. (1987). Characterization of *Pales* spermatocyte spindles, with reference to an MTOC-associated protein. *J. Cell Sci.,* **87,** 431–438.

Bastmeyer, M., Steffen, W., and Fuge, H. (1986). Immunostaining of spindle components in tipulid spermatocytes using a serum against pericentriolar material. *Eur. J. Cell Biol.* **42,** 305–310.

Blair Zajdel, M., and Blair, G. E. (1988). The intracellular distribution of the transformation-association protein p53 in adenovirus-transformed rodent cells. *Onogene* **2,** 579–584.

Blose, S. H., and Bushnell, A. (1982). Observations on the vimentin-10-nm filaments during mitosis in BHK-21 cells. *Exp. Cell Res.* **142,** 57–62.

Bloom, G. S., and Brashear, T. A. (1989). A novel 58-kDa protein associated with the Golgi apparatus and microtubules. *J. Biol. Chem.,* **264,** 16083–16092.

Bonifacino, J. S., Klausner, R. D., and Sandoval, I. V. (1985). A widely distributed nuclear protein immunologically related to the microtubule-associated protein MAP1 is associated with the mitotic spindle. *Proc. Natl. Acad. Sci. U.S.A.* **82,** 1146–1150.

Bornens, M. (1977). Is the centriole bound to be the nuclear membrane? *Nature (London)* **270,** 80–82.

Bornens, M. Paintrand, M., Berges, J., Marty, M. -C., and Karsenti, E. (1987). Structural and chemical characterization of isolated centrosomes. *Cell Motil. Cytosk.* **8,** 238–249.

Brenner, S. L., and Brinkley, B. R. (1982). Tubulin assembly sites and the organization of microtubule arrays in mammalian cells. *Cold Spring Harbor Symp. Quant. Biol.* **46,** 241–254.

Buendia, B., Antony, C., Verde, F., Bornens, M., and Karsenti, E. (1990). A centrosomal antigen localized on intermediate filaments and mitotic spindle poles. *J. Cell Sci.* **97,** 259–271.

Calarco-Gillam, P. D., Siebert, M. C., Hubble, R., Mitchison, T., and Kirschner, M. (1983). Centrosome development in early mouse embryos as defined by an autoantibody against pericentriolar material. *Cell* **35,** 621–629.

Centonze, V. E., and Borisy, G. G. (1990). Nucleation of microtubules from mitotic centrosomes is modulated by phosphorylated epitope. *J. Cell Sci.* **95,** 405–411.

Chaly, N., Bladon, T., Setterfield, G., Little, J. E., Kaplan, J. G., and Brown, D. L. (1984). Changes in distribution of nuclear matrix antigens during the mitotic cell cycle. *J. Cell Biol.* **99,** 661–671.

Chaput, M., Claes, V., Portetelle, D., Cludts, I., Cravador, A., Burny, A., Gras, H., and Tartar, A. (1988). The neurotrophic factor neuroleukin is 90% homologous with phosphohexose isomerase. *Nature (London)* **332,** 454–455.

Clayton, L., Black, C. M., and Lloyd, C. W. (1985). Microtubule nucleating sites in higher plant cells identified by an auto-antibody against percentriolar material. *J. Cell Biol.* **101,** 319–324.

Connolly, J. A., and Kalnins, V. I. (1978). Visualization of centrioles and basal bodies by fluorescent staining with nonimmune rabbit sera. *J. Cell Biol.* **79,** 526–532.

Courvalin, J. -C., Hernandez-Verdun, D., Gosti-Testu, F., Marty, M. -C., Maunoury, R., and Bornens, M. (1986). A protein of M_r 80,000 is associated with the nucleolus organizer of human cell lines. *Chromosoma (Berlin)* **94,** 353–361.

Dales, S., and Wallace, A. C. (1985). Nuclear pore complexes deposited in the glomerular basement membrane are associated with autoantibodies in a case of membranous nephritis. *J. Immunol.* **134,** 1588–1593.

Davis, F. M., Tsao, T. Y., Fowler, S. K., and Rao, P. N. (1983). Monoclonal antibodies to mitotic cells. *Proc. Natl. Acad. Sci. U.S.A.* **80,** 2926–2930.

Davis, F. M., Wright, D. A., Penkala, J. E., and Rao, P. N. (1989). Mitosis-specific monoclonal antibodies block cleavage in amphibian embryos. *Cell Struct. Funct.* **14,** 271–277.

De Camilli, P., Moretti, M., Donini, S. D., Walter, U., and Lohmann, S. M. (1986). Heterogeneous distribution of the cAMP receptor protein RII in the nervous system: Evidence for its intracellular accumulation on microtubules, microtubule-organizing centers, and in the area of the golgi complex. *J. Cell Biol.* **103,** 189–203.

De Mey, J., Aerts, F., De Raeymaeker, M., Daneels, G., Moeremans, M., De Wever, B., Vandré, D. D., Vallee, R. B., Borisy, G. G., and De Brabander, M. (1987). Microtubule associated protein (MAP) 1A: A phosphoprotein associated with spindle fibers and microtubule organizing centres. *In* "Nature and Function of Cytoskeletal Proteins in Motility and Transport: Progress in Zoology" (K. E. Wolhfarth-Botterman, ed.) Vol. 34, pp. 187–205. Fischer, New York.

Doxsey, S., Calarco, P., Seibert, P., Evans, L., Stein, P., and Kirschner, M. (1990). A centrosome protein involved in meiotic spindle formation. *J. Cell Biol.* **111,** 179a (*Abstract*).

Engle, D. B., Doonan, J. H., and Morris, N. R. (1988). Cell-cycle modulation of MPM-2 specific spindle pole body phosphorylation in *Aspergillus nidulans. Cell Motil. Cytosk.* **10,** 432–437.

Enos, A. P., and Morris, N. R. (1990). Mutation of a gene that encodes a kinesin-like protein blocks nuclear division in *A. nidulans. Cell* **60,** 1019–1027.

Frash, M., Glover, D. M., and Saumweber, H. (1986). Nuclear antigens follow different pathways into daughter nuclei during mitosis in early *Drosophila* embryos. *J. Cell Sci.,* **82,** 155–172.

Fung, B. P., and Kasamatsu, H. (1985). Immuno-electron-microscopic localization of a centriole-related antigen in ciliated cells. *Cell Tissue Res.* **239,** 43–50.

Geetha-Habib, M., Noiva, R., Kaplan, H. A., and Lennarz, W. J. (1988). Glycosylation site binding protein, a component of oligosaccharyl transferase, is highly similar to three other 57 kd luminal proteins of the ER. *Cell* **54,** 1053–1060.

Gonzalez, C. Saunders, R. D. C., Casal, J., Molina, I., Camena, M., Ripoll, P., and Glover, D. M. (1990). Mutations at the *asp* locus of *Drosophila* lead to multiple free centrosomes in syncytial embryos, but restrict centrosome duplication in larval neuroblasts. *J. Cell Sci.* **96,** 605–616.

Gosti, F., Marty, M.-C., Courvalin, J. C., Maunoury, R., and Bornens, M. (1987). Centrosomal proteins and lactate dehydrogenase possess a common epitope in human cell line. *Proc. Natl. Acad. Sci. U.S.A.* **84,** 1000–1004.

Gould, R. R. and Borisy, G. G. (1977). The pericentriolar material in Chinese hamster ovary cells nucleates microtubule formation. *J. Cell. Biol.* **73,** 601–615.

Hagan, I. M., and Yanagida, M. (1990). Novel potential mitotic motor protein encoded by fission yeast *cut7+* gene. *Nature (London)* **347,** 563–566.

Hamaguchi, M. S., Hamaguchi, Y., and Hiramoto, Y. (1986). Microinjected polystyrene beads move along astral rays in sand dollar eggs. *Dev. Growth Differ.* **28,** 461–470.

Harris, P. (1975). The role of membranes in the organization of the mitotic apparatus. *Exp. Cell Res.* **94,** 409–425.

Heath, I. B. (1981). Nucleus-associated organelles in fungi. *Int. Rev. Cytol.* **69,** 191–221.

Hecht, R. M., Berg-Zabelshansky, M., Rao, P. N., and Davis, F. M. (1987). Conditional absence of mitosis-specific antigens in a temperature-sensitive embryonic-arrest mutant of *Caenorhabditis elegans. J. Cell Sci.* **87,** 305–314.

Hiraoka, L., Golden, W., and Magnuson, T. (1989). Spindle-pole organization during early mouse development. *Dev. Biol.* **133,** 24–36.

Ho, W. C., Allan, V. J., van Meer, G., Berger, E. G., and Kreis, T. E. (1989). Reclustering of scattered golgi elements occurs along microtubules. *Eur. J. Cell Biol.* **48,** 250–263.

Hoffman, J. C., and Mullins, J. M. (1990). A nuclear and mitotically enhanced epitope. *Cell Motil. Cytosk.* **16,** 68–79.

Houliston, E., Peckering, S. J., and Maro, B. (1987). Redistribution of microtubules and pericentriolar material during the development of polarity in mouse blastomeres. *J. Cell Biol.* **104,** 1299–1308.

Huang, B., Watterson, D. M., Lee, V. D., and Schibler, M. J. (1988a). Purification and characterization of a basal body-associated Ca^{2+}-binding protein. *J. Cell Biol.* **107,** 121–131.

Huang, B., Mengersen, A., and Lee, V. D. (1988b). Molecular cloning of cDNA for caltractin, a basal body-associated Ca^{2+}-binding protein: Homology in its protein sequence with calmodulin and the yeast *CDC31* gene product. *J. Cell Biol.* **107,** 133–140.

Hurt, E. C. (1988). A novel nucleoskeletal-like protein located at the nuclear periphery is required for the life cycle of *Saccharomyces cerevisiae. EMBO J.* **7,** 4323–4334.

Joachim, S., and Schwoch, G. (1988). Immunoelectron microscopic localization of catalytic and regulatory subunits of cAMP-dependent protein kinases in the parotid gland. *Eur. J. Cell Biol.* **46,** 491–498.

Joswig, G., and Petzelt, C. (1990). The centrosomal cycle: Visualization in PtK cells by a monoclonal antibody to a centrosomal 32 kd protein. *Cell Motil. Cytosk.* **15,** 181–192.

Joswig, G., Petzelt, C., and Werner, D. (1990). Murine cDNAs coding for the centrosomal antigen centrosomin A. *J. Cell Sci.,* **98,** 37–43.

Kasamatsu, H., and Nehorayan, A. (1979). Intracellular localization of viral polypeptides during Simian virus 40 infection. *J. Virol.* **32,** 648–660.

Kasamatsu, H., Shyamala, M., and Lin, W. (1980). Host antigens in the centriolar region are induced in SV40-infected TC7 cells: SV40 small-T-function requirement. *Cold Spring Harbor Symp. Quant. Biol.* **44,** 243–252.

Kellogg, D. R., Field, C. M., and Alberts, B. M. (1989). Identification of microtubule-associated proteins in the centrosome, spindle, and kinetochore of the early *Drosophila* embryo. *J. Cell Biol.* **109,** 2977–2991.

Keryer, G., Davis, F. M., Rao, P. N., and Beisson, J. (1987). Protein phosphorylation and dynamics of cytoskeletal structures associated with basal bodies in *Paramecium. Cell Motil. Cytosk.* **8,** 44–54.

Keryer, G. De Loubresse, N. G., Bordes, N., and Bornens, M. (1989). Identification of a spindle-associated protein in ciliate micronuclei. *J. Cell Sci.* **93,** 287–298.

Kikuchi, Y., Shimatake, H., and Kikuchi, A. (1988). A yeast gene required for the G_1-to-S transition encodes a protein containing an A-kinase target site and GTPase domain. *EMBO J.* **7,** 1175–1182.

Kilmartin, J. V., Wright, B., and Milstein, C. (1982). Rat monoclonal antitubulin antibodies derived by raising a new nonsecreting rat cell line. *J. Cell Biol.* **93,** 576–582.

Kilmartin, J. V., and Rout, M. P. (1990). Yeast spindle body components. *In* "ASCB & EMBO Joint Meeting: The Cell Cycle." Airlie, Virginia.

Klotz, C. Bordes, N., Laine, M. C., Sandoz, D., and Bornens, M. (1986). A protein of 175,000 daltons associated with striated rootlets in ciliated epithelia, as revealed by a monoclonal antibody. *Cell Motil. Cytosk.* **6,** 56–67.

Kohler, G., and Milstein, C. (1976). Derivation of specific antibody-producing tissue culture and tumor lines by cell fusion. *Eur. J. Immunol.* **6,** 511–519.

Komesli, S., Tournier, F., Paintrand, M., Margolis, R. L., Job, D., and Bornens, M. (1989). Mass isolation of calf thymus centrosomes: Identification of a specific configuration. *J. Cell Biol.* **109,** 2869–2878.

Koonce, M. P., and McIntosh, J. R. (1990). Identification and immunolocalization of cytoplasmic dynein in *Dictyostelium. Cell Motil. Cytosk.* **15,** 51–62.

Kreis, T. E. (1990). Role of microtubules in the organization of the golgi apparatus. *Cell Motil. Cytosk.* **15,** 67–70.

Kronebusch, P. J., and Singer, S. J. (1987). The microtubule-oganizing complex and the Golgi apparatus are co-localized around the entire nuclear envelope of interphase cardiac myocytes. *J. Cell Sci.* **88,** 25–34.

Kuang, J. Zhao, J. -Y., Wright, D. A., Saunders, G. F., and Rao, P. N. (1989). Mitosis-specific monoclonal antibody MPM-2 *Xenopus* oocyte maturation and depletes maturation-promoting activity. *Proc. Natl. Acad. Sci. U.S.A.* **86,** 4982–4986.

Kuriyama, R. (1984). Activity and stability of centrosomes in Chinese hamser ovary cells in nucleation of microtubules *in vitro. J. Cell Sci.* **66,** 277–295.

Kuriyama, R. (1989). 225-kilodalton phosphoprotein associated with mitotic centrosomes in sea urchin eggs. *Cell Motil. Cytosk.* **12,** 90–103.

Kuriyama, R., and Borisy, G G. (1981a). Centriole cycle in Chinese hamster ovary cells as determined by whole-mount electron microscopy. *J. Cell Biol.* **91,** 814–821.

Kuriyama, R., and Borisy, G. G. (1981b). Microtubule-nucleating activity of centrosomes in Chinese hamster ovary cells is independent of the centriole cycle but coupled to the mitotic cycle. *J. Cell Biol.* **91,** 822–826.

Kuriyama, R., Sato, C., Fukui, Y., and Nishibayashi, S. (1982). *In vitro* nucleation of microtubules from microtubule-oganizing center prepared from cellular slime mold. *Cell Motil.* **2,** 257–272.

Kuriyama, R., Keryer, G., and Borisy, G. G. (1984). The mitotic spindle of Chinese hamster ovary cells isolated in taxol-containing medium. *J. Cell Sci.* **66,** 265–272.

Kuriyama, R., and Borisy, G. G. (1985). Identification of molecular components of the centrosphere in the mitotic spindle of sea urchin eggs. *J. Cell Biol.* **101,** 524–530.

Kuriyama, R., Rao, P. N., and Borisy, G. G. (1990a). Immunocytological evidence for centrosomal phosphoproteins in mitotic sea urchin eggs. *Cell Struct. Funct.* **15,** 13–20.

Kuriyama, R. Savereide, P., Lefebvre, P., and Dasgupta, S. (1990b). The predicted amino acid sequence of a centrosphere protein in dividing sea urchin eggs is similar to elongation factor (EF-1α). *J. Cell Sci.* **95,** 231–236.

Lechtreck, K.-F., McFadden, G. I., and Melkonian, M. (1989). The cytoskeleton of the naked green flagellate *Spermatozopsis similis:* Isolation, whole mount electron microscopy, and preliminary biochemical and immunological characterization. *Cell Motil. Cytosk.* **14,** 522–561.

Le Guen, P., and Crozet, N. (1989). Microtubule and centrosome distribution during sheep fertilization. *Eur. J. Cell Biol.* **48,** 239–249.

Lenstra, J. A., and Bloemendal, H. (1983). The major proteins form HeLa cells. Identification and intracellular localization. *Eur. J. Biochem.* **130,** 419–426.

Leslie, R., and Wilson, L. (1987). A human autoimmune serum recognizes an 80 kd protein in sea urchin mitotic spindle poles. *J. Cell Biol.* **105,** 284a (*Abstract*).

Lin, W., Fung, B., Shyamala, M., and Kasamatsu, H. (1981). Identification of antigenically related polypeptides at centrioles and basal bodies. *Proc. Natl. Acad. Sci. U.S.A.* **78,** 2373–2377.

Luca, F. C., Bloom, G. S., and Vallee, R. B. (1986). A monoclonal antibody that cross-reacts with phosphorylated epitopes on two microtubule-associated proteins and two neurofilament polypeptides. *Proc. Natl. Acad. Sci. U.S.A.* **83,** 1006–1010.

Lydersen, B. K., and Pettijohn, D. E. (1980). Human-specific nuclear protein that associates with the polar region of the mitotic apparatus: Distribution in a human/hamster hybrid cell. *Cell* **22,** 489–499.

McCarty, G. A., Valencia, D. W., and Fritzler, M. J. (1984). Antibody to the mitotic spindle apparatus: Immunologic characteristics and cytologic studies. *J. Rheumatol.* **11,** 213–218.

Maekawa, T., Leslie, R., and Kuriyama, R. (1991). Identification of a minus end-specific microtubule-associated protein located at the mitotic poles in cultured mammalian cells. *Eur. J. Cell Biol.* **54,** 255–267.

Maekawa, T., Leslie, R., and Kuriyama, R. (1990). Centrosomal antigens are differentially distributed in mitotic PtK_1 cells. *J. Cell Biol.* **111,** 180a (*Abstract*).

Mao, P. L., and Wang, E. (1988). Molecular characterization of statin, a protein marker for non-proliferating cells. *J. Cell Biol.* **107,** 76a (*Abstract*).

Maro, B., Paintrand, M., Sauron, M.-E., Paulin, D., and Bornens, M. (1984). Vimentin filaments and centrosomes. Are they attached? *Exp. Cell Res.* **150,** 452–458.

Maro, B. Howlett, S. K., and Webb, M. (1985). Non-spindle microtubule organizing centers in metaphase II-arrested mouse oocytes. *J. Cell Biol.* **101,** 1665–1672.

Masuda, H., Hirano, T., Yanagida, N., and Cande, W. Z. (1990). *In vitro* reactivation of spindle elongation in fission yeast *nuc2* mutant cells. *J. Cell Biol.* **110,** 417–425.

Maunoury, R. (1978). Localisation immunocytochimique de la centrosphere de cellules tumorales humaines par utilisation d' anticorps naturels de lapin. *C. R. Seances Acad. Sci. Ser. III* **286,** 503–506.

Maunoury, R. (1979). Mise en évidence de la centrosphère de cellules humaines par la technique d'immunoperoxydase avec des anticorps naturels. *Biol. Cell.* **36,** 91–94.

Millar, S. E., Freeman, M., and Glover, D. M. (1987). The distribution of a 'mitosis-specific' antigen during *Drosophila* development. *J. Cell Sci.* **87,** 95–104.

Mitchison, T., and Kirschner, M. (1984). Microtubule assembly nucleated by isolated centrosomes. *Nature* (*London*) **312,** 232–237.

Moudjou, M. Lanotte, M., and Bornens, M. (1989a). The fate of the centrosome–microtubule network in monocyte-derived giant cells. *J. Cell Sci.* **94,** 237–244.

Moudjou, M., Viguès, B., and Bornens, M. (1989b). Centrosome and calcium: Identification of a 62–64 kDa centrin homologue protein in human centrosomes. *J. Cell Biol.* **109,** 345a (*Abstract*).

Nadezhdina, E. S., Gais, D., and Chentsov, Y. S. (1979). On the association of centrioles with the interphase nucleus. *Eur. J. Cell Biol.* **19,** 109–115.

Neighbors, B. W., Williams, R. C., Jr., and McIntosh, J. R. (1988). Localization of kinesin in cultured cells. *J. Cell Biol.* **106,** 1193–1204.

Nenci, I., and Marchetti, E. (1978). Concerning the localization of steroids in centrioles and basal bodies by immunofluorescence. *J. Cell Biol.* **76,** 255–260.

Nigg, E. A., Schafer, G., Hilz, H., and Eppenberger, H. M. (1985). Cyclic-AMP-dependent protein kinase type II is associated with the Golgi complex and with centrosomes. *Cell* **41,** 1039–1051.

Nislow, C., Sellitto, C., Kuriyama, R., and McIntosh, J. R. (1990). A monoclonal antibody to a mitotic microtubule-associated protein blocks mitotic progression. *J. Cell Biol.* **111,** 511–522.

Numata, O., Takemasa, T., Takagi, I., Hirano, M., Hirano, H., Chiba, J., and Watanabe, Y. (1991). *Tetrahymena* 114-nm filament-forming protein has citrate synthase activity. *Biochem. Biophys. Res. Commun.* **174,** 1028–1034.

Oakley, B. R., Oakley, C. E., Yoon, Y., and Jung, M. K. (1990). γ-Tubulin is a component of the spindle pole body that is essential for microtubule function in *Aspergillus nidulans. Cell* **61,** 1289–1301.

Ohta, K., Toriyama, M., Endo, S., and Sakai, H. (1988a). Localization of mitotic apparatus-associated 51-kDa protein in unfertilized and fertilized sea urchin eggs. *Cell Motil. Cytosk.* **10,** 496–505.

Ohta, K. Toriyama, M., Hamaguchi, Y., Maekawa, S., Endo, S., and Sakai, H. (1988b). Characterization of monoclonal antibodies against the mitotic apparatus associated 51-kD protein and the effect of the microinjection on cell division in sand dollar eggs. *Protoplasma* [*Suppl*] **1,** 113–121.

Ohta, K. Toriyama, M., Miyazaki, M. Murofushi, M., Hosoda, S., Endo, S., and Sakai, H. (1990). The mitotic apparatus-associated 51-kDa protein from sea urchin egg is a GTP-binding protein and is immunologically related to yeast polypeptide elongation factor 1 alpha. *J. Biol. Chem.* **265,** 3240–3247.

Oliver, J. M., Osborne, W. R. A., Pfeiffer, J. R., Child, F. M., and Berlin, R. D. (1981). Purine nucleoside phosphorylase is associated with centrioles and basal bodies. *J. Cell. Biol.* **91,** 837–847.

Pepper, D. A., and Brinkley, B. R. (1977). Localization of tubulin in the mitotic apparatus of mammalian cells by immunofluorescence and immunoelectron microscopy. *Chromosoma* (*Berlin*) **60,** 223–235.

Pepper, D. A., and Brinkley, B. R. (1979). Microtubule in[illegible]ation at kinetochores and centrosomes in lysed mitotic cells. *J. Cell Biol.* **82,** 585–591.

Piatigorsky, J., and Wistow, G. J. (1989). Enzyme/crystallins: Gene sharing as an evolutionary strategy. *Cell* **57,** 197–199.

Pickett-Heaps, J. D. (1969). The evolution of the mitotic apparatus: An attempt at comparative ultrastructural cytology in dividing plant cells. *Cytobios* **3,** 257–280.

Pines, J., and Hunter, T. (1989). The involvement of human cyclins in progression through the cell cycle. *J. Cell Biol.* **109,** 283a (*Abstract*).

Price, C. M., McCarty, G. A., and Pettijohn, D. M. (1984). NuMA protein is a human autoantigen. *Arth. Rheum.* **27,** 774–779.

Raff, J. W., and Glover, D. M. (1988). Nuclear and cytoplasmic mitotic cycles continue in *Drosphila* embryos in which DNA synthesis is inhibited with aphidicolin. *J. Cell Biol.* **107,** 2009–2019.

Raff, J. W., and Glover, D. M. (1989). Centrosomes, and not nuclei, initiate pole cell formation in *Drosophila* embryos. *Cell,* **57,** 611–619.

Rao, P. N., Zhao, J. -Y., Ganju, R. K., and Ashorn, C. L. (1989). Monoclonal antibody against the centrosome. *J. Cell Sci.,* **93,** 63–69.

Rattner, J. B., and Phillips, S. G. (1973). Independence of centriole formation and DNA synthesis. *J. Cell Biol.* **57,** 359–372.

Rattner, J. B., Lew, J., and Wang, J. H. (1990). $p34^{cdc2}$ kinase is localized to distinct domains within the mitotic apparatus. *Cell Motil. Cytosk.* **17,** 227–235.

Riabowol, K, Draetta, G., Brizuela, L., Vandré, D., and Beach, D. (1989). The cdc2 kinase is a nuclear protein that is essential for mitosis in mammalian cells. *Cell* **57,** 393–401.

Rieder, C. L., and Borisy, G. G. (1982). The centrosome cycle in PtK_2 cells: Asymmetric distribution and structural changes in the pericentriolar material. *Biol. Cell* **44,** 117–132.

Ring, D., Hubble, R., Caput, D., and Kirschner, M. (1980). Isolation of microtubule organizing centers from mouse neuroblastoma cells. *In* "Microtubules and Microtubule Inhibitors" (M. De Brabander and J. De Mey, eds.), pp. 377–310. Elsevier/North Holland, Amsterdam.

Roobol, A., Havercroft, J. C., and Gull, K. (1982). Microtubule nucleation by the isolated microtubule-organizing centre of *Physarum polycephalum* myxamoebae. *J. Cell Sci.* **55,** 365–381.

Roos, U. -P. (1975). Fine structure of an organelle associates with the nucleus and cytoplasmic microtubules in the cellular slime mould *Polysphondylium violaceum. J. Cell Sci.* **18,** 315–326.

Rout, M. P., and Kilmartin, J. V. (1990). Components of the yeast spindle and spindle pole body. *J. Cell Biol.* **111,** 1913–1927.

Sager, P. R., Rothfield, N. L., Oliver, J. M., and Berlin, R. D. (1986). A novel mitotic spindle pole component that originates from the cytoplasm during prophase. *J. Cell Biol.* **103,** 1863–1872.

Sakai, H., Shimoda, S., and Hiramoto, Y. (1977). Mass isolation of mitotic apparatus using a glycerol/Mg^{2+}/Triton X-100 medium. *Exp. Cell Res.* **104,** 457–461.

Salisbury, J. L., Baron, A. T., Coling, D. E., Martindale, V. E., and Sanders, M. A. (1986). Calcium-modulated contractile proteins associated with the eucaryotic centrosome. *Cell Motil. Cytosk.* **6,** 193–197.

Salisbury, J. L., Sanders, M. A., and Harpst, L. (1987). Flagellar root contraction and nuclear movement during flagellar regeneration in *Chlamydomonas reinhardtii. J. Cell Biol.* **105,** 1799–1805.

Salisbury, J. L., Baron, A. T., and Sanders, M. A. (1988). The centrin-based cytoskeleton of *Chlamydomonas reinhardtii:* Distribution in interphase and mitotic cells. *J. Cell Biol.* **107,** 635–641.

Sandoval, I. V., Bonifacino, J. S., Klausner, R. D., Henkart, M., and Wehland, J. (1984). Role of microtubules in the organization and localization of the Golgi apparatus. *J. Cell Biol.* **99,** 113s–118s.

Sato, C., Nishizawa, K., Nakamura, H., Komagoe, Y., Shimada, K., Ueda, R., and Suzuki, S. (1983). Monoclonal antibody against microtubule associated protein-1 produces immunofluorescent spots in the nucleus and centrosome of cultured mammalian cells. *Cell Struct. Funct.* **8,** 245–254.

Sauron, M. E., Marty, M. C., Maunoury, R., Courvalin, J. C., Maro, B., and Bornens, M. (1986). Centiole-associated proteins in human cell lines. *J. Submicrosc. Cytol.* **16,** 133–135.

Schatten, G., Simerly, C., and Schatten, H. (1985). Microtubule configurations during fertilization, mitosis, and early development in the mouse and the requirement for egg microtubule-mediated motility during mammalian fertilization. *Proc. Natl. Acad. Sci. U.S.A.* **82,** 4152–4156.

Schatten, H., Walter, M., Mazia, D., Biessmann, H., Paweletz, N., Coffe, G., and Schatten, G. (1987). Centrosome detection in sea urchin eggs with a monoclonal antibody against *Drosophila* intermediate filament proteins: Characterization of stages of the division cycle of centrosomes. *Proc. Natl. Acad. Sci. U.S.A.* **84,** 8488–8492.

Segarini, P. R., Shyamala, M., Atcheson, C. L., and Kasamatsu, H. (1983). The centriolar antigen expression in Tc7 cells is dependent on growth conditions and occurs at a particular time point in G_1. *J. Cell Physiol.*

Sellitto, C., and Kuriyama, R. (1987). Analysis of MTOC-components with monoclonal antibodies. *J. Cell Biol.* **105,** 282a (*Abstract*).

Sellitto, C., and Kuriyama, R. (1988a). Distribution of a matrix component of the midbody during the cell cycle in Chinese hamster ovary cells. *J. Cell Biol.* **106,** 431–439.

Sellitto, C., and Kuriyama, R. (1988b). Distribution of pericentriolar material in multipolar spindles induced by Colcemid treatment in Chinese hamster ovary cells. *J. Cell Sci.* **89,** 57–65.

Sellitto, C., and Kuriyama, R. (1988c). Phosphorylated domains shared among microtubule-related structures as revealed by a monoclonal anticentrosomal antibody. *J. Cell Biol.* **107,** 240a (*Abstract*).

Senécal, J. -L., Oliver, J. M., and Rothfield, N. (1985). Anticytoskeletal autoantibodies in the connective tissue diseases. *Arth. Rheum.* **28,** 889–898.

Senécal, J. -L., and Rauch, J. (1988). Hybridoma lupus autoantibodies can bind major cytoskeletal filaments in the absence of DNA-binding activity. *Arth. Rheum.* **31,** 864–875.

Sherline, P., and Mascardo, R. N. (1982a). Epidermal growth factor induces rapid centrosomal separation in HeLa and 3T3 cells. *J. Cell Biol.* **93,** 507–511.

Sherline, P., and Mascardo, R. N. (1982b). Epidermal growth factor-induced centrosomal preparation: Mechanism and relationsip to mitogenesis. *J. Cell Biol.* **95,** 316–322.

Shyamala, M., Atcheson, C. L., and Kasamatsu, H. (1982). Stimulation of host centriolar antigen in Tc7 cells by simian virus 40: Requirement for RNA and protein synthesis and an intact simian virus 40 small-t gene function. *J. Virol.* **43,** 721–729.

Slobin, L. I. (1980). The role of eucaryotic elongation factor Tu in protein synthesis. The measurement of the elongation factor Tu content of rabbit reticulocytes and other mammalian cells by a sensitive radioimmunoassay. *Eur. J. Biochem.* **110,** 555–563.

Snyder, J. A., and McIntosh, J. R. (1975). Initiation and growth of microtubules from mitotic centers in lysed mammalian cells. *J. Cell Biol.* **67,** 744–760.

Snyder, M. (1989). The *SPA2* protein of yeast localizes to sites of cell growth. *J. Cell Biol.* **108,** 1419–1429.

Snyder, M., and Davis, R. W. (1988). *SPA1:* A gene important for chromosome segregation and other mitotic functions in *S. cerevisiae. Cell* **54,** 743–754.

Steffen, W., and Linck, R. W. (1989). Tektins in ciliary and flagellar microtubules and their association with other cytoskeletal system. *In* "Cell Movement. Kinesin, Dynein, and Microtubule Dynamics" (F. D. Warner and J. R. McIntosh, eds.), Vol. 2, pp. 67–81. Liss, New York.

Steinert, P. M., and Roop, R. (1988). Molecular and cellular biology of intermediate filaments. *Ann. Rev. Biochem.* **57,** 593–625.

Sterner, J. M., and Olmsted, J. B. (1989). Localization of microinjected IAF-derivatized MAP 2 domains. *J. Cell Biol.* **109,** 191a (*Abstract*).

Suprenant, K. A., Tempero, L. B., and Hammer, L. E. (1989). Association of ribosomes with *in vitro* assembled microtubules. *Cell Motil. Cytosk.* **14,** 401–415.

Tassin, A. -M., Maro, B., and Bornens, M. (1985). Fate of microtubule-organizing centers during myogenesis in vitro. *J. Cell Biol.* **100,** 35–46.

Tong, P. Y., Tollefsen, S. E., and Kornfeld, S. (1988). The cation-independent mannose 6-phosphate receptor binds insulin-like growth factor II. *J. Biol. Chem.* **263,** 2585–2588.

Toriyama, M.,Ohta, K., Endo, S., and Sakai, H. (1988). 51-kd protein, a component of microtubule-organizing granules in the mitotic apparatus involved in aster formation in vitro. *Cell Motil. Cytosk.* **9,** 117–128.

Tucker, R. W., Pardee, A. B., and Fujiwara, K. (1979). Centriole ciliation is related to quiescence and DNA snythesis in 3T3 cells. *Cell* **17,** 527–535.

Tuffanelli, D. L., McKeon, F., Kleinsmith, D. M., Burnham, T. K., and Kirschner, M. (1983). Anticentromere and anticentriole antibodies in the scleroderma spectrum. *Arch. Dermatol.* **119,** 560–566.

Turksen, K., Aubin, J. E., and Kalnins, V. I. (1982). Identification of a centiole-associated protein by antibodies present in normal rabbit sera. *Nature (London)* **298,** 763–765.

Uchida, T., and Endo, T. (1988). Immunoelectron microscopic demonstration of S-100b protein-like in centriole, cilia, and basal body. *J. Histochem. Cytochem.* **36,** 693–696.

Vale, R. D., Reese, T. S., and Sheetz, M. P. (1985). Identification of a novel force-generating protein, kinesin, involved in microtubule-based motility. *Cell* **42,** 39–50.

Vandré, D. D., Davis, F. M., Rao, P. N., and Borisy, G. G. (1984). Phosphoproteins are components of mitotic microtubule organizing centers. *Proc. Natl. Acad. Sci. U.S.A.* **81,** 4439–4443.

Vandré, D. D., Davis, F. M., Rao, P. N., and Borisy, G. G. (1986). Distribution of cytoskeletal proteins sharing a conserved phosphorylated epitope. *Eur.J. Cell Biol.* **41,** 72–81.

Vandré, D. D., and Borisy, G. G. (1989). Anaphase onset and dephosphorylation of mitotic phosphoproteins occur concommitantly. *J. Cell Sci.* **94,** 245–248.

Van Ness, J., and Pettijohn, D. E. (1984). Specific attachment of nuclear-mitotic apparatus protein to metaphase chromosomes and mitotic spindle poles: Possible function in nuclear reassembly. *J. Mol. Biol.* **171,** 175–205.

Viel, A., Dje, M. K., Mazabraud, A., Denis, H., and le Marie, M. (1987). Thesaurin a, the major protein of *Xenopus laevis* previtellogenic oocytes, present in the 42S particles, is homologus to elongation factor EF-1α. *FEBS Lett.* **223,** 232–236.

Vorobjev, I. A., and Chentsov, Y. S. (1982). Centrioles in the cell cycle. I. Epithelial cells. *J. Cell Biol.* **93,** 938–949.

Wadsworth, P. (1987). Microinjected carboxylated beads move predominantly poleward in sea urchin eggs. *Cell Motil. Cytosk.* **8,** 293–301.

Wang, E. Connolly, J. N., Kalnins, V. I., and Choppin, P. W. (1979). Relationship between movement and aggregation of centrioles in cyncytia and formation of microtubule bundles. *Proc. Natl. Acad. Sci. U.S.A.* **76,** 5719–5723.

Wang, Y. -L., and Taylor, D. L. (1979). Distribution of fluorescently labeled actin in living sea urchin eggs during early development. *J. Cell Biol.* **82,** 672–679.

Whitfield, W. G., Millar, S. E., Saumweber, H., Frasch, M., and Glover, D. M. (1988). Cloning of a gene encoding an antigen associated with the centrosome in *Drosophila. J. Cell Sci.* **89,** 467–480.

Wick, S. M. (1985). Immunofluorescence microscopy of tubulin and microtubule arrays in plant cells. *Cell Biol. Int. Rep.* **9,** 357–371.

Willingham, M. C., and Pastan, I. (1985). Ultrastructural immunocytochemical localization of the transferrin receptor using a monoclonal antibody in human KB cells. *J. Histochem. Cytochem.* **33,** 59–64.

Wistow, G. W., and Piatigorsky, J. (1988). Lens crystallins: The evolution and expression of proteins for a highly specialized tissue. *Ann. Rev. Biochem.* **57,** 479–504.

Wordeman, L., and Cande, W. Z. (1986). Kinase(s) association with functional mitotic spindles from the diatom *Stephanopyxis turris* phosphorylate a variety of spindle polypeptides. *J. Cell Biol.* **103,** 411a (*Abstract*).

Wright, R. L., Adler, S. A., Spanier, J. G., and Jarvik, J. W. (1989). Nucleus-basal body connector in *Chlamydomonas:* Evidence for a role in basal body segregation and against essential roles in mitosis or in determining cell polarity. *Cell Motil. Cytosk.* **14,** 516–526.

Zieve, G. W., Heidemann, S. R., and McIntosh, J. R. (1980). Isolation and partial characterization of a cage of filaments that surround the mammalian mitotic spindle. *J. Cell Biol.* **87,** 160–169.

Zimmerman, A. M. (1960). Physico-chemical analysis of the isolated mitotic apparatus. *Exp. Cell Res.* **20,** 529–547.

6

Role of Centrin in Spindle Pole Dynamics

ANDRE T. BARON AND JEFFREY L. SALISBURY

Laboratory for Cell Biology
Department of Biochemistry and Molecular Biology
Mayo Clinic Foundation
Rochester, Minnesota 55905

I. THE CENTROSOME AS THE DIVISION CENTER OF THE CELL

All cells must equipartition heritable information into daughter cells. In eukaryotes this task is directed by the centrosome. Boveri (1901) formally defined the centrosome as the larger of two differentiated bodies that reside at mitotic poles; the smaller bodies found within the centrosome were called centrioles. Although variable in size and detail from one cell type to another, the centrosome was seen by early light microscopists as a homogeneous sphere from which rays—now known to be microtubules—took their origin (see Wilson, 1925). The

Copyright © 1992 Academic Press, Inc.
All rights of reproduction in any form reserved.

centrioles were seen as minute granules or rods, often double, after staining with iron hematoxylin or crystal violet. We refer the reader to contemporary reviews that deal with the centrosome (Peterson and Berns, 1980; McIntosh, 1983; Bornens and Karsenti, 1984; Brinkley, 1985; Vandré and Borisy, 1989) and the centriole individually (De Harven, 1968; Fulton, 1971; Wolfe, 1972; Wheatley, 1982). Boveri's review (1901), "Über der Natur der Centrosomen," however, is of historic significance since it established the centrosome as the "reproductive organ" of the cell and provided a theoretical explanation for the distributive mechanism of inheritance. To quote Boveri (1887): "The centrosome is an independent permanent cell-organ, which, exactly like the chromatic elements (chromosomes), is transmitted by division to the daughter cells. The centrosome represents the dynamic center of the cell" and "it coordinates nuclear and cytoplasmic division" (Boveri, 1901). The primary function of the centrosome—to equipartition heritable information to daughter cells—and its status as the division center of the cell was thus clearly recognized late in the 19th century.

II. FLAGELLAR BASAL APPARATUS AND CENTROSOME AS FUNCTIONAL, STRUCTURAL, AND MOLECULAR HOMOLOGS

The flagellar basal apparatus of flagellate unicells and the centrosome of metazoan cells are functional equivalents; this generalization was first clearly appreciated by Henneguy and von Lenhossek in the late 1800s (see Wilson, 1925; Mazia, 1961). The flagellar apparatus and centrosome are the division centers of these respective cell types. Once in each cell cycle, the flagellar apparatus or centrosome doubles, divides, and separates; each newly replicated division center migrates to an opposite side of the nucleus during prophase to become a spindle pole and participate in chromosome segregation. In addition to their primary role as division centers, the centrosome and flagellar apparatus function secondarily as microtubule-organizing centers (MTOCs), a concept recognized by Porter (1966) and later formalized by Pickett-Heaps (1969). As MTOCs, the centrosome and flagellar apparatus play an active role in determining the number, stability, and polarity of cytoplasmic microtubules (reviewed by McIntosh, 1983, 1987; Bornens and Karsenti, 1984; Tucker, 1984; Brinkley, 1985; McIntosh and Koonce, 1989; Mitchison, 1989). Consequently, the centrosome, flagellar apparatus, and spindle poles help define the final form of both the interphase and mitotic microtubule arrays (Mazia, 1984; Mitchison and Kirschner, 1984).

As a consequence of evolution and their role as division centers and MTOCs, the flagellar apparatus and centrosome share structural components. At ultrastructural resolution, we find that both organelle complexes harbor cylindrical organelles composed of nine microtubule triplets; these are the basal bodies or centrioles—the nonflagellate equivalent of a basal body—that were observed by early light microscopists (reviewed by De Harven, 1968; Fulton, 1971; Wolfe, 1972; Wheatley, 1982). The centrosome and flagellar apparatus also serve as focal points for cytoplasmic microtubules (reviewed by Peterson and Berns, 1980; McIntosh, 1983; Bornens and Karsenti, 1984; Brinkley, 1985; Grain, 1986; Sandoz *et al.,* 1988). The electron microscope has further revealed that basal bodies of flagellate cells, and often centrioles, are associated with a variety of well-formed fibrous elements, which include basal feet (striated feet), connecting fibers (distal and proximal fibers), transition fibers (alar sheets), and striated roots of various types (reviewed by De Harven, 1968; Wolfe, 1972; Pitelka, 1974; Melkonian, 1980; Wheatley, 1982; Grain, 1986). In addition, centrioles of metazoan cells are generally enmeshed in a fibrous reticulum consisting of pericentriolar matrix (pericentriolar cloud or halo) and pericentriolar satellites (pericentriolar dense bodies)(reviewed by Peterson and Berns, 1980; McIntosh, 1983; Bornens and Karsenti, 1984; Brinkley, 1985). All these fibrillar components of division centers constitute pericentriolar material (PCM).

The functional homology of the centrosome and flagellar apparatus as division centers and MTOCs, as well as their structural similarity, predict that these organelle complexes share molecular components. The combined techniques of sodium dodecyl sulfate–polyacrylamide gel electrophoresis (SDS-PAGE), isoelectric focusing gel electrophoresis, cell fractionation, immunocytochemistry, and Western blotting have identified many protein components of both the centrosome and flagellar apparatus over the last two decades (see subsequent text). Consistent with the prediction just made, many studies demonstrate collectively that the flagellar apparatus and centrosome of divergent organisms harbor common antigenic determinants (Connolly and Kalnins, 1978; Kasamatsu *et al.,* 1980; Ring *et al.,* 1980; Lin *et al.,* 1981; Oliver *et al.,* 1981; Calarco-Gillam *et al.,* 1983; Moroi *et al.,* 1983; Vandré *et al.,* 1984; Salisbury *et al.,* 1984, 1986b; Fung and Kasamatsu, 1985; Maro *et al.,* 1985; Bastmeyer *et al.,* 1986; Klotz *et al.,* 1986; Houliston *et al.,* 1987; Keryer *et al.,* 1987; Baron and Salisbury, 1988; Steffen and Linck, 1988; Hiraoka *et al.,* 1989; La Claire and Goddard, 1989; Rao *et al.,* 1989). In most instances, however, the molecular basis for these epitopes remains to be elucidated.

A. Protein Components of Division Centers

The earliest strategies for identifying proteins of division centers combined gel electrophoresis technology (Weber and Osborn, 1969; Laemmli, 1970; O'Farrell, 1975) with methods of cell fractionation and selective solubilization of indigenous organelles, such as basal bodies and striated roots. Gel analysis of the soluble proteins derived from a complex cellular fraction is, however, equivocal with respect to the cytological localization of the proteins identified. One can never be sure that the solubilization steps are selective specifically for the organelle of interest. The purity of the initial cellular fraction is also of concern. The use of immunologic reagents, advances in immunocytochemical methods (Cuello, 1983), and the development of Western blotting (Towbin *et al.,* 1979; Burnett, 1981) have in recent years dispelled many of these concerns. The identification and localization of specific proteins to the PCM of division centers at light and electron microscopic resolution is now routine.

Proteins common to the flagellar apparatus of unicells have been identified by SDS-PAGE of preparations enriched in striated roots (costa, kinetodesmal fibers, rhizoplasts). In addition to centrin (20K) of striated flagellar roots (SFRs) of *Tetraselmis striata* (Salisbury *et al.,* 1984), they include a 90K protein of the costa of *Trichomonas gigantea* (Amos *et al.,* 1979); 21K (Rubin and Cunningham, 1973) and 250K (Vaudaux *et al.,* 1977; Williams *et al.,* 1979) proteins of kinetodesmal fibers of *Tetrahymena pyriformis;* 240K (Gardiner *et al.,* 1981) and 170K (Dingle and Larson, 1981; Larson and Dingle, 1981) proteins of the rhizoplast of *Naegleria gruberi;* and 76K, 50K, and 31K proteins of basal body–flagellar root complexes of *Dunaliella bioculata* (Marano *et al.,* 1985). Using a rabbit antiserum produced against electrophoretically purified preparations of the 170K protein of *Naegleria,* Dingle and Larson (1981) confirmed the localization of this protein to the rhizoplast by indirect immunofluorescence microscopy and the apparent absence of the 170K protein from *Tetrahymena.* The immunolocalization of centrin to SFRs of *Tetraselmis* was also confirmed at both light and electron microscopic resolution (Salisbury *et al.,* 1984).

The ciliary basal apparatus of metazoan epithelial cells consists of multiple basal bodies, basal feet, transition fibers, and striated roots, as well as the classical cytoskeletal filaments and their constituent proteins, that is, microtubules, microfilaments, and intermediate filaments (reviewed by Sandoz *et al.,* 1988). An approach that has been successful in identifying protein components of division centers involves the preparation and isolation of deciliated basal apparatus of ciliated epithelial cells (ciliary cortex). Stephens (1975) isolated the ciliary cortex of gill epithelial cells of the bay scallop *Aequipecten irradians.* Solubilization of the constituent striated roots and analysis of the supernatants by SDS-PAGE revealed a doublet of 230K and 250K; several other protein components of lower molecular weight were also present. Anderson and Floyd (1980) demonstrated the presence of tubulin and proteins of 17K, 90K, and 180K in basal body-enriched ciliary cortices from chicken oviduct. It is generally agreed that basal bodies (centrioles), as microtubule-based structures, are composed, in part, of tubulin (Wolfe, 1972; Wheatley, 1982). A more recent study reports the isolation of ciliary cortices from quail oviduct epithelium and the generation of monoclonal antibodies (Klotz *et al.,* 1984). Four monoclonal reagents—referred to as 2E6, 2F12, 3C10, and 3A8—with distinct ciliary basal apparatus immunolocalization patterns were identified. Each monoclonal antibody also recognized a unique polypeptide(s) by immunoblot analysis (2E6, 58K; 2F12, 100K; 3C10, 80K; 3A8, 50K, 60K, 150K, and 400K). Another monoclonal antibody, CC-310, also produced against quail oviduct ciliary cortices, has identified and localized a 175K protein to striated roots of these epithelial cells (Klotz *et al.*, 1986). A common epitope of the 175K quail protein is present on proteins of 48K and 92K from frog palate and mussel gill striated roots, respectively (see Sandoz *et al.*, 1988); the centrosome of PtK_2 cells also labels with the CC-310 antibody. These observations indicate that a specific epitope is conserved on elements of PCM from different species.

Several studies have capitalized on the observation that certain nonimmune rabbit sera (Connolly and Kalnins, 1978; Maunoury, 1978; Sauron *et al.*, 1984) and autoimmune human sera (Moroi *et al.*, 1983; Senécal *et al.*, 1985) fortuitously contain antibodies to components of division centers. Using nonimmune rabbit sera, Turksen and co-workers (1982) identified a 50K protein that was common to the ciliary apparatus of the ciliate *Tetrahymena pyriformis* and chicken tracheal epithelial cells by immunoblot procedures. Immunofluorescence microscopy localized this protein closely to the basal bodies of *Tetrahymena* cortices. The relationship of this molecule to the 50K protein identified in quail oviduct ciliary apparatus by Klotz and co-workers (1984) is unknown. Another rabbit nonimmune serum, referred to as 0013, reacts strictly with centrosomes and soluble cytosolic components of primate cells (Maunoury, 1978). This serum was shown to recognize a complex mixture of polypeptides, including a 36K protein (Sauron *et al.*, 1984). Further study demonstrated that the 36K protein was a subunit of the enzyme lactate dehydrogenase (LDH), and that antibodies affinity purified against the 36K protein shared an epitope with a number of high molecular weight proteins (140K to 250K) associated with isolated centrosomes from human T lymphoblasts of the cell line KE37 (Gosti-Testu *et al.*, 1986; Gosti *et al.*, 1987). The molecular basis and functional relationship of the high molecular weight centrosomal proteins to the 36K subunit of LDH is presently unknown. Two polypeptides of 14K and 17K have been localized to the centrosome of mammalian cell lines of monkey, human, hamster, rat, and mouse origin, and to the ciliary apparatus of cat, chicken, and steer tracheal epithelial cells with normal and immune rabbit sera (Kasamatsu *et al.*, 1980; Lin *et al.*, 1981, Fung and Kasamatsu, 1985). Immunogold labeling at electron microscopic resolution demonstrates that these antigens are associated with the PCM that lies around the basal bodies and striated roots of cat tracheal epithelial cells. The relationship of this 17K polypeptide to that reported by Anderson and Floyd (1980) for ciliary apparatus of chicken oviduct epithelium has not been addressed. Finally, a human autoimmune serum from a patient with CREST (calcinosis, Raynaud's phenomenon, esophageal dysmotility, sclerodactyly, and telangiectasias) syndrome has localized proteins of

110K and 115K to the spindle poles of mitotic HeLa cells; the specificity of this localization was confirmed with affinity purified antibodies to the 110K and 115K proteins (Sager *et al.,* 1986).

Isolation of mitotic apparatus (Kuriyama and Borisy, 1985; Sellitto and Kuriyama, 1988; Toriyama *et al.,* 1988) and centrosomes (Bornens *et al.,* 1987; Kuriyama, 1989; Joswig and Petzelt, 1990) has also provided information about the protein composition of division centers (see Chapters 1 and 5). Two hybridomas, SU4 and SU5, produced against mitotic apparatus of eggs of the sea urchin *Strongylocentrotus purpuratus* identify proteins of 20K, 180K, and 190K, and 50K, respectively (Kuriyama and Borisy, 1985). A monoclonal antibody produced against centrosome fractions of PtK cells recognizes a 32K protein (Joswig and Petzelt, 1990). All three monoclonal reagents localize to the centrosphere of these respective cells and demonstrate cell-cycle-dependent changes in centrosome shape and pattern. Another hybridoma, referred to as CHO1, generated against mitotic apparatus of Chinese hamster ovary (CHO) cells is specific for two polypeptides of 95K and 105K in these cells (Sellitto and Kuriyama, 1988). Indirect immunofluorescence microscopy with CHO1 revealed staining of interphase centrosomes and nuclei. During mitosis, spindle staining was observed; as cells progressed through mitosis, labeling became concentrated at the equatorial region of the spindle and finally the midbody. A second hybridoma produced against mitotic apparatus of CHO cells, referred to as CHO3, reacts with a thiophosphoprotein of 225K in eggs of the sea urchin *Strongylocentrotus purpuratus* (Kuriyama, 1989). Immunogold labeling at electron microscopic resolution demonstrates that this protein localizes to material from which microtubules emanate. Toriyama and co-workers (1988) have isolated an aster-forming protein fraction by phosphocellulose chromatography from mitotic apparatus of eggs of the sea urchin *Hemicentrotus pulcherimus.* The major component of this aster-forming fraction is a protein of 51K. Antisera prepared against the 51K protein localize to the center of microtubule asters.

Monoclonal antibodies generated against extracts of mitotic HeLa cells also have proved useful in the biochemical dissection of division centers (Davis *et al.,* 1983; reviewed by David and Rao, 1987). Two hybridomas, MPM-1 and MPM-2, recognize numerous mitosis-specific phosphoproteins; the number of proteins recognized from different phylogenetic sources precludes a detailed discussion here. Indirect immunofluorescence microscopy demonstrates that some of these phosphoproteins are components of spindle poles, kinetochores, and midbodies of mitotic tissue culture cells (Vandré *et al.,* 1984; reviewed by Vandré and Borisy, 1985). In addition to spindle poles and centrosomes, MPM-1 and MPM-2 label flagellar and ciliary apparatus of phylogenetically diverse cells; one of the molecules identified corresponds to microtubule-associated protein 1 (MAP1) (Vandré *et al.,* 1986). Another monoclonal antibody, MPM-13, produced against HeLa cell extracts, identifies proteins of 43K and 56K from a variety of mammalian tissue culture cells and the ciliate *Tetrahymena* (Rao *et al.,* 1989). Immunofluorescence microscopy demonstrates MPM-13 staining of the centrosome of mammalian cells.

Other proteins common to division centers for which we have both biochemical and functional information include MAP1 (Sato *et al.,* 1983, 1985; De Mey *et al.,* 1987), kinesin (Neighbors *et al.,* 1988), purine nucleotide phosphorylase (Oliver *et al.,* 1981), cAMP-dependent protein kinase (Nigg *et al.,* 1985; De Camilli *et al.,* 1986; Joachim and Schwoch, 1988), calmodulin (Deery *et al.,* 1984), tektin (Steffen and Linck, 1988), p34^{cdc2} kinase of maturation promoting factor (MPF) (Bailly *et al.,* 1989; Riabowol *et al.,* 1989), and centrin (Salisbury *et al.,* 1984, 1986b; Baron and Salisbury, 1988). In all cases these molecules were discovered first, antibodies were produced next, and the proteins were localized to the centrosome and/or flagellar and ciliary apparatus last. This strategy has also identified a centrosome-associated 68K protein in eggs of the sea urchin *Strongylocentrotus* with a monoclonal antibody, Ah6, produced against *Drosophila* intermediate filament proteins (Schatten *et al.,* 1987). Bonifacino and co-workers (1985) have found that both monoclonal 8D12 and polyclonal antibodies produced against MAP1 identify an immunologically related molecule of 280K in mammalian tissue culture cells. This molecule localizes to the centrosome. A novel approach using microtubule affinity columns has recently identified more than 50 microtubule-associated proteins from *Drosophila* embryos (Kellogg *et al.,* 1989). Mouse polyclonal antibodies have been produced to 24 of these proteins; of these sera, 19 label the centrosome or spindle poles with apparently unique patterns of localization.

The centrosome, ciliary apparatus, and flagellar apparatus are clearly heterogeneous in protein composition. Some proteins of these division centers appear to be identical. Others share immunologic epitopes, but are of disparate molecular weight. In the case of proteins reactive to monoclonal antibodies MPM-1 and MPM-2, the common epitopes correspond to sites of phosphorylation. For others, the molecular and functional basis for common immunoreactive epitopes is unknown. Molecular cloning will be useful in dissecting the immunologic relatedness of these proteins, and promises to shed light on the functional importance of phylogenetically conserved epitopes. In general, it is thought that the molecules of division centers regulate or contribute directly to their microtubule organizing activity or to their motile behavior.

III. STRIATED FLAGELLAR ROOTS AND THE PERICENTRIOLAR LATTICE AS STRUCTURAL AND MOLECULAR HOMOLOGS

SFRs of the alga *Tetraselmis striata* are an elaborately organized form of PCM (Fig. 1). These striated organelles consist of axially aligned 3–8-nm diameter filaments (reviewed by Salisbury, 1983). Fimbria composed of these filaments splay from this organelle and attach to basal bodies of the flagellar apparatus, to the nuclear envelope, and to the plasmalemma (Manton and Parke, 1965; Stewart *et al.*, 1974; Salisbury and Floyd, 1978; Melkonian, 1979; Robenek and Melkonian, 1979; Salisbury *et al.*, 1981).

Our approach to understanding the biochemical nature of *Tetraselmis* SFRs involved their isolation and separation from other cellular organelles (Salisbury *et al.*, 1984). Following cell disruption, SFRs were separated by differential centrifugation through gradients of sucrose and flotation on gradients of colloidal silica. We discovered by SDS-PAGE that these particular SFRs are predominantly composed of an acidic 20K calcium-binding phosphoprotein that consists of two isoforms, α (p*I* 4.9) and β (p*I* 4.8). The ubiquitous occurrence of this protein in centrosomes and flagellar apparatus led to the name centrin (reviewed by Melkonian *et al.*, 1988; Salisbury 1989a,b; Bazinet *et al.*, 1990). The molecular mass of centrin among algal cells is consistently in the range of 20K (Salisbury *et al.*, 1984, 1986a,b, 1988; Wright *et al.*, 1985; McFadden *et al.*, 1987; Schulze *et al.*, 1987; Hohfeld *et al.*, 1988; Huang *et al.*, 1988b; Lechtreck *et al.*, 1989).

We have previously immunoprecipitated a protein having a molecular mass of 165K from [S^{35}] methionine and [S^{35}] cysteine-labeled whole cell lysates of the *Potorous tridactylis* kidney epithelial cell line PtK_2 with a rabbit anticentrin polyclonal antiserum, designated 08/28 (Baron and Salisbury, 1988). Recent immunoprecipitation experiments, using three newly developed rabbit anticentrin polyclonal antisera and mouse anticentrin monoclonal antibodies, however, identify a protein of 20K in PtK_2 and other higher eukaryotic cells (Baron, Greenwood, Nemeth, and Salisbury; in preparation). In PtK_2 cells, centrin localizes to components of PCM, e.g., basal feet, pericentriolar matrix, and pericentriolar satellites. Ultrastructural features of basal feet, pericentriolar matrix, and pericentriolar satellites can be seen in Fig. 2. Basal feet and pericentriolar satellites are, in part, composed of 3–8-nm diameter filaments; these filaments also connect pericentriolar satellites to one another. Together, these elements of the PCM are components of a fibrous three-dimensional centrosomal matrix which we refer to as the pericentriolar lattice (PCL).

Fig. 1. Striated flagellar roots are contractile organelles. (A) Thin-sectioned *Tetraselmis* cell fixed in the absence of added calcium. One of the two fully extended striated flagellar roots (SFR) of this cell is visible. (B) Thin-sectioned *Tetraselmis* cell fixed within 30 s of a calcium shock. Both flagellar roots have contracted (CFR), resulting in deflagellation, displacement of the flagellar apparatus (FA) toward the nucleus (N), and inpocketing of the plasmalemma (S) away from the theca (T). V, vacuole; P, pyrenoid; F, flagella. Bar: 1 μm. [Reproduced from Salisbury *et al.*, *Journal of Cell Biology* (1984), **99,** 962–970, by copyright permission of the Rockefeller University Press.] Contraction of striated flagellar roots is mediated by filament supercoiling (arrows). (C) Filaments 3–8 nm in diameter (D) bend and twist, forming (E) electron-dense bodies. Bar: 0.25 μm. [Reproduced from Salisbury, *Journal of Submicroscopic Cytology* (1983), **15,** 105–110, with permission.]

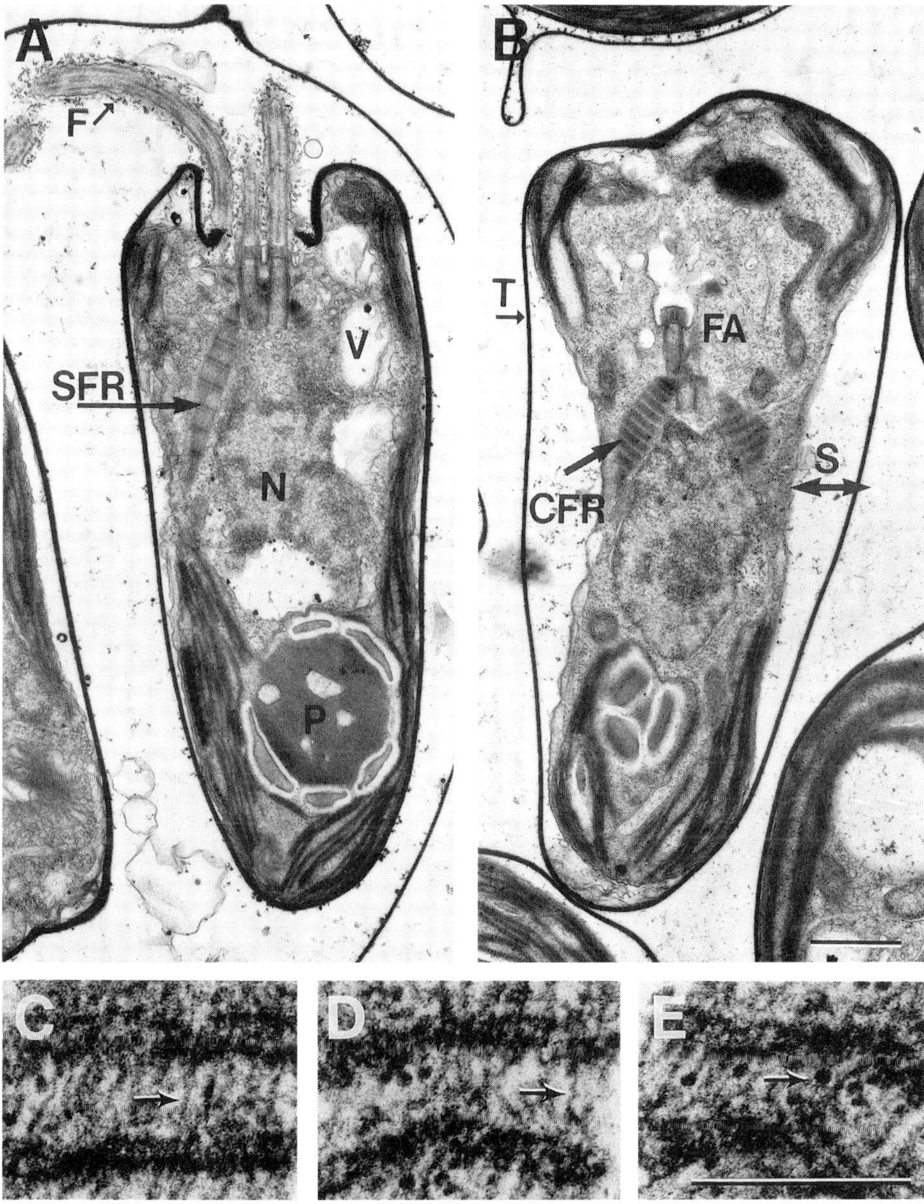
A
F
SFR
V
N
P
B
T
FA
CFR
S
C
D
E

The PCL shares structural features with SFRs and other centrin-based fiber systems such as the nucleus–basal body connectors (NBBCs) and distal fibers of *Chlamydomonas reinhardtii* (Wright *et al.,* 1985; Salisbury *et al.,* 1986) and *Spermatozopsis similis* (McFadden *et al.,* 1987). Ultrastructural examination has revealed that SFRs, NBBCs, distal fibers, pericentriolar satellites, and basal feet are all characterized by 3–8-nm diameter filaments (Salisbury, 1982, 1983; Wright *et al.,* 1985; Baron and Salisbury, 1988; Salisbury *et al.,* 1988). The centrin-based stellate-pattern fibers in the transition zone of *Chlamydomonas* are also of similar diameter (Sanders and Salisbury, 1989). The striated appearance of basal feet resembles that of SFRs and distal fibers. In addition, pericentriolar satellites are similar in appearance to the electron-dense bodies that form the wide cross-striations of contracted SFRs (cf., Fig. 1C–E, Fig. 2B); these electron-dense bodies represent supercoiled filaments (reviewed by Salisbury, 1983). Although the PCL

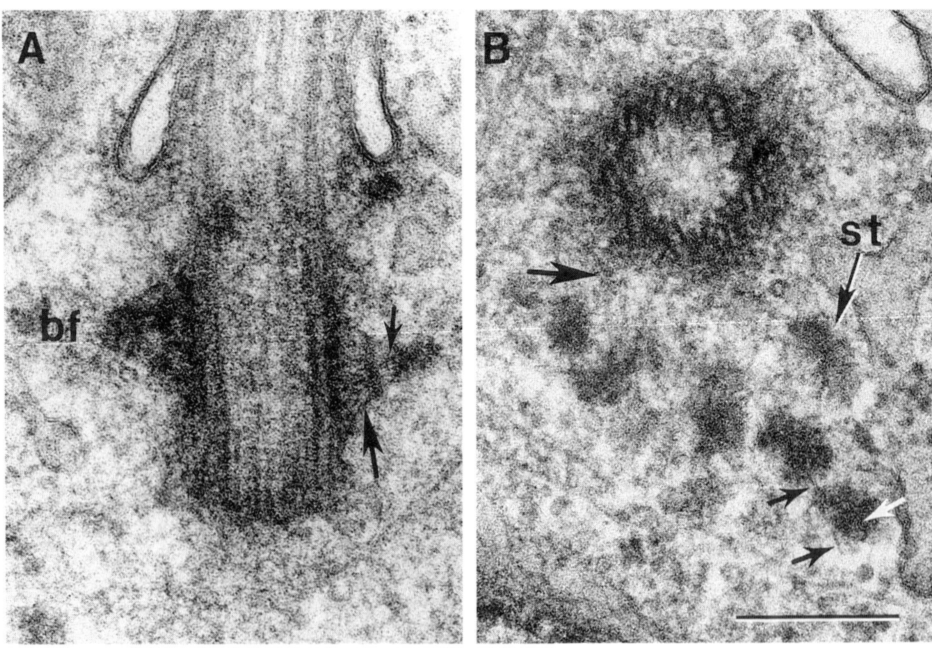

Fig. 2. Basal feet and pericentriolar satellites of PtK_2 cells are composed of filaments. (A) Basal feet (bf) have a striated morphology and are composed of filaments that have diameters of 3–8 nm (small arrow) and 7–11 nm (large arrow). (B) Pericentriolar satellites (st) are indirectly connected to centrioles through the pericentriolar matrix (large arrow) and are composed of and interconnected by 3–8-nm diameter filaments (small arrows). Bar: 0.25 μm. [Reproduced from Baron and Salisbury, *Journal of Cell Biology* (1988), **107,** 2669–2678, by copyright permission of the Rockefeller University Press.]

is a less-ordered three-dimensional arrangement of PCM compared with the *Tetraselmis* SFRs, it is nevertheless a molecular and structural homolog of these SFRs and other centrin-based fiber systems.

IV. CENTRIN-CONTAINING FIBER SYSTEMS AS CALCIUM-MODULATED CONTRACTILE ORGANELLES

Striated flagellar roots of *Tetraselmis* were first shown to be calcium-modulated contractile organelles by Salisbury and Floyd (1978); these organelles undergo cycles of contraction and extension in the presence of ATP. Since the initial identification and characterization of centrin as the major protein component of these SFRs (Salisbury *et al.,* 1984), numerous studies have shown that centrin-containing cytoskeletal systems are characterized by calcium-modulated contractile behavior (Wright *et al.,* 1985; McFadden *et al.,* 1987; Salisbury *et al.,* 1987, 1988; Hohfeld *et al.,* 1988; Koutoulis *et al.,* 1988; Sanders and Salisbury, 1989).

Unlike the sliding filament mechanism of muscle contraction (Huxley and Hanson, 1954; Huxley, 1969), centrin-based fiber systems contract via a mechanism that involves filament supercoiling (Salisbury and Floyd, 1978; reviewed by Salisbury, 1983). This is best seen in SFRs, in which extended filaments twist and bend to form electron-dense bodies as they contract (Fig. 1C–D). Cytochemical studies demonstrate that substantial amounts of calcium ions are bound to contracted SFRs (Salisbury, 1982). Alkaline urea gels run in the presence of calcium or EGTA show that centrin exhibits a decrease in mobility when calcium ions are included in the buffer system, thus indicating that calcium-binding directly affects the conformational state of centrin (Salisbury *et al.,* 1984). Cytochemical studies also demonstrate that the cross-striated regions of SFRs exhibit ATPase activity (Salisbury, 1983). Labeling of *Tetraselmis* cells with $^{32}PO_4$, followed by two-dimensional gel analysis and autoradiography, shows that the β isoform of centrin is phosphorylated (Salisbury *et al.,* 1984; Martindale and Salisbury, 1990). Analysis of contracted versus extended SFRs indicates that contraction decreases the proportion of the phosphorylated β isoform relative to the α isoform. ATP is required to potentiate the centrin-based NBBC for calcium-modulated contraction in detergent-extracted cell models of *Chlamydomonas* (Salisbury *et al.,* 1987). Additional studies on flagellar excision, a centrin-based motility behavior, in detergent-extracted cell models of *Chlamydomonas* indicate that excision occurs when free calcium levels are increased to 10^{-7} *M* or higher (Sanders and Salisbury, 1989). These observations suggest that force production and contraction, that is, supercoiling of centrin-based filaments is mediated by calcium binding, dephosphorylation, and conformational alterations of centrin subunit structure.

The centrosome of PtK_2 cells harbors a filamentous reticulum—the PCL—that shares several structural features with other centrin-based fiber systems (Baron and

Salisbury, 1988). To address whether the PCL displays calcium-modulated contractile behavior, we elevated transiently (within 30 sec) the intracellular free calcium level of interphase PtK_2 cells from a resting level of 1×10^{-8} *M* to 2×10^{-7} *M* with the calcium ionophore ionomycin. This caused a rapid (within 2 min) aggregation of centrin-based pericentrosomal spots toward the cell center. As free calcium returned to a basal level (within 5 min), the pericentrosomal spots became redispersed throughout the cytoplasm (Baron, Greenwood, Nemeth, and Salisbury, unpublished observations). We interpret these observations to indicate a calcium-modulated contraction and subsequent reextension of the PCL to and from the cell center. At electron microscopic resolution, elevated levels of free calcium promote conversion of the pericentriolar matrix into electron dense pericentriolar satellites around the centrioles (Baron, Greenwood, and Salisbury, unpublished observations). Structurally, pericentriolar satellites resemble a ball of tangled string. Our observations suggest that pericentriolar satellites of interphase PtK_2 cells represent the supercoiled, that is, contracted, state of a filamentous pericentriolar matrix. The PCL therefore shares with other centrin-based fiber systems, such as *Tetraselmis* SFRs, the conserved property of calcium-modulated contractile behavior.

V. CELL CYCLE-DEPENDENT CHANGES IN THE CYTOPLASMIC ORGANIZATION OF CENTRIN

Chlamydomonas (Johnson and Porter, 1968; Cavalier-Smith, 1974; Coss, 1974) and the kidney epithelial cell line PtK (Roos, 1973; Snyder and McIntosh, 1975; Rattner and Berns, 1976; Berns and Richardson, 1977; Berns *et al.,* 1977; Rieder and Borisy, 1982; Snyder *et al.,* 1982) have rich histories as model systems for the study of centrosome behavior. To date, only *Chlamydomonas reinhardtii* (Salisbury *et al.,* 1988), and the cell line PtK_2 (Baron *et al.,* 1991) have been thoroughly studied with respect to the behavior of centrin during the cell cycle.

A. Centrin Behavior during Mitosis in *Chlamydomonas reinhardtii*

The interphase centrin-based cytoskeleton of *Chlamydomonas* consists of a stellate pattern of filaments in the transition zone of the flagellum (Sanders and Salisbury, 1989), a distal fiber of axially aligned filaments that links the paired basal bodies of the flagellar apparatus, and a NBBC (Wright *et al.,* 1985; Salisbury *et al.,* 1987, 1988; Huang *et al.,* 1988b; reviewed by Salisbury, 1988). Immunofluorescence micrographs of the interphase and mitotic centrin-based cytoskeleton of *Chlamydomonas* are shown in Fig. 3. The distal fiber appears as a bar across the top of the NBBC of interphase cells. The NBBC, in turn, consists of two major

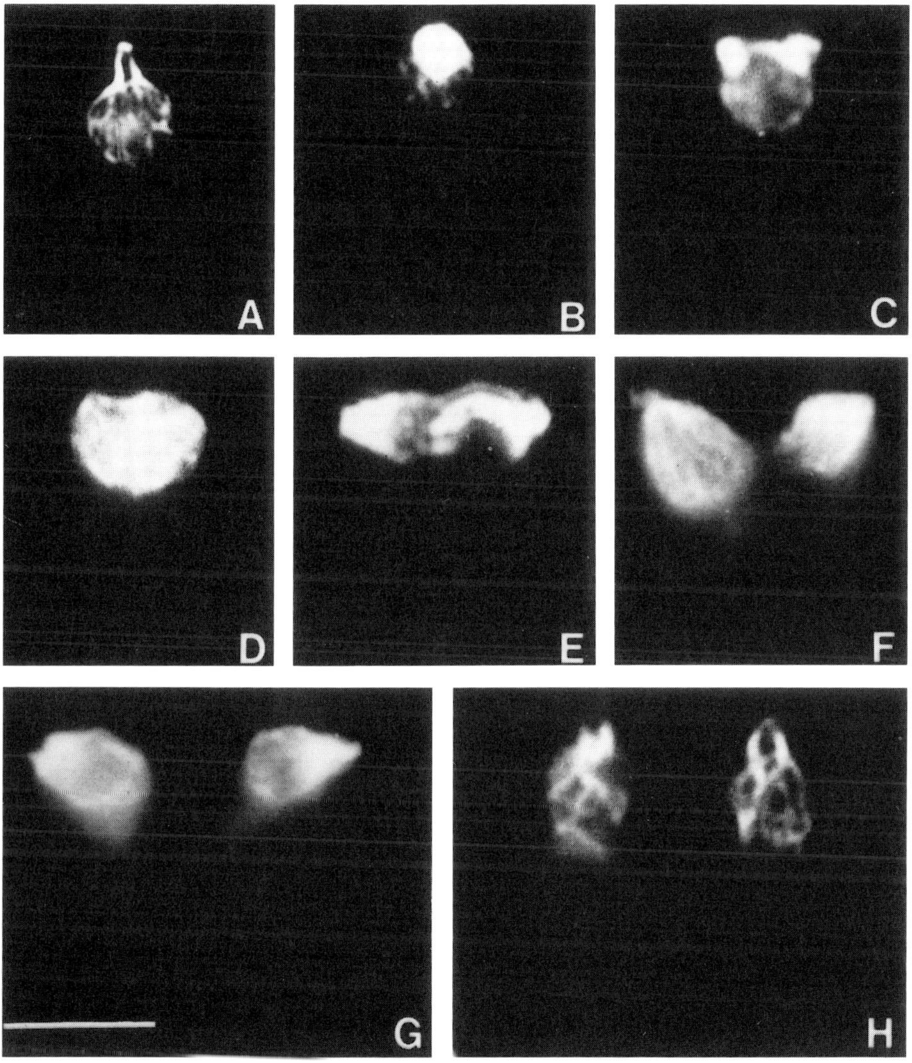

Fig. 3. The centrin-based nucleus–basal body connector (NBBC) of *Chlamydomonas* goes through a sequence of contractions and re-extensions during the mitotic cycle. The cells were labeled for indirect immunofluorescence microscopy with anticentrin monoclonal antibody 17E10. The anterior of the cell is at the top of each micrograph; the entire cell volume is included in each micrograph. (A) Interphase centrin-based NBBC. (B) Preprophase contraction of the NBBC into one focus. (C) Prophase separation of centrin foci and beginning of re-extension. (D) Metaphase spindle outlined by a re-extended centrin-based fiber system. (E) Metaphase/anaphase boundary illustrating a second transient contraction of each daughter NBBC. (F, G) Telophase re-extension of each daughter NBBC within the mother cell. (H) Interphase cells after the completion of cytokinesis. Bar: 5 μm. [Reproduced from Salisbury *et al.*, *Journal of Cell Biology* (1988), **107,** 635–641, by copyright permission of the Rockefeller University Press.]

descending fibers or arms that originate from each basal body and branch into 8–16 fimbria; these fimbria surround the nucleus and embrace the outer nuclear envelope. The NBBC therefore links the flagellar apparatus or division center to the nucleus. Analogous to SFRs, the NBBC can undergo cycles of calcium-modulated contraction and extension, both *in vitro* and in living cells (Wright *et al.,* 1985; Salisbury *et al.,* 1987).

Preprophase (late G_2) in *Chlamydomonas* is marked by contraction of a duplicated NBBC and displacement of the nucleus toward the basal bodies (Salisbury *et al.,* 1988). This event is followed by division, separation, and reextension of each newly formed sister NBBC during prophase and prometaphase; a network of centrin-based fibers now extends from each spindle pole and surrounds the karyoplasm. Each NBBC appears to maintain an attachment to the outer nuclear envelope throughout mitosis and therefore connects the karyoplasm to each division center. The NBBCs remain in an extended or relaxed state throughout metaphase. Anaphase onset is characterized by a second transient contraction of each sister NBBC; the timing of chromosome separation is coincident with this event. Re-extension of each sister NBBC occurs during telophase. By the time of cytokinesis, each daughter cytoplast has re-established an interphase-like array of centrin-based fibers.

B. Centrin Behavior during Mitosis in PtK_2 Cells

The PCL of PtK_2 cells is seen with anticentrin antibodies and the epifluorescence microscope as a constellation of pericentrosomal spots (Baron and Salisbury, 1988). Cells in late G_1, S, and early G_2 phase of the cell cycle are characterized by a dense centrosomal focus of spots; additional spots are radially dispersed throughout the cytoplasm (Baron *et al.,* 1991). Depending on the size and shape of the cell, the peripheral spots may be 30–50 μm from the cell center. These spots may be induced to move toward the cell center when intracellular free calcium levels are elevated (Baron, Greenwood, Nemeth, and Salisbury, unpublished observations).

During late G_2 in PtK_2 cells, the pericentrosomal spots dispersed throughout the cytoplasm aggregate at the cell center; one bright centrosomal focus of superimposed spots results. As the cells proceed through prophase, this single focus divides into two foci, that is, two PCLs that move toward opposite sides of the nucleus; this behavior is contemporaneous with reorganization of the interphase microtubule complex and assembly of a bipolar spindle (Fig. 4). Prometaphase is characterized by dispersion of the pericentrosomal spots from the spindle poles. At metaphase, the spots are distributed throughout each half-spindle (Fig. 5). Early in anaphase, the pericentrosomal spots lie between the poles and chromosomes. As the sister chromatids move poleward, some spots are seen beside and behind the chromatids; other spots are clustered at the poles. At telophase, the pericentrosomal spots form a dispersed pattern from the spindle poles; they occupy the cytoplasm that surrounds each mass of chromatin. This dispersed state is maintained throughout early G_1 phase of the cell cycle. Finally, as the cells progress into late G_1 they reestablish the interphase labeling pattern.

C. Analogous Behavior of NBBC and PCL during Mitosis: Interpretations

The most striking behavioral feature shared by the NBBC of *Chlamydomonas* and the PCL of PtK_2 cells is the contraction (aggregation) that occurs at preprophase (late G_2). This contracted (aggregated) state is maintained throughout prophase as each of these respective division centers divides and separates with the nascent spindle poles. During prometaphase, the NBBC and PCL re-extend (disperse). They remain in an extended (dispersed) state throughout metaphase. Anaphase in *Chlamydomonas* is marked by a second transient contraction of the NBBC coincident with karyokinesis. We also observe a cluster of pericentrosomal spots at the spindle poles during anaphase in PtK_2 cells. Telophase and G_1 are characterized by an extended (dispersed) state of both the NBBC and the PCL. These observations suggest a transient increase in intracellular free calcium during prophase and at the anaphase/metaphase boundary. We interpret our observations to indicate that both the NBBC and the PCL go through a sequence of calcium-modulated contractions and re-extensions during mitosis.

VI. CENTRIN AS A COMPONENT OF THE CYTOMATRIX

Our present view of cytoplasm is that of a structured, dynamic, and kinetic gel (dynakinetic) that is referred to as the cytomatrix (reviewed by Porter, 1976, 1984; Porter *et al.,* 1979, 1983). The cytomatrix is a collective structure consisting of the cell's major cytoskeletal filament systems and their associated proteins (Wolosewick and Porter, 1979; Schliwa and van Blerkom, 1981; Gershon *et al.,* 1985; Stossel *et al.,* 1985; Pollard and Cooper, 1986). It is, in large part, responsible for the "living" properties of cytoplasm. Interactions between the principal cytoskeletal filament systems and their associated proteins combine to give the cytomatrix contractile, elastic, viscous, and sieving properties (reviewed by Marsland, 1942; Frey-Wyssling, 1953; Allen, 1961; Taylor and Condeelis,1979; Taylor and Fechheimer, 1982; Pollard, 1984; Porter, 1984; Stossel *et al.,* 1985; Luby-Phelps *et al.,* 1988). Dynakinetic changes in the state of cytomatrix contractility, elasticity, viscosity, and sieving properties manifest themselves in sol–gel transitions that convey direction to motile cellular phenomena and anisometry to cell form. Although we know much about the microtubule, microfilament, and intermediate

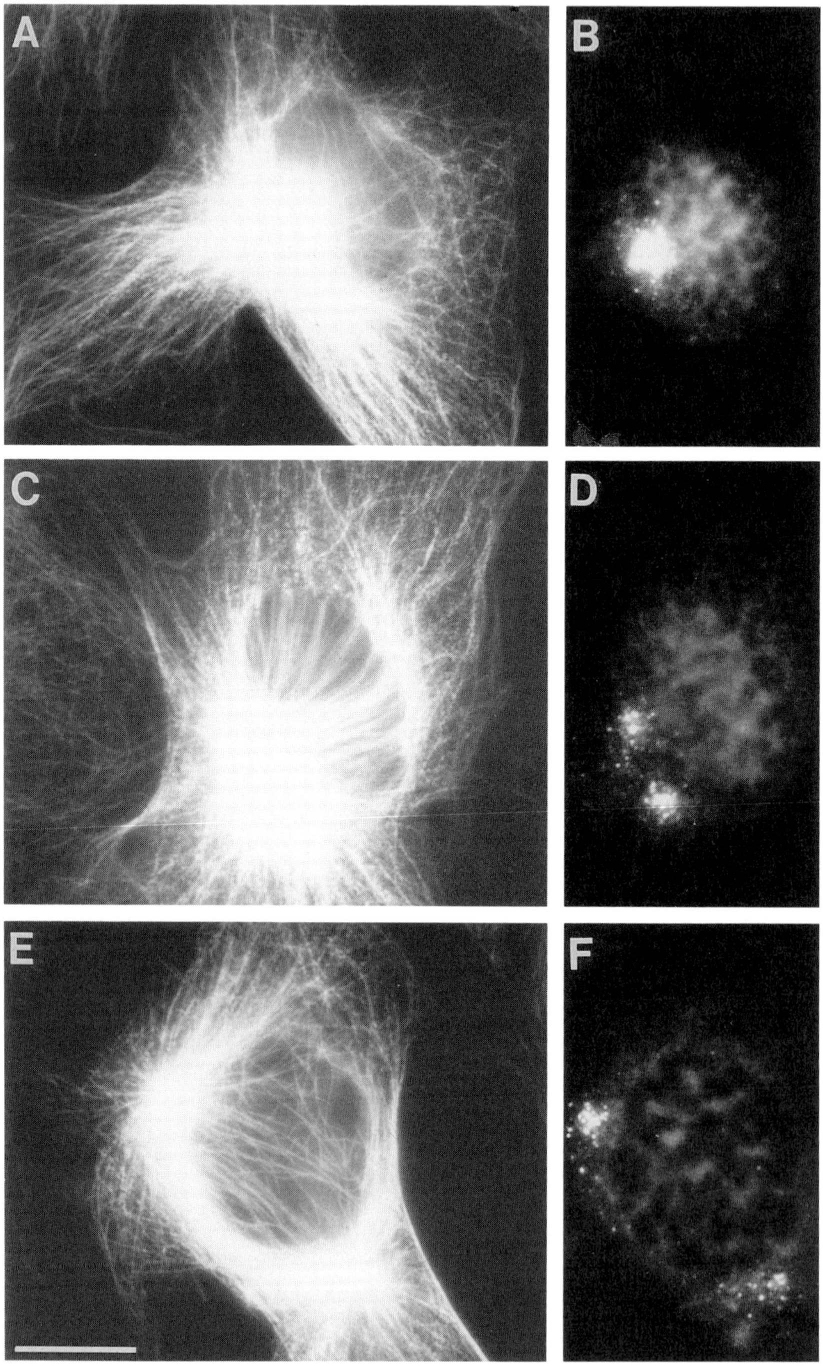
A
B
C
D
E
F

filament systems of cells individually, we do not understand fully how these filament systems and their associated proteins function collectively as the cytomatrix. Furthermore, we know very little regarding the cytomatrix constituents referred to as the pericentriolar matrix and the spindle matrix.

Centrin is a major component of the algal cytoskeleton. Of the unicellular green algae, 26 taxa incorporate centrin into a cytoskeletal fiber system that connects the nucleus and basal bodies and is referred to as the NBBC (Schulze *et al.*, 1987). Centrin also has been localized to a number of other filamentous structures including distal fibers of basal bodies (*Chlamydomonas reinhardtii, Spermatozopsis similis, Dunaliella bioculata, Monostroma grevillei, Enteromorpha sp.*, and *Polytomella parva;* McFadden *et al.*, 1987; Melkonian *et al.*, 1988; Salisbury *et al.*, 1988), paraxonemal fibers of transverse flagella (*Peridinium inconspicuum;* Hohfeld *et al.*, 1988), stellate fibers of basal body transition zones (*Chlamydomonas reinhardtii, Enteromorpha sp.;* Melkonian *et al.*, 1988; Sanders and Salisbury, 1989), and fibrous bundles of spine scales (*Apedinella radians;* Koutoulis *et al.*, 1988). In addition, centrin is a component of the mitotic apparatus of *Chlamydomonas* (Salisbury *et al.*, 1988). In the cell line PtK_2, centrin localizes to basal feet, pericentriolar matrix, and pericentriolar satellites (Baron and Salisbury, 1988). Observations further indicate that centrin is not confined strictly to the cytoplasm immediately surrounding the centrioles, but that this molecule may be 30–50 μm from the cell center. Finally, centrin is a component of the mitotic spindle matrix (Baron *et al.*, 1991). In broad terms, therefore, centrin is a dynamic component of the cytomatrix.

VII. SPECULATIONS ON THE FUNCTION OF CENTRIN DURING MITOSIS

Centrin-based fibers are involved in a variety of calcium-modulated contractile phenomena and therefore serve a variety of functions. The precise role of these fibers is defined by how they are incorporated into the cytoarchitecture of the cell; this is exemplified by the NBBC, distal fiber, paraxonemal fiber, fibrous bundles, and stellate pattern fibers. Contraction of the NBBC in *Chlamydomonas* moves the nucleus and basal bodies toward one another (Wright *et al.*, 1985; Salisbury *et al.*, 1987, 1988). Nuclear movement occurs in response to flagellar excision and

Fig. 4. The pericentriolar lattice of PtK_2 cells undergoes division and separation during prophase of mitosis. Cells were double labeled with (A,C,E) antitubulin and (B,D,F) polyclonal anticentrin 08/28 antibodies. (A,B) A single focus of superimposed pericentrosomal spots (C,D) divides into two foci (E,F), which move toward opposite sides of the nucleus. This behavior is coincident spatially and temporally with division, separation, and reorganization of the spindle poles and cytoplasmic microtubule complex. (B,D,F) Condensed chromosomes are seen as dark objects within a background of nuclear autofluorescence. Bar: 10 μm. [Reproduced from Baron *et al.*, *Cell Motility and the Cytoskeleton* (1991), by copyright permission of John Wiley & Sons, Inc.]

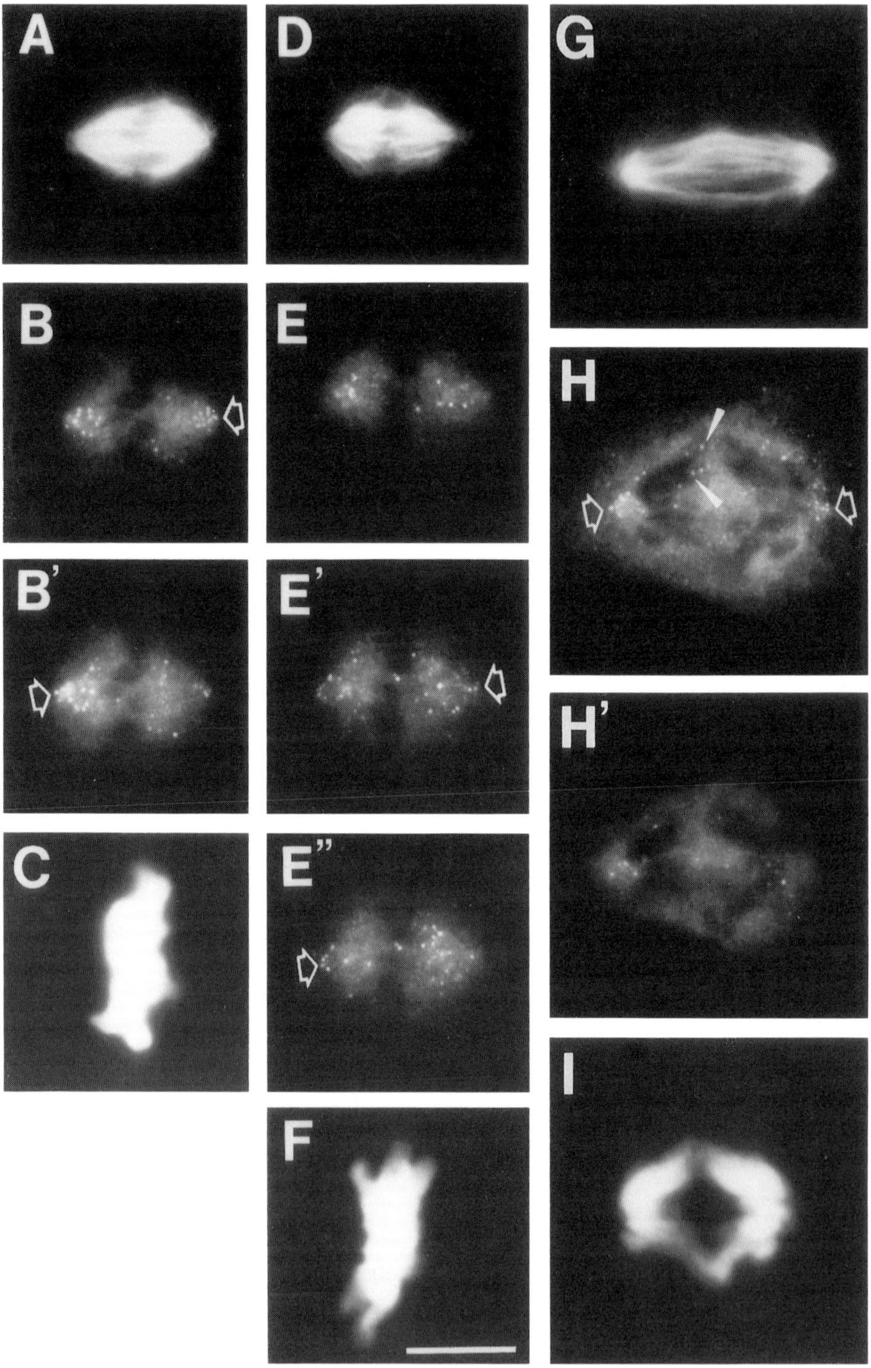
A
B
B'
C
D
E
E'
E''
F
G
H
H'
I

includes a change in nuclear shape from spherical to pyriform (Salisbury *et al.*, 1987). It has been proposed that these motile events may be causally related to signaling and induction of flagellar precursor genes or to the transport of flagellar precursors or their messages to sites of synthesis and assembly near the flagellar apparatus. The role of the NBBC in gene induction during flagellar regeneration is not yet resolved fully and is a subject of controversy (see Wright *et al.*, 1989). Reversal of swimming direction in *Spermatozopsis* is accompanied by contraction of the distal fiber and results in a 180° change in angle of the paired basal bodies to one another (McFadden *et al.*, 1987). Extreme contraction of the paraxonemal fiber of *Peridinium* causes supercoiling of the transverse flagellum; more moderate contractions of the paraxonemal fiber are implicated in changing the pattern of flagellar beat (Hohfeld *et al.*, 1988). In the marine unicell *Apedinella,* centrin-based fibrous bundles are thought to reorient extracellular spine scales during a change in swimming direction (Koutoulis *et al.*, 1988). Finally, contraction of the stellate pattern fibers in the transition zone of *Chlamydomonas* affects flagellar excision by severing the nine axonemal microtubule doublets; the flagellum is thus cast off (Sanders and Salisbury, 1989).

The precise function(s) of centrin during mitosis is uncertain at the present time. Nonetheless, we predict that a functional role will involve calcium-modulated contractions and re-extensions of filamentous components of the cytomatrix. We will speculate, here, on the possible function(s) that centrin-based filaments may play in mitotic progression.

In preparation for mitosis, the centrosome or division center goes through a replication cycle (reviewed by Vandré and Borisy, 1989). Functionally, each centrosome contains two reproductive elements, called polar organizers, whose identity remains unknown. Although these elements are associated with centrioles, these organelles are not themselves the polar organizers (Sluder and Rieder, 1985; Sluder *et al.*, 1989, see Chapter 9). Once during each cell cycle, the polar organizers split, duplicate, and separate (Mazia *et al.*, 1960). These events are manifest

---

Fig. 5. The pericentriolar lattice of PtK_2 cells is a component of the spindle matrix. Cells were triple labeled with (A,D,G) monoclonal antitubulin antibodies (B,B′,E,E′,E″,H,H′) polyclonal anticentrin 08/28 antibodies, and (C,F,I) the DNA-binding fluorochrome DAPI. The anticentrin images in each column represent different focal planes through the same cell. (A–F) Two metaphase cells which have their chromosomes aligned at the cell equator (B,B′,E,E′,E″). Although both spindles are positioned at an angle relative to the plane of focus, careful examination reveals that the pericentrosomal spots occupy the limits of the spindle and extend from the poles (open arrows) to the chromosomes. (G–I) The cell shown has nearly completed anaphase A; the sister chromatids have almost reached the spindle poles (cf. C,F,I). The kinetochore microtubules have greatly shortened and spindle elongation has progressed (cf. A,D,G). A cluster of pericentrosomal spots demarcate the spindle poles (open arrows); additional spots (solid arrowheads) are seen beside and behind the trailing chromatid arms. Bar: 10 μm. [Reproduced from Baron *et al., Cell Motility and the Cytoskeleton* (1991), by copyright permission of John Wiley & Sons, Inc.]

on the cellular level by the splitting, duplication, and separation of the centrosome or flagellar apparatus and establishment of the spindle poles, which in turn determine the bipolarity of the forthcoming mitotic division.

The major MTOC and division center of the yeast *Saccharomyces* is the spindle pole body (SPB) (Robinow and Marak, 1966; Moens and Rapport, 1971; Peterson and Ris, 1976; see Chapter 7). Cells carrying temperature-sensitive alleles of the *CDC31* gene are defective in SPB reproduction and do not progress through mitosis (Byers, 1981). When shifted to the restrictive temperature, these cells develop one SPB that is twice as large as SPBs of other arrested CDC-mutants. This larger SPB nucleates approximately twice the number of microtubules as a normal SPB. The *CDC31* defect thus prevents splitting of the polar organizers. As a consequence, the SPB, although nucleating twice the number of microtubules, remains functionally singular as a division center. Molecular cloning of the centrin gene from the alga *Chlamydomonas* has shown that centrin shares sequence homology (50% identity) with the deduced amino acid sequence of the *CDC31* gene product (Baum *et al.,* 1986; Huang *et al.,* 1988a). These observations implicate centrin and the *CDC31* gene product in the splitting and separation of division centers. This leads us to question whether the *CDC31* gene product and centrin are components of the elusive polar organizers. Observations that the centrin-based NBBC of *Chlamydomonas* (Salisbury *et al.,* 1988; Wright *et al.,* 1989) and the PCL of PtK_2 cells (Baron *et al.,* 1991) undergo duplication, splitting, and separation during normal mitosis are consistent with this hypothesis.

Centrin plays a role in the appropriate segregation of basal bodies into daughter cells during cell division in *Chlamydomonas* (Wright *et al.,* 1985, 1989). The variable flagellar number mutants *vfl-1, vfl-2,* and *vfl-3* of *Chlamydomonas* harbor structurally defective distal fibers and/or NBBCs (Wright *et al.,* 1983, 1985, 1989; Adams *et al.,* 1985). As a consequence of these defects, the *vfl* mutants possess from 0 to 10 flagella per cell—the result of aberrant basal body segregation during cytokinesis. It appears that the physical association established by a normal NBBC between the basal bodies and nucleus is required for the appropriate segregation of the basal bodies relative to chromosomes during mitosis.

Our observations indicate that progression through mitosis involves a sequence of contractions and re-extensions of the NBBC and PCL. In light of the evidence that anaphase chromosome movement is modulated by calcium (Cande, 1979, 1981, 1982; Kiehart, 1981; Izant, 1983; Hepler, 1985; Keith *et al.,* 1985; Poenie *et al.,* 1985, 1986; Ratan and Shelanski, 1986; Ratan *et al.,* 1986, 1988; Tsien and Poenie, 1986; Hepler and Callaham, 1987; Tombes and Borisy, 1989), it is possible that force production by contraction of the NBBC and the PCL of *Chlamydomonas* and PtK_2 cells, respectively, contributes to or modulates karyokinesis. This proposal is consistent with the contractile/elastic matrix model of chromosome motility, which suggests that a spring or contractile filamentous spindle matrix moves chromosomes poleward (Pickett-Heaps *et al.,* 1982, 1984; Pickett-Heaps, 1986). Another possible motile mechanism suggested by the involvement of centrin-based fibers in severing axonemal microtubules is that centrin actively severs spindle microtubules (see Sanders and Salisbury, 1989). The generation of free microtubule ends in the spindle would conceivably influence spindle microtubule dynamics and, hence, chromosome segregation.

VIII. CONCLUSIONS

Several generalizations are currently possible with respect to centrin. (1) These molecules are associated with 3–8-nm diameter filament systems. (2) Centrin-based motility phenomena are characterized by calcium-modulated contractile behavior. (3) Centrin performs a variety of functions associated with division centers. (4) Finally, the precise cytoarchitecture of the division center determines centrin's cellular function(s).

ACKNOWLEDGMENTS

We thank David Estervig and John Harper for a critical reading of this manuscript and helpful discussions. This work was supported by NIH grant GM 35258.

REFERENCES

Adams, G. M. W., Wright, R. L., and Jarvik, J. W. (1985). Defective temporal and spatial control of flagellar assembly in a mutant of *Chlamydomonas reinhardtii* with variable flagellar number. *J. Cell Biol.* **100,** 955–964.

Allen, R. D. (1961). Ameboid movement. *In* "The Cell: Biochemistry, Physiology, Morphology" (J. Brachet and A. E. Mirsky, eds.), Vol. 2, pp. 135–216. Academic Press, New York.

Amos, W. B., Grimstone, A. V., Rothschild, L. J., and Allen, R. D. (1979). Structure, protein composition and birefringence of the costa: A motile flagellar root fiber in the flagellate *Trichomonas. J. Cell Biol.* **35,** 139–164.

Anderson, R. G. W., and Floyd, A. K. (1980). Electrophoretic analysis of basal body (centriole) proteins. *Biochemistry* **19,** 5625–5631.

Bailly, E., Doree, M., Nurse, P., and Bornens, M. (1989). $p34^{cdc2}$ is located in both nucleus and cytoplasm; part is centrosomally associated at G_2/M and enters vesicles at anaphase. *EMBO J.* **8(13),** 3985–3995.

Baron, A. T., and Salisbury, J. L. (1988). Identification and localization of a novel, cytoskeletal, centrosome-associated protein in PtK_2 cells. *J. Cell Biol.* **107,** 2669–2678.

Baron, A. T., Greenwood, T. M., and Salisbury, J. L. (1991). The centrin-related 165,000 M_r protein of PtK_2 cells is a component of the cytomatrix. *Cell Motil, Cytosk.* **18,** 1–14.

Bastmeyer, M., Steffen, W., and Fuge, H. (1986). Immunostaining of spindle components in tipulid spermatocytes using a serum against pericentriolar material. *Eur. J. Cell Biol.* **42,** 305–310.

Baum, P., Furlong, C., and Byers, B. (1986). Yeast gene required for spindle pole body duplication: Homology of its product with Ca^{2+}-binding proteins. *Proc. Natl. Acad. Sci. U.S.A.* **83,** 5512–5516.

Bazinet, C. W., Baron, A. T., and Salisbury, J. L. (1990). Centrin: A calcium-binding protein associated with centrosomes. *In* "Stimulus Response Coupling: The Role of Intracellular Calcium-binding Proteins" (V. L. Smith and J. R. Dedman, eds.) pp. 39–56. CRC Press, Inc., Baco Raton.

Berns, M. W., Rattner, J. B., and Brenner, S. (1977). The role of the centriolar region in animal cell mitosis: A laser microbeam study. *J. Cell Biol.* **72,** 351–367.

Berns, M. W., and Richardson, S. M. (1977). Continuation of mitosis after selective laser microbeam destruction of the centriolar region. *J. Cell Biol.* **75,** 977–982.

Bonifacino, J. S., Klausner, R. D., and Sandoval, I. V. (1985). A widely distributed nuclear protein immunologically related to the microtubule-associated protein MAP1 is associated with the mitotic spindle. *Proc. Natl. Acad. Sci. U.S.A.* **82,** 1146–1150.

Bornens, M., and Karsenti, E. (1984). The centrosome. *In* "Membrane Structure and Function" (E. E. Bittar, ed.), Vol. 6, pp. 99–171. John Wiley & Sons, New York.

Bornens, M., Paintrand, M. Berges, J., Marty, M. -C., and Karsenti, E. (1987). Structural and chemical characterization of isolated centrosomes. *Cell Motil. Cytosk.* **8,** 238–249.

Boveri, T. (1887). Über die Befruchtung der Eier von *Ascaris megalocephala. Sitz.-Ber. Ges. Morph. Phys. München* **3,** 153.

Boveri, T. (1901). "Zellen-studien IV. Uber der Natur der Centrosomen." Fischer, Jena.

Brinkley, B. R. (1985). Microtubule organizing centers. *Ann. Rev. Cell Biol.* **1,** 145–172.

Burnett, W. N. (1981). "Western blotting." Electrophoretic transfer of proteins from SDS-polyacrylamide gels to unmodified nitrocellulose and radiographic detection with antibody and radioiodinated protein A. *Anal. Biochem.* **112,** 195–203.

Byers, B. (1981). Multiple roles of the spindle pole bodies in the life cycle of *Saccharomyces. In* "Molecular Genetics in Yeast" (D. Von Wettstein, J. Friis, M. Keilland-Brandt, and A. Stenderup, eds.), Vol. 17, pp. 119–131. Munksgaard, Copenhagen.

Calarco-Gillam, P. D., Siebert, M. C., Hubble, R., Mitchison, T., and Kirschner, M. (1983). Centrosome development in early mouse embryos as defined by an autoantibody against pericentriolar material. *Cell* **35,** 621–629.

Cande, W. Z. (1979). Anaphase chromosome movement: Studies using lysed mitotic cells. *In* "Cell Motility: Molecules and Organization" (S. Hatano, H. Ishikawa, and H. Sato, eds.), pp. 593–608. University of Tokyo Press, Tokyo.

Cande, W. Z. (1981). Physiology of chromosome movement in lysed cell models. *In* "International Cell Biology" (H. G. Schweiger, ed.), pp. 382–391. Springer-Verlag, Berlin.

Cande, W. Z. (1982). Permeabilized cell models for studying chromosome movement in dividing PtK_1 cells. *Meth. Cell Biol.* **25,** 57–68.

Cavalier-Smith, T. (1974). Basal body and flagellar development during vegetative cell cycle and the sexual cycle of *Chlamydomonas reinhardtii. J. Cell Sci.* **16,** 529–556.

Connolly, J. A., and Kalnins, V. I. (1978). Visualization of centrioles and basal bodies by fluorescent staining with nonimmune rabbit sera. *J. Cell Biol.* **79,** 526–532.

Coss, R. A. (1974). Mitosis in *Chlamydomonas reinhardtii* basal bodies and the mitotic apparatus. *J. Cell Biol.* **63,** 325–329.

Cuello, A. (1983). "Immunohistochemistry." Wiley & Sons, Chichester.

Davis, F. M., Tsao, T. Y., Fowler, S. K., and Rao, P. N. (1983). Monoclonal antibodies to mitotic cells. *Proc. Natl. Acad. Sci. U.S.A.* **80,** 2926–2930.

Davis, F. M., and Rao, P. N. (1987). Antibodies to mitosis-specific phosphoproteins. *In* "Molecular Regulation of Nuclear Events in Mitosis and Meiosis" (R. A. Schlegel, M. S. Halleck, and P. N. Rao, eds.), pp. 259–293. Academic Press, Orlando, Florida.

De Camilli, P., Moretti, M., Donini, S. D., Walter, U., and Lohmann, S. M. (1986). Heterogeneous distribution of the cAMP receptor protein RII in the nervous system: Evidence for its intracellular accumulation on microtubules, microtubule-organizing centers, and in the area of the golgi complex. *J. Cell Biol.* **103,** 189–203.

Deery, W. J., Means, A. R., and Brinkley, B. R. (1984). Calmodulin-microtubule association in cultured mammalian cells. *J. Cell Biol.* **98,** 904–910.

De Harven, E. (1968). The centriole and the mitotic cycle. *In* "The Nucleus" (A. J. Dalton and F. Haquenae, eds.), pp. 197–227. Academic Press, New York.

De Mey, J., Aerts, F., De Raeymaeker, M., Daneels, G., Moeremans, M., De Wever, B., Vandre, D. D., Vallee, R. B., Borisy, G. G., and De Brabander, M. (1987). Microtubule associated protein (MAP) 1A: A phosphoprotein associated with spindle fibers and microtubule organizing centers. *In* "Nature and Function of Cytoskeletal Proteins in Motility and Transport: Progress in Zoology" (K. E. Wolhfarth-Botterman, ed.), Vol. 34, pp. 187–205. Gustav Fisher Verlag, Stuttgart.

Dingle, A. D., and Larson, D. E. (1981). Structure and protein composition of the striated flagellar rootlets of some protists. *Biosystems* **14,** 345–358.

Frey-Wyssling, A. (1953). "Submicroscopic Morphology of Protoplasm," 2d Ed. Elsevier, Amsterdam.

Fulton, C. (1971). Centrioles. *In* "Origin and Continuity of Cell Organelles" (J. Reinert and H. Uspring, eds.), pp. 170–221. Springer-Verlag, Berlin.

Fung, B. P., and Kasamatsu, H. (1985). Immuno-electron-microscopic localization of a centriole-related antigen in ciliated cells. *Cell Tissue Res.* **239,** 43–50.

Gardiner, P. R., Miller, R. H., and Marsh, M. C. P. (1981). Studies on the rhizoplast from *Naegleria gruberi. J. Cell Sci.* **47,** 277–293.

Gershon, N. D., Porter, K. R., and Trus, B. L. (1985). The cytoplasmic matrix: Its volume and surface area and the diffusion of molecules through it. *Proc. Natl. Acad. Sci. U.S.A.* **82,** 5030–5034.

Gosti, F., Marty, M. -C., Courvalin, J. C., Maunoury, R., and Bornens, M. (1987). Centrosomal proteins and lactase dehydrogenase possess a common epitope in human cell lines. *Proc. Natl. Acad. Sci. U.S.A.* **84,** 1000–1004.

Gosti-Testu, F., Marty, M. -C., Berges, J., Maunoury, R., and Bornens, M. (1986). Identification of centrosomal proteins in a human lymphoblastic cell line. *EMBO J.* **5(10),** 2545–2550.

Grain, J. (1986). The cytoskeleton in protists: Nature, structure, and functions. *Int. Rev. Cytol.* **104,** 153 219.

Hepler, P. K. (1985). Calcium restriction prolongs metaphase in dividing *Tradescantia* stamen hair cells. *J. Cell Biol.* **100,** 1363–1368.

Hepler, P. K., and Callaham, D. A. (1987). Free calcium increases during anaphase in stamen hair cells of *Tradescantia. J. Cell Biol.* **105,** 2137–2143.

Hiraoka, L., Golden, W., and Magnuson, T. (1989). Spindle-pole organization during early mouse development. *Devel. Biol.* **132,** 24–36.

Hohfeld, I., Otten, J., and Melkonian, M. (1988). Contractile eukaryotic flagella: Centrin is involved. *Protoplasma* **147,** 16–24.

Houliston, E., Pickering, S. J., and Maro, B. (1987). Redistribution of microtubules and pericentriolar material during the development of polarity in mouse blastomeres. *J. Cell Biol.* **104,** 1299–1308.

Huang, B., Mengersen, A., and Lee, V. D. (1988a). Molecular cloning of cDNA for caltractin, a basal body-associated Ca^{2+}-binding protein: Homology in its protein sequence with calmodulin and the yeast CDC31 gene product. *J. Cell Biol.* **107,** 133–140.

Huang, B., Watterson, D. M., Lee, V. D., and Schibler, M. J. (1988b). Purification and characterization of a basal body-associated Ca^{2+}-binding protein. *J. Cell Biol.* **107,** 121–131.

Huxley, H. E. (1969). The mechanism of muscle contraction. *Science (Washington, D.C.)* **164,** 1356–1365.

Huxley, H. E., and Hanson, J. (1954). Changes in the cross-striations of muscle during contraction and stretch and their structural interpretation. *Nature (London)* **173,** 973–976.

Izant, J. C. (1983). The role of calcium ions during mitosis. Calcium participates in the anaphase trigger. *Chromosoma (Berlin)* **88,** 1–10.

Joachim, S., and Schwoch, G. (1988). Immunoelectron microscopic localization of catalytic and regulatory subunits of cAMP-dependent protein kinases in the parotid gland. *Eur. J. Cell Biol.* **46,** 491–498.

Johnson, U. W., and Porter, K. R. (1968). Fine structure of cell division in *Chlamydomonas reinhardtii. J. Cell Biol.* **38,** 403–425.

Joswig, G., and Petzelt, C. (1990). The centrosome cycle: Visualization in PtK cells by a monoclonal antibody to a centrosomal 32 kd protein. *Cell Motil. Cytosk.* **15,** 181–192.

Kasamatsu, H., Shyamala, M., and Lin, W. (1980). Host antigens in the centriolar region are induced in SV40-infected TC7 cells: SV40 small T-function requirement. *Cold Spring Harbor Symp. Quant. Biol.* **44,** 243–252.

Keith, C. H., Ratan, R., Maxfield, F. R., Bajer, A., and Shelanski, M. L. (1985). Local cytoplasmic calcium gradients in living mitotic cells. *Nature (London)* **316,** 848–850.

Kellogg, D. R., Field, C. M., and Alberts, B. M. (1989). Identification of microtubule-associated proteins in the centrosome, spindle, and kinetochore of the early *Drosophila* embryo. *J. Cell Biol.* **109(6),** 2977–2991.

Keryer, G., Davis, F. M., Rao, P. N., and Beisson, J. (1987). Protein phosphorylation and dynamics of cytoskeletal structures associated with basal bodies in *Paramecium. Cell Motil. Cytosk.* **8,** 44–54.

Kiehart, D. P. (1981). Studies on the *in vivo* sensitivity of spindle microtubules to calcium ions and evidence for a vesicular calcium-sequestering system. *J. Cell Biol.* **88,** 604–617.

Klotz, C., Bordes, N., Laine, M. C., Sandoz, D., and Bornens, M. (1984). Study of cytoskeleton proteins of ciliated cells by the use of monoclonal antibodies. *J. Submicrosc. Cytol.* **16(1),** 115–117.

Klotz, C., Bordes, N., Laine, M. C., Sandoz, D., and Bornens, M. (1986). A protein of 175,000 daltons associated with striated rootlets in ciliated epithelia, as revealed by a monoclonal antibody. *Cell Motil. Cytosk.* **6,** 56–67.

Koutoulis, A., McFadden, G. I., and Wetherbee, R. (1988). Spine-scale reorientation in *Apedinella radians (Pedinellales, Chrysophyceae):* The microarchitecture and immunocytochemistry of the associated cytoskeleton. *Protoplasma* **147,** 25–41.

Kuriyama, R. (1989). 225-kilodalton phosphoprotein associated with mitotic centrosomes in sea urchin eggs. *Cell Motil. Cytosk.* **12,** 90–103.

Kuriyama, R., and Borisy, G. G. (1985). Identification of molecular components of the centrosphere in the mitotic spindle of sea urchin eggs. *J. Cell Biol.* **101,** 524–530.

La Claire, J. W., II, and Goddard, R. H. (1989). Immunolocalization of pericentriolar material in algal mitotic spindle. *Cell Motil. Cytosk.* **13,** 225–238.

Laemmli, U. K. (1970). Cleavage of structural proteins during the assembly of the head of bacteriophage T4. *Nature (London)* **227,** 680–685.

Larson, D. E., and Dingle, A. D. (1981). Isolation, ultrastructure, and protein composition of the flagellar rootlet of *Naegleria gruberi. J. Cell Biol.* **89,** 424–432.

Lechtreck, K. -F., McFadden, G. I., and Melkonian, M. (1989). The cytoskeleton of the naked green flagellate *Spermatozopsis similis:* Isolation, whole mount electron microscopy, and preliminary biochemical and immunological characterization. *Cell Motil. Cytosk.* **14,** 552–561.

Lin, W., Fung, B., Shyamala, M., and Kasamatsu, H. (1981). Identification of antigenically related polypeptides at centrioles and basal bodies. *Proc. Natl. Acad. Sci. U.S.A.* **78(4),** 2372–2377.

Luby-Phelps, K., Lanni, F., and Taylor, D. L. (1988). The submicroscopic properties of cytoplasm as a determinant of cellular function. *Ann. Rev. Biophys. Biophys. Chem.* **17,** 369–396.

Manton, I., and Parke, M. (1965). Observations on the fine structure of two species of *Platymonas* with special reference to flagellar scales and the mode of origin of the theca. *J. Mar. Biol. Assoc. U.K.* **45,** 743–754.

Marano, F., Santa-Maria, A., and Krishnawamy, S. (1985). The flagellar apparatus of *Dunaliella:* Isolation of basal body-flagellar root complex. *Protoplasma* **127,** 82–92.

Maro, B., Howlett, S. K., and Webb, M. (1985). Non-spindle microtubule organizing centers in metaphase II-arrested mouse oocytes. *J. Cell Biol.* **101,** 1665–1672.

Marsland, D. A. (1942). Protoplasmic streaming in relation to gel structure in the cytoplasm. *In* "The Structure of Protoplasm: A Monograph of the American Society of Plant Physiologists" (W. Seifriz, ed.), pp. 127–160. Iowa State College Press, Ames.

Martindale, V. E., and Salisbury, J. L. (1990). Phosphorylation of algal centrin is rapidly responsive to changes in the external milieu. *J. Cell Sci.* **96,** 395–402.

Maunoury, R. (1978). Localization immunocytochimique de la centrosphere de cellules tumorales humaines par utilisation d'anticorps naturels de lapin. *C. R. Acad. Sci. (Paris)* **286,** 503–506.

Mazia, D. (1961). Mitosis and the physiology of cell division. *In* "The Cell: Biochemistry, Physiology, Morphology" (J. Brachet and A. E. Mirsky, eds.), Vol. 3, pp. 77–412. Academic Press, New York.

Mazia, D. (1984). Centrosomes and mitotic poles. *Exp. Cell Res.* **153,** 1–15.

Mazia, D., Harris, P. J., and Bibring, T. (1960). The multiplicity of the mitotic centers and the time-course of their duplication and separation. *J. Biophys. Biochem. Cytol.* **7(1),** 1–20.

McFadden, G. I., Schulze, D., Surek, B., Salisbury, J. L., and Melkonian, M. (1987). Basal body reorientation mediated by a Ca^{2+}-modulated contractile protein. *J. Cell Biol.* **105,** 903–912.

McIntosh, J. R. (1983). The centrosome as an organizer of the cytoskeleton. *Mod. Cell Biol.* **2,** 115–142.

McIntosh, J. R. (1987). Progress in research on mitosis. *In* "Biomechanics of Cell Division" (N. Akkas, ed.), pp. 123–143. Plenum Press, New York.

McIntosh, J. R., and Koonce, M. P. (1989). Mitosis. *Science (Washington, D.C.)* **246,** 622–628.

Melkonian, M. (1979). An ultrastructural study of the flagellate *Tetraselmis cordiformis* (Stein) (Chlorophyceae) with emphasis on the flagellar apparatus. *Protoplasma* **98,** 139–151.

Melkonian, M. (1980). Ultrastructural aspects of basal body associated fibrous structures in green algae: A critical review. *Biosystems* **12,** 85–104.

Melkonian, M., Schulze, D., McFadden, G. I., and Robenek, H. (1988). A polyclonal antibody (anticentrin) distinguishes between two types of fibrous flagellar roots in green algae. Protoplasma **144,** 56–61.

Mitchison, T. J. (1989). Mitosis: Basic concepts. *Curr. Opn. Cell Biol.* **1,** 67–74.

Mitchison, T., and Kirschner, M. (1984). Microtubule dynamics and cellular morphogenesis. *In* "Molecular Biology of the Cytoskeleton" (G. G. Borisy, D. W. Cleveland, and D. B. Murphy, eds.), pp. 27–44. Cold Spring Harbor Laboratory Press, Cold Spring Harbor, New York.

Moens, P. B., and Rapport, E. (1971). Spindles, spindle plaques, and meiosis in the yeast *Saccharomyces cerevisiae* (Hansen). *J. Cell Biol.* **50,** 344–361.

Moroi, Y., Murata, I., Takeuchi, A., Kamatani, N., Tanimoto, K., and Yokohari, R. (1983). Human anticentriole autoantibody in patients with scleroderma and Raynaud's phenomenon. *Clin. Immunol. Immunopathol.* **29,** 381–390.

Neighbors, B. W., Williams, R. C., Jr., and McIntosh, J. R. (1988). Localization of kinesin in cultured cells. *J. Cell Biol.* **106,** 1193–1204.

Nigg, E. A., Schafer, G., Hilz, H., and Eppenberger, H. M. (1985). Cyclic-AMP-dependent protein kinase type II is associated with the golgi complex and with centrosomes. *Cell* **41,** 1039–1051.

O'Farrell, P. H. (1975). High resolution two-dimensional electrophoresis of proteins. *J. Biol. Chem.* **250(10),** 4007–4021.

Oliver, J. M., Osborne, W. R. A., Pfeiffer, J. R., Child, F. M., and Berlin, R. D. (1981). Purine nucleoside phosphorylase is associated with centrioles and basal bodies. *J. Cell Biol.* **91,** 837–847.

Peterson, J. B., and Ris, H. (1976). Electron-microscopic study of the spindle and chromosome movement in the yeast *Saccharomyces cerevisiae. J. Cell Sci.* **22,** 219–242.

Peterson, S. P., and Berns, M. W. (1980). The centriolar complex. *Int. Rev. Cytol.* **64,** 81–106.

Pickett-Heaps, J. D. (1969). The evolution of the mitotic apparatus: An attempt at comparative ultrastructural cytology in dividing plant cells. *Cytobios.* **3,** 257–280.

Pickett-Heaps, J. D. (1986). Mitotic mechanisms: An alternative view. *Trends Biochem. Sci.* **11,** 504–507.

Pickett-Heaps, J. D., Tippit, D. H., and Porter, K. R. (1982). Rethinking mitosis. *Cell* **29,** 729–744.

Pickett-Heaps, J. D., Spurch, T., and Tippit, D. H. (1984). Chromosome motion and the spindle matrix. *J. Cell Biol.* **99,** 137s–143s.

Pitelka, D. R. (1974). Basal bodies and root structures. *In* "Cilia and Flagella " (M. A. Sleigh, ed.), pp. 437–464. Academic Press, New York.

Poenie, M., Alderton, J., Tsien, R. Y., and Steinhardt, R. A. (1985). Changes of free calcium levels with stages of the cell division cycle. *Nature (London)* **315,** 147–149.

Poenie, M., Alderton, J., Steinhardt, R., and Tsien, R. (1986). Calcium rises abruptly and briefly throughout the cell at the onset of anaphase. *Science (Washington, D.C.)* **233,** 886–889.

Pollard, T. D. (1984). Molecular architecture of the cytoplasmic matrix. *In* "White Cell Mechanics: Basic Science and Clinical Aspects" (H. J. Meiselman, M. A. Lichtman, and P. L. LaCelle, eds.), pp. 75–86. Liss, New York.

Pollard, T. D., and Cooper, J. A. (1986). Actin and actin-binding proteins. A critical evaluation of mechanisms and functions. *Ann. Rev. Biochem.* **55,** 987–1035.

Porter, K. R. (1966). Cytoplasmic microtubules and their function. *In* "Principles of Biomolecular Organization" (G. E. W. Wolstenholme and M. O'Connor, eds.). pp. 308–357. Churchill, London.

Porter, K. R. (1976). Introduction: Motility in cells. *In* "Cell Motility" (R. Goldman, T. Pollard, and J. Rosenbaum, eds.), Vol 3, pp. 1–28. Cold Spring Harbor Laboratory Press, Cold Spring Harbor, New York.

Porter, K. R. (1984). The cytomatrix: A short history of its study. *J. Cell Biol.* **99(1),** 3s–12s.

Porter, K. R., Byers, H. R., and Ellisman, M. H. (1979). The cytoskeleton. *In* "The Neurosciences: Fourth Study Program" (F. O. Schmitt and F. C. Warden, eds.), pp. 703–722. MIT Press, Cambridge.

Porter, K. R., Beckerle, M., and McNiven, M. (1983). The cytoplasmic matrix. *Mod. Cell Biol.* **2,** 259–302.

Rao, P. N., Zhao, J. -Y., Ganju, R. K., and Ashorn, C. L. (1989). Monoclonal antibody against the centrosome. *J. Cell Sci.* **93,** 63–69.

Ratan, R. R., and Shelanski, M. L. (1986). Calcium and the regulation of mitotic events. *Trends Biochem. Sci.* **11,** 456–459.

Ratan, R. R., Shelanski, M. L., and Maxfield, F. R. (1986). Transition from metaphase to anaphase is accompanied by local changes in cytoplasmic free calcium in PtK_2 kidney epithelial cells. *Proc. Natl. Acad. Sci. U.S.A.* **83,** 5136–5140.

Ratan, R. R., Maxfield, F. R., and Shelanski, M. L. (1988). Long-lasting and rapid calcium changes during mitosis. *J. Cell Biol.* **107,** 993–999.

Rattner, J. B., and Berns, M. W. (1976). Centriole behavior in early mitosis of rat kangaroo cells (PtK_2). *Chromosoma (Berlin)* **54,** 359–395.

Riabowol, K., Draetta, G., Brizuela, L., Vandre, D., and Beach, D. (1989). The cdc2 kinase is a nuclear protein that is essential for mitosis in mammalian cells. *Cell* **57,** 393–401.

Rieder, C. L., and Borisy, G. G. (1982). The centrosome cycle in PtK_2 cells: Asymmetric distribution and structural changes in the pericentriolar material. *Biol. Cell* **44,** 117–132.

Ring, D., Hubble, R., Caput, D., and Kirschner, M. (1980). Isolation of microtubule organizing centers from mouse neuroblastoma cells. *In* "Microtubules and microtubule Inhibitors" (M. De Brabander and J. De Mey, eds.), pp. 297–310. Elsevier North-Holland Biomedical Press, Amsterdam.

Robenek, H., and Melkonian, M. (1979). Rhizoplast-membrane associations in the flagellate *Tetraselmis cordiformis* Stein (Chlorophyceae) revealed by freeze-etching and thin sections. *Arch. Protistenkd.* **122,** 340–351.

Robinow, C. F., and Marak, J. (1966). A fiber apparatus in the nucleus of the yeast cell. *J. Cell Biol.* **29,** 129–151.

Roos, U. -P. (1973). Light and electron microscopy of rat kangaroo cells in mitosis. I. Formation and breakdown of the mitotic apparatus. *Chromosoma (Berlin)* **40,** 43–82.

Rubin, R. W., and Cunningham, W. P. (1973). Partial purification and phosphotungstate solubilization of basal bodies and kinetodesmal fibers from *Tetrahymena pyriformis. J. Cell Biol.* **57,** 601–612.

Sager, P. R., Rothfield, N. L., Oliver, J. M., and Berlin, R. D. (1986). A novel mitotic spindle pole component that originates from the cytoplasm during prophase. *J. Cell Biol.* **103,** 1863–1872.

Salisbury, J. L. (1982). Calcium-sequestering vesicles and contractile flagellar roots. *J. Cell Sci.* **58,** 433–443.

Salisbury, J. L. (1983). Contractile flagellar roots: The role of calcium. *J. Submicrosc. Cytol.* **15,** 105–110.

Salisbury, J. L. (1988). The lost neuromotor apparatus of *Chlamydomonas:* Rediscovered. *J. Protozool.* **35(4),** 574–577.

Salisbury, J. L. (1989a). Algal centrin: Calcium-sensitive contractile organelles. *In* "Algae as Experimental Systems" (A. W. Coleman, L. J. Goff, and J. R. Stein-Taylor, eds.), pp. 19–37. Liss, New York.

Salisbury, J. L. (1989b). Centrin and the algal flagellar apparatus. *J. Phycol.* **25,** 201–206.

Salisbury, J. L., and Floyd, G. L. (1978). Calcium-induced contraction of the rhizoplast of a quadriflagellate green alga. *Science (Washington, D.C.)* **202,** 975–977.

Salisbury, J. L., Swanson, J. A., Floyd, G. L., Hall, R., and Maihle, N. J. (1981). Ultrastructure of the flagellar apparatus of the green alga *Tetraselmis subcordiformis,* with special consideration given to the function of the rhizoplast and rhizanchora. *Protoplasma* **107,** 1–11.

Salisbury, J. L., Baron, A. T., Surek, B., and Melkonian, M. (1984). Striated flagellar roots: Isolation and partial characterization of a calcium-modulated contractile organelle. *J. Cell Biol.* **99,** 962–970.

Salisbury, J.L., Aebig, K. W., and Coling, D. E. (1986a). Isolation of the calcium-modulated contractile protein of striated flagellar roots. *Meth. Enzymol.* **134,** 408–414.

Salisbury, J. L., Baron, A. T., Coling, D. E., Martindale, V. E., and Sanders, M. A. (1986b). Calcium-modulated contractile proteins associated with the eukaryotic centrosome. *Cell Motil. Cytosk.* **6,** 193–197.

Salisbury, J. L., Sanders, M. A., and Harpst, L. (1987). Flagellar root contraction and nuclear movement during flagellar regeneration in *Chlamydomonas reinhardtii. J. Cell Biol.* **105,** 1799–1805.

Salisbury, J. L., Baron, A. T., and Sanders, M. A. (1988). The centrin-based cytoskeleton of *Chlamydomonas reinhardtii:* Distribution in interphase and mitotic cells. *J. Cell Biol.* **107,** 635–641.

Sanders, M. A., and Salisbury, J. L. (1989). Centrin-mediated microtubule severing during flagellar excision in *Chlamydomonas reinhardtii. J. Cell Biol.* **108,** 1751–1760.

Sandoz, D., Chailley, B., Boisvieux-Ulrich, E., Lemullois, M., Laine, M. -C., and Bautista-Harris, G. (1988). Organization and functions of cytoskeleton in metazoan ciliated cells. *Biol. Cell* **63,** 183–193.

Sato, C., Nishizawa, K., Nakamura, H., Komagoe, Y., Shimada, K., Ueda, R., and Suzuki, S. (1983). Monoclonal antibody against microtubule associated protein-1 produces immunofluorescent spots in the nucleus and centrosome of cultured mammalian cells. *Cell Struct. Funct.* **8,** 245–254.

Sato, C., Tanabe, K., Nishizawa, K., Nakayama, T., Kobayashi, T., and Nakamura, H. (1985). Localization of 350K molecular weight and related proteins in both the cytoskeleton and nuclear flecks that increase during G_1 phase. *Exp. Cell Res.* **160,** 206–220.
Sauron, M. E., Marty, M. C., Maunoury, R., Courvalin, J. C., Maro, B., and Bornens, M. (1984). Centriole-associated proteins in human cell lines. *J. Submicrosc. Cytol.* **16(1),** 133–135.
Schatten, H., Walter, M., Mazia, D., Biessmann, H. Paweletz, N., Coffe, G., and Schatten, G. (1987). Centrosome detection in sea urchin eggs with a monoclonal antibody against *Drosophila* intermediate filament proteins: Characterization of stages of the division cycle of centrosomes. *Proc. Natl. Acad. Sci. U.S.A.* **84,** 8488–8492.
Schliwa, M., and van Blerkom, J. (1981). Structural interactions of cytoskeletal components. *J. Cell Biol.* **90,** 222–235.
Schulze, D., Robenek, H., McFadden, G. I. and Melkonian, M. (1987). Immunolocalization of a Ca^{2+}-modulated contractile protein in the flagellar apparatus of green algae: The nucleus-basal body connector. *Eur. J. Cell Biol.* **45,** 51–61.
Sellitto, C., and Kuriyama, R. (1988). Distribution of a matrix component of the midbody during the cell cycle in Chinese hamster ovary cells. *J. Cell Biol.* **106,** 431–439.
Senécal, J. -L., Oliver, J. M., and Rothfield, N. (1985). Anticytoskeletal autoantibodies in the connective tissue diseases. *Arthr. Rheum.* **28(8),** 889–898.
Sluder, G., and Rieder, C. L. (1985). Centriole number and the reproductive capacity of spindle poles. *J. Cell Biol.* **100,** 887–896.
Sluder, G., Miller, F. J., and Rieder, C. L. (1989). Reproductive capacity of sea urchin centrosomes wihout centrioles. *Cell Motil. Cytosk.* **13,** 264–273.
Snyder, J. A., and McIntosh, J. R. (1975). Initiation and growth of microtubules from mitotic centers in lysed mammalian cells. *J. Cell Biol.* **67,** 744–760.
Snyder, J. A., Hamilton, B. T., and Mullins, J. M. (1982). Loss of mitotic centrosomal microtubule initiation capacity at the metaphase–anaphase transition. *Eur. J. Cell Biol.* **27,** 191–199.
Steffen, W., and Linck, R. W. (1988). Evidence for tektins in centrioles and axonemal microtubules. *Proc. Natl. Acad. Sci. U.S.A.* **85,** 2643–2647.
Stephens, R. E. (1975). The basal apparatus: Mass isolation from the molluscan ciliated gill epithelium and a preliminary characterization of striated rootlets. *J. Cell Biol.* **64,** 408–420.
Stewart, K. D., Mattox, K. R., and Chandler, C. D. (1974). Mitosis and cytokinesis in *Platymonas subcordiformis,* a scaly green monad. *J. Phycol.* **10,** 65–79.
Stossel, T. P., Chapponier, C., Ezzell, R. M., Hartwig, J. H., Janmey, P. A., Kwiatkowski, D. J., Lind, S. E., Smith, D. B., Southwick, F. S., Yin, H. L., and Zaner, K. S. (1985). Nonmuscle actin-binding proteins. *Ann. Rev. Cell Biol.* **1,** 353–402.
Taylor, D. L., and Condeelis, J. S. (1979). Cytoplasmic structure and contractility in ameboid cells. *Int. Rev. Cytol.* **56,** 57–144.
Taylor, D. L., and Fechheimer, M. (1982). Cytoplasmic structure and contractility: The solation–contraction coupling hypothesis. *Phil. Trans. R. Soc. Lond.* [*Biol.*] **299,** 185–187.
Tombes, R. M., and Borisy, G. G. (1989). Intracellular free calcium and mitosis in mammalian cells: Anaphase onset is calcium modulated, but is not triggered by a brief transient. *J. Cell Biol.* **109,** 627–636.
Toriyama, M., Ohta, K., Endo, S., and Sakai, H. (1988). 51-kd protein, a component of microtubule-organizing granules in the mitotic apparatus involved in aster formation *in vitro. Cell Motil. Cytosk.* **9,** 117–128.
Towbin, H., Stahelin, T., and Gordon, J. (1979). Electrophoretic transfer of proteins from polyacrylamide gels to nitrocellulose sheets. *Proc. Natl. Acad. Sci. U.S.A.* **76,** 4350–4354.
Tsien, R. Y., and Poenie, M. (1986). Fluorescence ratio imaging: A new window into intracellular ionic signaling. *Trends Biochem. Sci.* **11,** 450–455.

Tucker, J. B. (1984). Spatial organization of microtubule-organizing centers and microtubules. *J. Cell Biol.* **99(1),** 55s–62s.

Turksen, K., Aubin, J. E., and Kalnins, V. I. (1982). Identification of a centriole-associated protein by antibodies present in normal rabbit sera. *Nature (London)* **298,** 763–765.

Vandré, D. D., Davis, F. M., Rao, P. N., and Borisy, G. G. (1984). Phosphoproteins are components of mitotic microtubule organizing centers. *Proc. Natl. Acad. Sci. U.S.A.* **81,** 4439–4443.

Vandré, D. D., and Borisy, G. G. (1985). The interphase–mitosis transformation of the microtubule network in mammalian cells. *In* "Cell Motility: Mechanism and Regulation" (H. Ishikawa, S. Hatano, and H. Sato, eds.), pp. 389–401. University of Tokyo Press, Tokyo.

Vandré, D. D., Davis, F. M., Rao, P. N., and Borisy, G. G. (1986). Distribution of cytoskeletal proteins sharing a conserved phosphorylated epitope. *Eur. J. Cell Biol.* **41,** 72–81.

Vandré, D. D., and Borisy, G. G. (1989). The centrosome cycle in animal cells. *In* "Mitosis: Molecules and Mechanisms" (J. S. Hyams and B. R. Brinkley, eds.), pp. 39–75. Academic Press, San Diego.

Vaudaux, P. E., Williams, N. E., Frankel, J., and Vaudaux, C. (1977). Inter-strain variability of structural proteins in *Tetrahymena. J. Protozool.* **24(3),** 453–458.

Weber, K., and Osborn, M. (1969). The reliability of molecular weight determinations by dodecyl sulfate-polyacrylamide gel electrophoresis. *J. Biol. Chem.* **244,** 4406–4412

Wheatley, D. N. (1982). "The Centriole: A Central Enigma of Cell Biology." Elsevier Biomedical Press, Amsterdam.

Williams, N. E., Vaudaux, P. E., and Skriver, L. (1979). Cytoskeletal proteins of the cell surface in *Tetrahymena.* I. Identification and localization of major proteins. *Exp. Cell Res.* **123,** 311–320.

Wilson, E. B. (1925). "The Cell in Development and Heredity," 3d Ed. MacMillan, New York.

Wolfe, J. (1972). Basal body fine structure and chemistry. *Adv. Cell Mol. Biol.* **2,** 151–192.

Wolosewick, J. J., and Porter, K. R. (1979). Microtrabecular lattice of the cytoplasmic ground substance: Artifact or reality. *J. Cell Biol.* **82,** 114–139.

Wright, R. L., Chojnacki, B., and Jarvik, J. (1983). Abnormal basal body number, location, and orientation in a striated fiber-defective mutant of *Chlamydomonas reinhardtii. J. Cell Biol.* **96,** 1697–1707.

Wright, R. L., Salisbury, J. L., and Jarvik, J. W. (1985). A nucleus–basal body connector in *Chlamydomonas reinhardtii* that may function in basal body localization or segregation. *J. Cell Biol.* **101,** 1903–1912.

Wright, R. L., Adler, S. A., Spanier, J. G., and Jarvik, J. W. (1989). Nucleus–basal body connector in *Chlamydomonas:* Evidence for a role in basal body segregation and against essential roles in mitosis or in determining cell polarity. *Cell Motil. Cytosk.* **14,** 516–526.

7

Spindle Pole Body of *Saccharomyces cerevisiae:* A Model for Genetic Analysis of the Centrosome Cycle

MARK WINEY[1] AND BRECK BYERS

Department of Genetics
University of Washington
Seattle, Washington 98195

I. INTRODUCTION

Throughout the evolutionary range of eukaryotes are vast differences in the size and complexity of centrosomes. One of the simplest is the spindle pole body (SPB) of the budding yeast, *Saccharomyces cerevisiae.* This discrete organelle, like centrosomes of greater complexity, acts as a microtubule organizing center (MTOC), serving not only as the focus of microtubule organization *in vivo*

[1]Current Address: Department of Molecular, Cellular and Developmental Biology, Porter Biosciences, University of Colorado at Boulder, Boulder, Colorado 80309.

Copyright © 1992 Academic Press, Inc.
All rights of reproduction in any form reserved.

(reviewed by Byers, 1981a) but also as a structure capable of nucleating microtubule assembly *in vitro* (Byers *et al.,* 1978; Hymans and Borisy, 1978). Implicit in its name, the SPB plays a crucial role in organizing the poles of both mitotic and meiotic spindles.

The life cycle of *S. cerevisiae* typically consists of both a haplophase and a diplophase. Haploid yeast cells of either mating type are capable of mitotic growth, but can mate with cells of the opposite mating type to form diploids. Mating begins with arrest of the haploids in G_1 and is followed by fusion, first of the cells and then of the nuclei (karyogamy). Karyogamy is initiated at the SPBs that are borne on either nucleus (Byers and Goetsch, 1975a). Diploids normally undergo mitotic growth in a manner similar to that of haploids, but will undergo meiosis and sporulation when starved for nitrogen compounds while being provided with a nonfermentable carbon source. The SPB in meiotic cells is crucial not only in controlling the formation of the meiotic spindles but also in serving as an initiation site for spore wall formation (reviewed by Byers, 1981a).

There are two predominant reasons to choose budding yeast as an organism for detailed analysis of centrosome-like organelles. First, the facility of genetic analysis in yeast makes SPB functions accessible to study by mutation. Second, the discrete nature of the SPB, as viewed by electron microscopy, enables one to specify the precise state of the organelle in any phase of the life cycle. It has therefore been possible, by analyzing both wild-type and mutant yeast strains, to define the requirements for various aspects of SPB behavior with respect to other features of the yeast life cycle. This chapter will focus on the results of such cytological and genetic studies of SPB behavior in mitosis, meiosis, and karyogamy.

II. ANATOMY OF SPINDLE POLE BODY

The yeast SPB was first visualized by electron microscopy (Robinow and Marak, 1966). This organelle, originally referred to as a "centriolar plaque," was seen as a disk of darkly staining material lying in the plane of the nuclear envelope. Microtubules of the mitotic spindle were clearly seen to have their ends embedded in the dense disk. More detailed analysis of this structure by Moens and Rapport (1971) and later by others (Peterson *et al.,* 1972; Byers and Goetsch, 1974, 1975a,b; Zickler and Olson, 1975; Peterson and Ris, 1976) revealed further aspects of its anatomy, as shown in Fig. 1.

Some substructural elements of the SPB are permanent features seen at all stages of the life cycle, whereas other features are transient, appearing only at specific stages. Prominent among the permanent components of the SPB is the darkly staining disk embedded in the nuclear envelope. This component, which is termed the central layer or central zone (Fig. 1a), has an area (on either flat surface) of

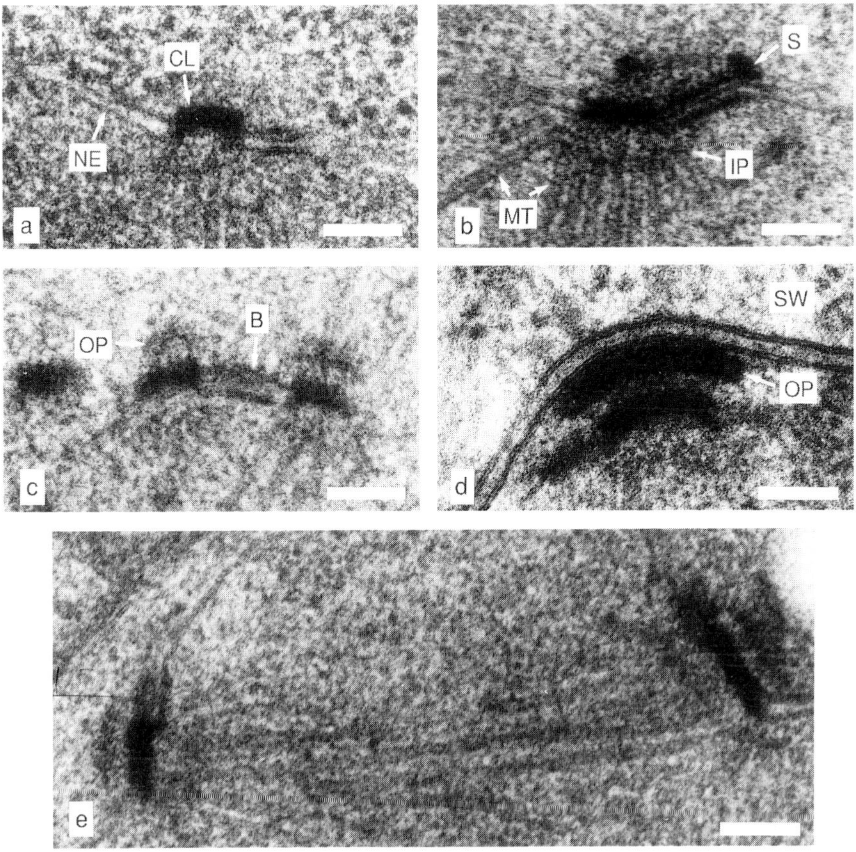

Fig. 1. Electron micrographs of the spindle pole body in *Saccharomyces cerevisiae* at various stages. The cells were prepared for electron microscopy as described by Byers and Goetsch (1974, 1975a). The stages shown include (a) a single spindle pole body, (b) a satellite-bearing spindle pole body, (c) the duplicated but unseparated (side-by-side) state, (d) a meiosis II spindle pole body with the modified outer plaque (OP), and (e) a mitotic spindle. The highlighted features of the spindle pole body are the central layer (CL), the inner plaque (IP), the satellite (S), the outer plaque (OP), and the bridge (B). Other structures indicated include the nuclear envelope (NE), microtubules (MT), and prospore wall (SW). Bars: 0.1 μm

about 0.017 μm^2 in haploid strains and of roughly twice that in diploids (Byers and Goetsch, 1974; Byers, 1981b). The nuclear envelope remains closed at its junction with the central layer of the SPB, creating a topological arrangement similar to the insertion of nuclear pore complexes in the envelope (Peterson *et al.,* 1972). Layered over either flat surface of the central zone are lightly staining disks termed the inner and outer plaques, on the nuclear and cytoplasmic faces, respectively (Fig. 1b,c). The central layer and the inner and outer plaques give the SPB its distinctive multilayered appearance. The plaques are probably critical to pole function because nuclear and cytoplasmic microtubules have their proximal ends associated with the inner and outer plaques, respectively. Indeed, the inner plaque is clearly a major site of microtubule nucleation *in vitro* (Byers *et al.,* 1978; Hymans and Borisy, 1978).

In addition to these permanent features of the SPB, other components appear transiently during the duplication cycle or in specific stages of the yeast life cycle. A notable transient feature is the bridge (Fig. 1c) that interconnects the two SPBs shortly after duplication has occurred. The bridge appears to consist of two membrane-like layers joined together at their edges, where they contact the central layer of either SPB. The membrane-like material seen here must differ from the more typical unit membrane of the nuclear envelope, because it stains much more darkly when prepared for electron microscopy. When the duplicated SPBs undergo separation to form either pole of the mitotic (or meiotic) spindle, there is an apparent scission at the middle of the bridge (midway between the two SPBs) so each separated SPB retains a remnant termed the "half-bridge" (Byers and Goetsch, 1974). The half-bridge persists and seems to play an important role in the next cycle of SPB duplication, as described next.

When the yeast cell enters the next cycle of nuclear division, the persistent half-bridge appears similar to that seen in the later phases of the preceding cycle. However, as the cell prepares for initiation of the new nuclear division cycle, a small tuft of darkly staining material similar in appearance to the central layer makes its appearance near the "free" edge of the half-bridge (distal to the juncture between the half-bridge and the spindle pole proper; Fig. 1b; Byers and Goetsch, 1974). This "satellite" persists until a *bona fide* daughter SPB arises at the same site. No intermediate structures between the satellite and the new SPB have been recognized, so the transition is presumed to occur rapidly. During mating, the transition is delayed until after karyogamy, so satellite-bearing SPBs persist in cells treated with mating pheromones.

Another transient feature of SPBs that is seen only in meiosis is an extensive modification of the outer plaque that occurs during the second meiotic division (Fig. 1d; Moens and Rapport, 1971; Peterson *et al.,* 1972; Zickler and Olson, 1975). The outer plaque becomes enlarged; its staining properties are intensified. This remarkable structural transformation of the outer plaque appears to be accompanied by a functional alteration, because prospore wall formation is initiated on the outer surface of this structure (Moens and Rapport, 1971). The absence of prospore wall formation adjacent to SPBs that fail to undergo this modification under certain experimental conditions indicates that the modified outer plaque is an essential prerequisite for spore wall development (Davidow *et al.*, 1980).

As discussed earlier, the ends of microtubules are associated with the inner and outer plaques of the SPB. Because SPBs will nucleate microtubules *in vitro,* it is assumed that the proximal ends are the minus ends of the microtubules. Decoration of these microtubules in isolated nuclei with excess tubulin (Heidemann and McIntosh, 1980) supports this interpretation (L. Goetsch and B. Byers, unpublished results). As in the case of centrosomes, the nature of microtubule association with this MTOC is difficult to discern because the proximal ends of the microtubules are obscured by the densely staining material from which they emanate. A clearer view is attained by examining this association in cell lysates, with which one can use negative staining (uranyl acetate) to examine the association of SPBs with either endogenous microtubules or microtubules polymerized *in vitro* (Figs. 2,3; Byers *et al.,* 1978; Hymans and Borisy, 1978). In either case, it is possible to effect a partial disruption of the microtubule/plaque junctions by prolonged exposure to the negative stain, thereby revealing an unusual structure at the formerly attached proximal ends of the microtubules (Fig. 3a; Byers *et al.,* 1978). These microtubule ends have a "closed" structure; that is, the lateral walls of each microtubule appear to be continuous with a hemispheric terminal wall of similar thickness (Fig. 3b; Byers *et al.,* 1978). By contrast, the distal ends of these same microtubules have the more typical "open" appearance (Fig. 3c). The role of the closed ends is still unknown, but it seems probable that they play a role in microtubule nucleation or stabilization.

Several central issues in cell biology are being elucidated by addressing them genetically in both *S. cerevisiae* and the fission yeast *Schizosaccharomyces pombe.* Using both yeasts has proven valuable because of the great evolutionary distance evident between these two organisms. A comparative approach might eventually yield critical insight into the SPB as well, but *S. pombe* has proven rather intractable to cytological analysis. Nevertheless, some success in characterization of the SPB has been attained recently by analyzing freeze-substituted cells (Tanaka and Kanbe, 1986; Kanbe *et al.,* 1989). The SPBs seen by this method appear similar to those in *S. cerevisiae;* each one is seen as a dark staining structure embedded in the nuclear envelope and associated with both cytoplasmic and nuclear microtubules. Detailed description of its structure, as well as its behavior during meiotic division and karyogamy, will have to await further study.

III COORDINATION OF SPINDLE POLE BODY CYCLE WITH CELLULAR BEHAVIOR

A. Mitosis

Duplication of the SPB in mitotic yeast must be integrated with other activities if a functional bipolar spindle is to be formed. Distinctive features of the SPB duplication cycle, as first described by Moens and Rapport (1971) and later elaborated by Byers and Goetsch (1975a), are shown in Fig. 4. Unbudded cells in early stages of G_1 exhibit an unmodified single SPB, whereas a satellite appears

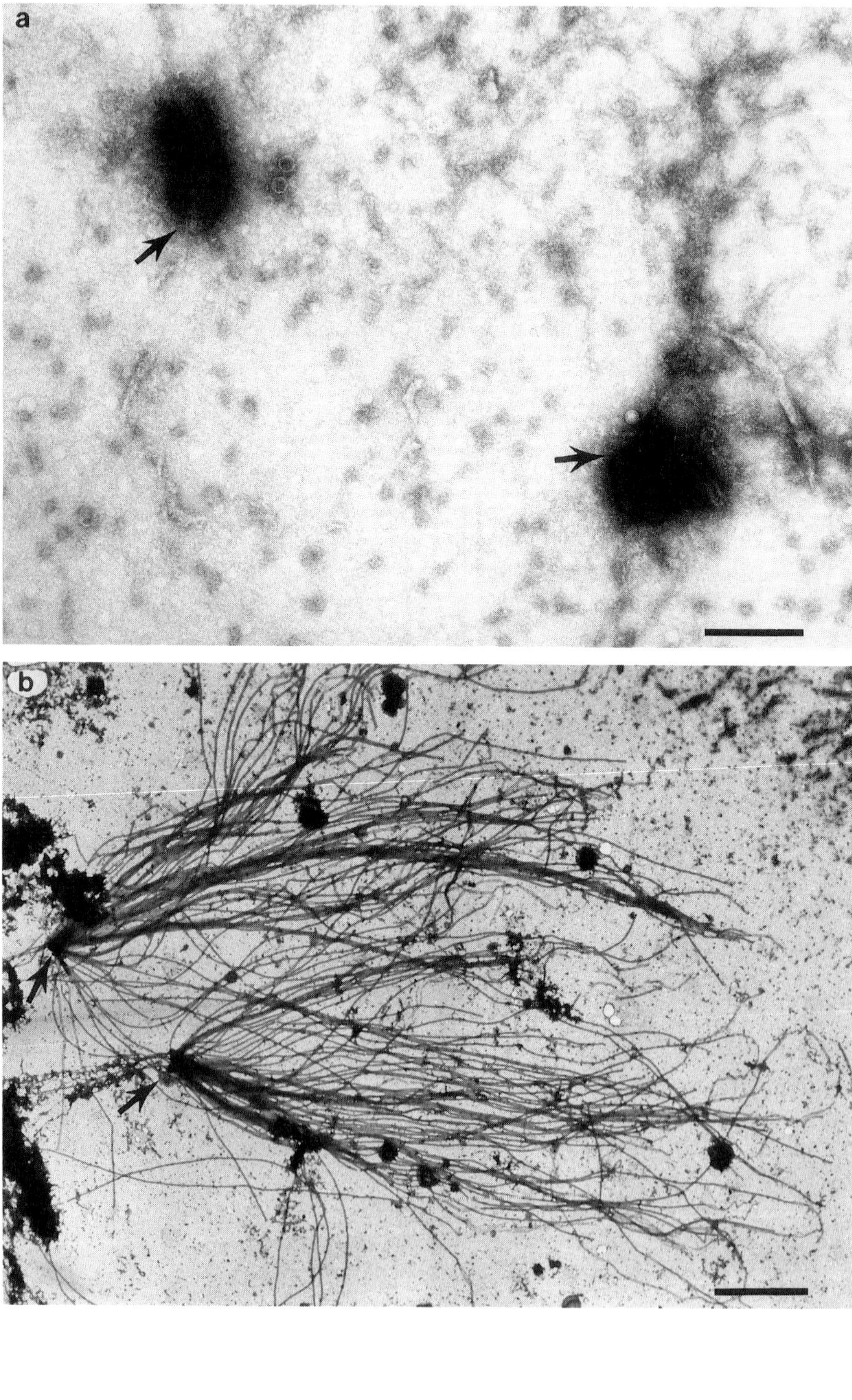

on the half-bridge in the latter part of G_1. Formation of a satellite-bearing pole is thought to be a requisite step in the duplication process, for the satellite appears to be an essential precursor to the new pole. Pole duplication per se has been found to occur near the time of transition from G_1 to S phase, nearly coincident with bud emergence (Fig. 5; Moens and Rapport, 1971; Byers and Goetsch, 1975a). Once duplication has occurred, the poles are found in a side-by-side configuration, interconnected by the bridge. As bud enlargement proceeds, the poles eventually separate from one another by a mechanism that seems to involve scission of the bridge, yielding a half-bridge associated with each pole. The separated poles then interact via their nuclear microtubules to form a short spindle (similar to that in Fig. 1e). This spindle rapidly achieves a length approximately equal to the diameter of the nucleus, further elongation being delayed until the end of G_2. During the stage preceding spindle elongation, the nucleus becomes deformed by its migration into the neck of the bud. At this point, the spindle often lies in the bud neck but is too short to reach the outer edges of the dumbbell-shaped nucleus. Later, when DNA replication and other mitotic functions have been completed, spindle elongation is renewed, and the poles (with attached nuclear material) are driven to the distal margins of the mother and daughter cell. The ensuing nuclear division and cytokinesis events leave one pole of the spindle in each cell.

Integration of these functions with other aspects of the cell cycle have been elucidated, in part, by studies of cell division cycle (*cdc*) mutants (Goetsch and Byers, 1974; 1975a). Temperature-sensitive mutations in the *cdc* genes were isolated by Hartwell *et al.*, (1973) and found to cause cells to achieve uniform arrest in characteristic states of the cell division cycle. Electron microscopy revealed that the SPB cycle was similarly subject to uniform states of arrest (Byers and Goetsch, 1974). Correlations between these various features have suggested dependent relationships between stages of spindle pole duplication and other landmark events in the cell cycle (Fig. 4; reviewed by Pringle and Hartwell, 1981).

One of the earliest states of arrest in the cell cycle is that caused either by the mating pheromones or by the so-called "START" mutations, such as *cdc28.* These cells remain unbudded and fail to initiate DNA replication (G_1 arrest). Their SPBs are arrested prior to duplication but after appearance of the satellite. Accordingly, one concludes that arrest prior to the crucial cell-cycle-initiating function termed "START" includes arrest of SPB development at the state immediately preceding its duplication. Another set of mutant genes, including *cdc4* and *cdc34,* causes a failure to initiate DNA replication, but the arrested cells undergo multiple cycles

Fig. 2. Spindle pole bodies (arrows) isolated from *Saccharomyces cerevisiae* and negatively stained for electron microscopy after incubation in buffer to permit depolymerization of attached microtubules, shown (a) before and (b) after addition of exogenous tubulin to permit nucleation of microtubule assembly *in vitro.* Bar: (a) 0.2 μm; (b) 1.0 μm. [Reproduced from Byers *et al., Journal of Cell Science* (1978), **30,** 331–352, by copyright permission of The Company of Biologists, Limited.]

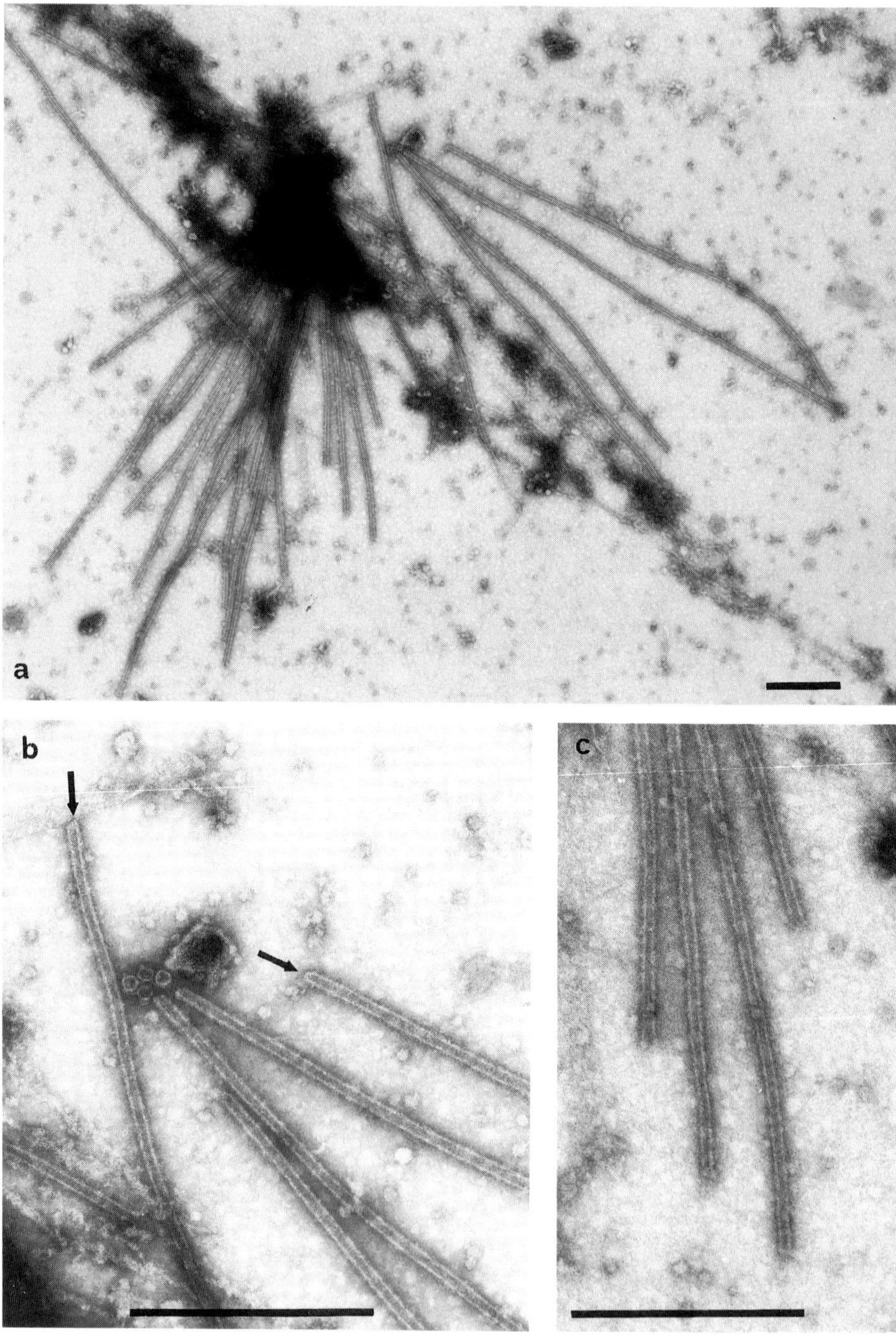
a
b
c

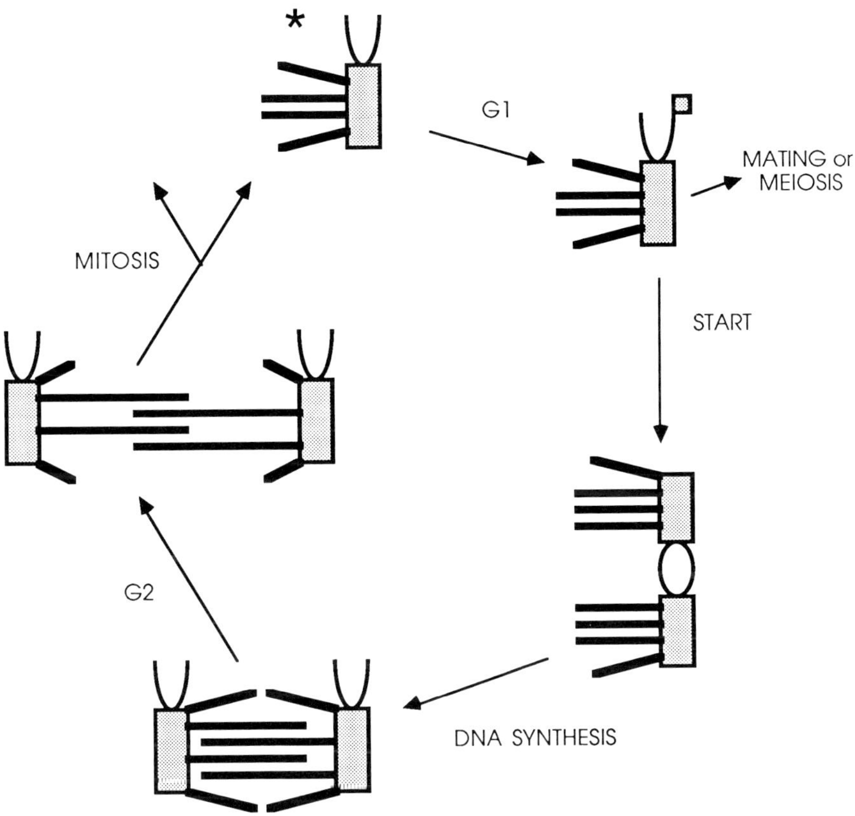

Fig. 4. Schematic representation of spindle pole body stages throughout the mitotic cell cycle of *Saccharomyces cerevisiae.* The stage diagrammed at the top (*) is found both in stationary phase cells and in the earlier portion of the ongoing cell cycle, before the satellite appears on the half-bridge (G_1 stage). Mating (in haploids) and meiosis (in diploids) may be initiated at this stage. Duplication coincides with start. The side-by-side stage is found during the period of DNA synthesis. Scission of the bridge permits the two daughter spindle pole bodies to become poles of the short spindle present during medial division. Spindle elongation (mitosis) immediately precedes cell division.

Fig. 3. (a) Partial disruption of endogenous spindle microtubules from an isolated spindle pole body by increased exposure to negative stain reveals, at higher magnification, that (b) the proximal ends of the microtubules (arrows) have a "closed" structure, whereas (c) the distal ends display the "open" configuration usually seen on the ends of microtubules. Bars: 0.5 μm. [Reproduced from Byers *et al., Journal of Cell Science* (1978), **30,** 331–352, by copyright permission of The Company of Biologist, Limited.]

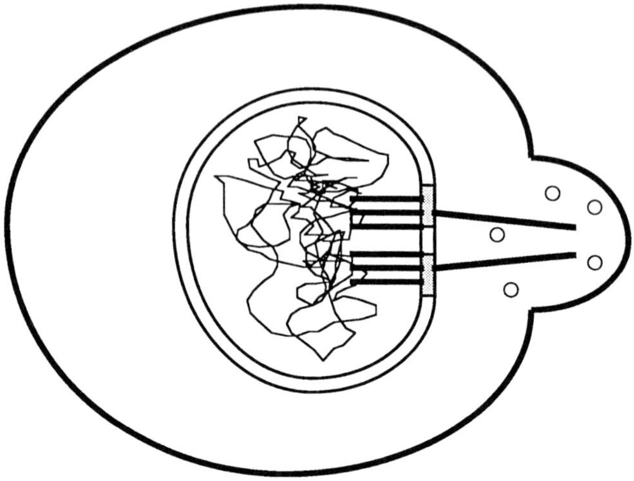

Fig. 5. Schematic diagram of a yeast cell that has just duplicated its spindle pole bodies as previously described (Moens and Rapport, 1971; Byers and Goetsch, 1975a). Bud emergence occurs adjacent to the newly formed side-by-side spindle pole bodies. Typically, cytoplasmic microtubules extend from the spindle pole bodies into the base of the new bud, where vesicles implicated in bud growth are concentrated.

of bud emergence. Here, the SPBs have undergone duplication but are unable to separate from one another to form a mitotic spindle. On the other hand, arrest at a later stage in G_1 by the *cdc7* mutation is accompanied by separation of the duplicated SPBs to form a mitotic spindle (Byers and Goetsch, 1974). Similarly, mitotic spindles of various lengths are found in the remaining *cdc* mutants that arrest the cell cycle in S phase, G_2, or mitosis.

B. Meiosis

The cytology of meiosis in *S. cerevisiae* has been described by Moens and Rapport (1971) and by Zickler and Olson (1975). As in mitotic cells, meiotic pole duplication proceeds to the same sort of intermediate stage, in which side-by-side poles are interconnected by the complete bridge; the poles remain in this state throughout meiotic prophase. Eventually, these SPBs undergo separation to form either pole of the meiosis I spindle. Each pole of the MI spindle duplicates in turn to yield two side-by-side poles, which then separate from one another to form a meiosis II spindle. Modification of the outer plaque on the MII poles (see Section II) may arise before separation of the duplicated poles, but initiation of the prospore wall does not seem to occur until after the MII spindles have formed.

The study of *cdc* mutants in meiosis has demonstrated that each mutation usually causes arrest at a characteristic stage of meiosis, but the stage of meiotic arrest often differs from that occurring in mitosis. For instance, *cdc28* mutants, which become arrested in mitosis with a satellite-bearing pole, are arrested in meiosis after premeiotic S phase has occurred but before the duplicated poles have separated to form a spindle for meiosis I (Shuster and Byers, 1989). A similar state of meiotic arrest is seen also for mutants in *cdc2, cdc7, cdc8,* and *cdc21,* which are defective in DNA synthesis both in mitosis and meiosis (Schild and Byers, 1978). Strikingly, the mitotic arrests caused by these mutations are characterized by the formation of a complete spindle (Byers and Goetsch, 1974). Similarly, inhibition of DNA synthesis by addition of hydroxyurea results in the same respective states of arrest: a complete spindle in mitotic cells and side-by-side SPBs in meiotic cells (Schild and Byers, 1978). One must conclude that scission of the bridge and spindle formation are not intrinsically dependent on DNA replication, but are coupled to it by mechanisms specific to the mitotic or meiotic pathways.

C. Karyogamy

Mating of *S. cerevisiae* entails cellular fusion and subsequent karyogamy (nuclear fusion). Mating is initiated when cells of opposite mating types sense the proximity of potential mating partners by detecting the mating pheromones they emit (reviewed by Cross *et al.,* 1988). The cells respond by arresting in the G_1 phase of the cell cycle; each cell displays a single satellite-bearing SPB. This type of pole persists throughout the stages of cellular fusion: adhesion of the mating partners, dissolution of the intervening cell walls, and mixing of their cytoplasms. Once cytoplasmic mixing has occurred, the cytoplasmic microtubules emanating outward from either SPB are seen to interact in a manner that appears important for the repositioning of the two nuclei in preparation for karyogamy (Byers and Goetsch, 1974, 1975a).

At this point, as in the fertilization of an egg by a sperm cell, two critical functions must be performed before the diplophase is fully established: karyogamy (fusion of the two haploid genomes) and the establishment of a single centrosome equivalent (in this case, the SPB). The SPB of the new diploid might conceivably arise by any of several mechanisms, such as sole inheritance from one parent or the other or action of each of the parental SPBs as one pole in the initial mitosis of the diploid. Instead, the two SPBs become fused into a single larger one, which subsequently undergoes a typical cycle of duplication as the zygote enters its first mitotic cycle (Byers and Goetsch, 1975). Although the specific mechanism of SPB fusion remains obscure, it appears that the two organelles approach one another at a roughly perpendicular orientation and fusion is initiated on one edge of either body, so the bridges and their satellites become fused before the SPBs themselves are joined. Fusion of the SPBs in this manner serves as the initial stage of the subsequent fusion of the nuclei upon which they are borne. Therefore, karyogamy in yeast is mechanistically dependent on fusion of the centrosome equivalents.

Experimental evidence that proper karyogamy depends on a specific program of mating-related events (rather than simply the admixture of two nuclei in a common cytoplasm) is provided by the work of Rose *et al.*, (1986), who showed that when spheroplasts of cells of the same mating type are artificially fused, their nuclei undergo fusion rather inefficiently. There is, however, a substantial enhancement in the frequency of successful karyogamy in these experiments if the two haploid cultures are previously subjected to treatment with mating pheromone. This facilitation of karyogamy by mating factor depends, in part, on arrest of the SPB cycle at the appropriate stage but must include other effects as well, because simply arresting the cells at a morphologically similar G_1 state by use of the *cdc28* mutation does not yield a similar enhancement of karyogamy.

IV. GENETIC ANALYSIS OF THE SPINDLE POLE BODY CYCLE

A. Spindle Pole Body Fusion

Some of the genetic mechanisms responsible for SPB fusion have been revealed in part by the characterization of mutants defective in karyogamy. The so-called karyogamy-defective (*kar*) mutants (Conde and Fink, 1976) remain capable of responding to mating factor and conjugating with a mating partner. The nuclei enter a common cytoplasm, but nuclear fusion fails and a heterokaryon is formed. Mutations causing defects in karyogamy may be either unilateral or bilateral. A unilateral *kar* mutation need only be carried in one of the two parental strains for karyogamy to fail to occur (Conde and Fink, 1976); bilateral *kar* phenotypes require its presence in both mating partners (Trueheart *et al.*, 1987). The favored interpretation of these different phenotypes is that unilateral *kar* mutants are defective in gene functions that have been expressed before cell fusion, so wild-type gene product from the mating partner arrives too late to rescue the defective nucleus (Dutcher and Hartwell, 1983). Two of the original *kar* mutants (*KAR1* and *KAR2*) fall into this category (Conde and Fink, 1976; Polaina and Conde, 1982). Bilateral karyogamy gene functions, in contrast, must remain reparable by the introduction of wild-type gene product after cell fusion.

The bilateral karyogamy phenotype was first described for *bik1* (bilateral karyogamy; Trueheart *et al.*, 1987) but the function of this gene remains unknown. More readily interpretable are the bilateral karyogamy defects of some *tub2* mutations, which cause defects in β-tubulin (Huffaker *et al.*, 1988). The bilateral nature of their defects confirms that microtubules play a crucial role in karyogamy and can fulfill that role even if they are unable to form normally until after cytoplasms have mixed. Microtubules previously had been implicated in karyogamy on the basis of cytological analysis (Byers and Goetsch, 1975a) and the observed karyogamy defects caused by the microtubule-depolymerizing drug benomyl (Delgado and Conde, 1984).

The analysis of the *KAR3* gene is an interesting demonstration of the involvement of a diffusible gene product in karyogamy. The original allele, *kar3-1,* was isolated as a semidominant unilateral *kar* mutation (Polaina and Conde, 1982). However, when the gene was isolated and a disruption allele was constructed, this viable mutant did not act unilaterally, but displayed a strong bilateral defect (Meluh and Rose, 1990). These results suggest that the *KAR3* gene product is diffusible in heterokaryons, so unilateral matings (deletion with wild-type) can be cured, whereas *kar3-1* is a dominant negative mutation that produces a "poisonous" product.

This interpretation is supported by the molecular analysis of the *KAR3* gene and its product (Meluh and Rose, 1990). The inferred protein sequence suggests that the *KAR3* protein is similar to kinesin, a microtubule-binding motor (force-generating) protein. Indeed, the *KAR3* protein has been localized to the cytoplasmic microtubules in yeast. Analysis of the *kar3-1* gene has shown that the mutation lies within a putative ATP-binding site. An inference from the genetic analysis is that the mutant protein may bind to the microtubules but not function there, hence interfering with the function of any wild-type product that might be present. Although the normal function of the *KAR3* protein is suggested by its similarity to kinesin, there is an unusual twist in its organization. Whereas the globular ATP-binding domain is located at the amino terminus in kinesin, it resides at the carboxy terminus of the *KAR3* protein. How this difference is related to function remains unknown. Nonetheless, the behavior of *kar3* confirms the importance of cytoplasmic microtubules for karyogamy and suggests an active role for cytoplasmic microtubules in the associated nuclear movements.

Whereas some bilateral *kar* mutations like *KAR3* appear to identify microtubule-related functions necessary for nuclear fusion, one might hope that the class of unilateral *kar* mutants would include some mutants that encode defective components of the SPBs. Pertinent to this logic, when Dutcher and Hartwell (1982) reported karyogamy defects in some of the *cdc* mutants, including *cdc28, cdc4, cdc34,* and *cdc37,* they suggested that the observed unilateral phenotypes might reflect defects specific to the SPB. More recently, several of these genes have been isolated and their DNA sequence determined, revealing that *CDC28* encodes the *S. cerevisiae* functional homolog of the *cdc2* kinase (Beach *et al.,* 1982; Lorincz and Reed, 1984), *CDC4* encodes a protein with sequence similarities to β-transducin (Fong *et al.,* 1986; Yochem and Byers, 1987) and *CDC34* encodes a ubiquitin-conjugating enzyme (Goebl *et al.,* 1988). Whether or not any of these functions plays a direct role in control of the SPB is not yet evident. Furthermore, immunological localization of the gene products encoded by *CDC4* (S. Johnson and B. Byers, personal communication) and *CDC4* (M. Goebl and B. Byers, personal communication) to the entire nucleus, rather than exclusively to the SPB, suggests that the relationship between these gene functions and SPB behavior may be indirect. Similarly, the finding that *KAR2* encodes a 70K heat-shock protein that is the *S. cerevisiae* BiP/GRP78 homolog (Rose *et al.,* 1989), and is localized in the endoplasmic reticulum, also fails to provide evidence that this gene product is a structural component of the SPB. The *CDC37* gene remains uncharacterized. These gene products may well serve to prepare the nucleus for karyogamy (Dutcher and Hartwell, 1983), for example, by promoting induction of the karyogamy-competent state in response to mating factor (Rose *et al.,* 1986), rather than to participate in nuclear fusion.

KAR1, on the other hand, appears to be directly involved in the structure of the SPB (Rose and Fink, 1987). Whereas the original allele (*kar1-1*) is defective only in karyogamy, alleles isolated more recently have demonstrated an essential role for the gene in SPB duplication (see Section IV,B). Furthermore, molecular analysis demonstrates localization of the gene product to the SPB (L. Vallen and M. Rose, personal communication). The cloned gene was used to construct chimeric *KAR1::lacZ* genes that, when expressed in yeast, result in localization of the *lacZ* epitope to the vicinity of the SPB (L. Vallen and M. Rose, personal communication). This localization, in conjunction with the genetic evidence for an essential role for *KAR1* in SPB duplication, strongly indicates that the *KAR1* product serves as a crucial component of the SPB.

B. Spindle Pole Body Duplication

Another aspect of SPB behavior that demands genetic analysis is the mode of its duplication, which appears to follow similar pathways in mitotic and meiotic cells. In both situations, the duplication mechanism appears to be a conservative process, the existing pole remaining intact while a new pole is assembled adjacent to it. The satellite seen during G_1, as in cells arrested by treatment with mating pheromone, is thought to be the precursor of the new pole. Not only does the satellite make its appearance just prior to the new pole, but it also resides at the site at which the new pole will appear. Additional evidence for a conservative mechanism of SPB duplication derives from studies of the pole-specific localization of *KAR1::lacZ* fusion proteins (L. Vallen and M. Rose, personal communication; see Section IV,A). Their detailed analysis has revealed that only one SPB in each cell shows this mode of localization, suggesting that only one pole (presumably the newly forming pole) is capable of incorporating the chimeric *KAR1::lacZ* protein into its structure.

Further evidence for conservative duplication has been provided by cytological and genetic analysis of meiotic cells. When sporulating cells were subjected to incubation at 36.5°C, they became arrested in pachytene; release from this arrest yielded a high proportion of asci containing only two spores instead of the usual four (Davidow *et al.,* 1980). Genetic analysis of these spores showed that they were almost exclusively haploids containing nonsister centromeres. This segregation pattern could be explained if both meiotic divisions had occurred, but only one product of each of the MII divisions was packaged into a spore. Electron microscopic analysis of cells in meiosis under these conditions revealed that both MI and MII spindles were indeed formed, but the dense outer plaque of the MII SPBs was absent from one pole in each MII spindle. This outer plaque-free SPB failed to initiate spore wall formation adjacent to itself. It is apparent from these observations that spore morphogenesis is regulated at the level of SPB modification, a phenomenon limited to one pole per MII spindle. This differential modification of MII SPBs, like the differential labeling of mitotic SPBs by *KAR1*::*lacZ,* is consistent with a conservative duplication mechanism enabling the cell to distinguish one pole from the other.

The mechanism of pole duplication, conservative or not, might be resolved by identifying the components of the system and understanding how they function. Several workers have attempted to identify yeast mutants defective in the process and infer the functions of the genes from phenotypic analysis and determination of gene structure. One important class of conditional mutants contains those that fail to duplicate their SPBs at the nonpermissive temperature and therefore form monopolar spindles. These include *cdc31* (Byers, 1981b; Schild *et al.,* 1981) and *KAR1* (Rose and Fink, 1987). For both mutants, cells subjected to nonpermissive conditions become arrested in a state with a large bud but only one SPB. Mutations in these genes also cause defects in chromosome segregation and microtubule organization.

The formation of monopolar spindles was first analyzed in *cdc31* (Byers, 1981b). Here, the existing SPB was seen to grow larger during the arrest, suggesting that the defect was specific to duplication per se, and not the assembly of the materials that constitute the spindle pole. Kinetic analysis suggested that *CDC31* was required for satellite deposition, because no satellites were detected after the mutant cells were shifted to the nonpermissive condition. Conceivably, the failure to form satellites was directly responsible for the consequent failure to produce a daughter pole. Many of the cells remained able to undergo a monopolar mitosis, in which all the chromosomes segregated to the single pole to yield cells of twice the original ploidy (Schild *et al.,* 1981). Also, diploid strains homozygous for the *cdc31* mutation were found to fail in MII, yielding two diploid spores instead of the normal four haploid spores (Byers, 1981b). Therefore *CDC31* is required for SPB duplication in both mitotic and meiotic divisions.

Baum *et al.,* (1986) isolated the *CDC31* gene and found it to be essential for viability. Sequence analysis showed that it encodes a small Ca^{2+}-binding protein similar to calmodulin. A *CDC31*::*trpE* fusion protein purified from *Escherichia coli* was able to bind calcium, whereas trpE itself did not, so the calcium binding could be attributed to the CDC31 moiety (P. Baum and B. Byers, unpublished results). Recently, Huang *et al.,* (1988) isolated the gene encoding caltractin (also referred to as centrin), a small Ca^{2+}-binding protein found in the basal bodies of *Chlamydomonas,* and found that it displays significant similarity to CDC31. However, attempts to demonstrate functional homology by complementation of yeast *cdc31* mutations with the *Chlamydomonas* gene have failed (B. Huang, personal communication). The functional significance of the similarity between these two proteins, as well as the role calcium may play in spindle pole duplication, remains unclear.

Alleles of *KAR1* (discussed previously) were also found to display a similar defect in SPB duplication (Rose and Fink, 1987). Specifically, when strains containing a conditional allele (temperature-sensitive) were shifted to the nonpermissive temperature, they arrested as large budded cells with monopolar spindles, similar to those is shown for *cdc31* mutants. Interestingly, overproduction of the *KAR1* gene product by use of an inducible promoter in yeast leads to an arrest of the cell cycle similar to the arrest caused by the conditional alleles, suggesting that SPB duplication is sensitive to either elevation or reduction in the level of the *KAR1* gene product.

A cold-sensitive allele of the *NDC1* gene has several phenotypes in common with the monopole-forming mutants just described (Thomas and Botstein, 1986). Strains containing the *ndc1-1* mutation show diploidization after a transient shift to the nonpermissive condition in a manner similar to that shown for *cdc31.* An additional similarity is that diploids homozygous for *ndc1-1* often fail in meiosis II and yield diploid spores containing sister chromatids. However, *ndc1-1* strains incubated at the nonpermissive temperature and examined by immunofluorescence staining of microtubules appeared to have two SPBs, unlike the single SPB found in the monopolar mutants. This experiment showed, however, that some aspect of spindle formation was defective in *ndc1-1,* because the DNA in these cells, visualized by 4',6-diamidino-2-phenylindole (DAPI) staining, was associated with only one of the two apparent SPBs. This displacement of the DNA appears to reflect the asymmetric segregation defect leading to increased ploidy in *ndc1* mutants, similar to the increase in ploidy observed in the monopolar mutants. Thomas and Botstein (1986) suggested two models to explain the *ndc1-1* phenotypes. One is that NDC1 is required for kinetochore function, so loss of this function results in the failure of the sister chromosomes to separate from one another, leading to defects in spindle morphology secondarily. The second model suggests that the mitotic spindle is dysfunctional because of a defect in either the SPBs or the assembly or function of the spindle microtubules. Recent electron microscopic examination of *ndc1-1* strains in our laboratory supports the second model, in which SPB is seen to be aberrant, yielding a new SPB that is dysfunctional (M. Winey, L. Goetsch, and B. Byers, unpublished results).

A second class of pole duplication mutants is composed of those that cause the accumulation of extra SPBs at the nonpermissive condition. The mutation designated *esp1* (extra spindle pole bodies) was isolated from a screen for mutants that become diploid after a transient shift to the nonpermissive condition (Baum *et al.*, 1988). Although this selection was expected to yield new mutants that failed to duplicate their SPBs, electron microscopic analysis of arrested *esp1-1* strains showed cells with extra SPBs (some cells had as many as nine poles in a single nucleus). These extra poles appeared to be cytologically normal and all were associated with cytoplasmic and nuclear microtubules.

How, then, did this multipolar state arise? One model suggests that poles are duplicated normally, but are not segregated correctly. A second model is that negative regulation of pole duplication may simply be lost in the absence of *ESP1* gene function. Kinetic analysis of *esp1-1* strains suggests that the terminal phenotype results from both a pole segregation defect and a subsequent loss of negative control (P. Baum, L. Goetsch, and B. Byers, unpublished results). Analysis of cells progressing into the cell cycle immediately after loss of *ESP1* gene function (at the nonpermissive condition) has shown that a mitotic spindle of normal appearance is formed. However, when it should elongate (at anaphase), it collapses, and the two halves of the spindle become disconnected from one another. This event may result in the segregation of both poles to the daughter cell. In any event, duplication is henceforth deregulated, and the multipolar state ensues.

This aberrant pattern of SPB segregation has disastrous effects on chromosome segregation. The missegregation dependent on the *esp* defect is so severe that flow cytometry of *esp1-1* cells shifted to the nonpermissive temperature shows rapid decay from normal ploidy to a wide mixture of hypo- and hyperploid cells (P. Baum and B. Byers, personal communication). Similarly, in *S. pombe,* Hirano *et al.,* (1986) have reported isolation of mutations in 10 *cut* genes, all of which display a similar mitotic failure causing substantial lethality following aberrant chromosome segregation. Evidence for a more fundamental similarity between the *cut* mutants and *esp1-1* is found in the fact that the derived amino acid sequences of ESP1 and CUT1 are nearly the same length and exhibit substantial similarity (51% identity in a carboxy-terminal segment of 99 amino acids; Uzawa *et al.*, 1990). The specific functions of the *ESP1* and *cut1* genes remain unknown, but it is expected that further analysis of these types of mutants will lend insight into the mitotic mechanism and the control of spindle pole duplication.

The foregoing analyses of both monopolar and multipolar mutants have been directed toward gaining an understanding of the genes required for SPB duplication, but they also have proven instructive with regard to the integration of the SPB cycle with other cell cycle functions. In particular, we note that the cell cycle can proceed to the G_2/M boundary in the absence of pole duplication in a *cdc31* mutant, although pole duplication normally is accomplished early in the cell cycle, before the onset of DNA synthesis. This clearly demonstrates that pole duplication per se is not a requisite precondition for bud emergence or DNA synthesis (Byers, 1981b).

These studies also provide insight into the relationship between two SPB-related functions: fusion and duplication. It is interesting to note that the point in the pole duplication cycle at which poles can become committed to karyogamy is an intermediate stage in duplication. The satellite-bearing SPB present at this point may mark a state of the pole crucial for its commitment either to fusion or to duplication. This commitment may well involve *KAR1,* which is involved in both pathways (Rose and Fink, 1987). Additionally, the study of the *cdc* mutants in mitotic and meiotic cells suggests connections between these two aspects of the spindle pole life cycle. A complete elaboration of the SPB life cycle and its interaction with other aspects of the life cycle of the cell will require considerable further study.

V. SPINDLE POLE BODY COMPONENTS

A central issue regarding the genes implicated in the SPB cycle is the possible localization of their products. It seems likely that some of these products are themselves components of the SPB but exploration of this question has proven technically difficult to pursue. Among the genes with demonstrable roles in spindle pole duplication (see Section IV,B), evidence for such localization has been put forth only for the *KAR1* gene product (L. Vallen and M. Rose, personal communication). In this case, it was shown that the product of an overexpressed *KAR1::lacZ* chimeric gene could be localized by immunofluorescence staining with anti-LACZ antibodies. Although promising, this result requires confirmation both by detection of the endogenous *KAR1* gene product and by the use of immunoelectron microscopy. It remains to be established whether the products of the other genes—such as *CDC31, NDC1,* and *ESP1*—that play crucial roles in SPB behavior are also components of the pole.

Other proteins that may serve as components of the SPB have been identified by immunochemical methods. In one approach, yeast genes of potential interest were identified by screening for yeast proteins that cross-react with human autoimmune sera specific to centrosomes. The spindle pole antigen genes, *SPA1* and *SPA2,* were isolated from a yeast expression library by probing with these antibodies (Snyder and Davis, 1988; Snyder, 1989). When the antigens were localized in yeast cells, SPA1 was found, by cell fractionation, to reside in the nuclear envelope (Snyder and Davis, 1988) whereas SPA2 localized to sites of cell growth inconsistent with *SPA2* encoding a SPB component (Snyder, 1989). Disruption of the *SPA1* gene did not cause inviability, so the protein must not be essential to spindle pole function. However, the deletion strains were sick and showed increased levels of chromosome loss and binucleate cells, suggesting possible defects in the behavior of the SPB or mitotic spindle. Immunological staining of the *SPA1* gene product indicates that the protein does not reside entirely in the SPB but is found at several sites in the cell (M. Snyder, personal communication). Therefore, the role of *SPA1* remains obscure.

A second immunochemical approach has entailed partial purification of the SPB and use of these preparations to produce antibodies that might have specificity to its constituents. Initial attempts at purification by gradient fractionation, using microtubule-nucleation activity to identify SPBs, demonstrated substantial enrichment (as in Fig. 2; Byers *et al.,* 1978). There have also been attempts to affinity purify SPBs using antibodies against tubulin (Kilmartin and Fogg, 1982). Recently, Rout and Kilmartin (1990) identified two components (90K and 110K) of the SPB by use of monoclonal antibodies prepared against a sedimentation gradient fraction enriched 600-fold for SPBs. Further characterization of these components, which are found to be proteins of 90K and 110K, is eagerly awaited.

VI. PERSPECTIVES

Cytological analysis of the SPB in wild-type and mutant yeast strains has illuminated numerous features of the behavioral repertoire exhibited by this centrosome-like organelle. Although the molecular underpinnings of these behaviors remain obscure, the ability to couple genetic analysis with molecular biology and immunochemical approaches promises to provide specific information about the essential components of the SPB. The insights gained from studies in this simple eukaryote, in which molecular genetics is so accessible, should help frame the right questions in the search for related mechanisms in centrosomes of greater complexity. In addition, evolutionarily conserved mechanisms may be directly revealed by these studies. The use of complementary approaches involving biochemistry in higher eukaryotes in conjunction with genetics in yeast has proven invaluable to the study of many conserved biological phenomena. A notable recent example is the manner in which our understanding of cell cycle control has benefitted from the knowledge that a homolog of the yeast the *cdc2*/p34 kinase is an essential component of M-phase promoting factor (Nurse, 1990). Integrated approaches of this sort should prove equally useful in the analysis of centrosomes and equivalent organelles.

ADDENDUM

Molecular analysis of the *BIK1* (bilateral karyogamy defect) gene product has demonstrated that this protein colocalizes with microtubule structures in yeast (Berlin *et al.,* 1990). In addition, γ-tubulin has been shown to be a component of centrosomes and spindle pole bodies in a wide variety of organisms, including budding and fission yeasts (Zheng *et al.*, 1991; Stearns *et al.*, 1991). Finally, new mutants that give rise to defects in spindle pole body duplication have been isolated and found to identify distinct steps in the duplication pathway (Winey *et al.*, 1991).

ACKNOWLEDGMENTS

We would like to thank the members of our laboratory for discussions and comments on the manuscript. We are indebted to Loretta Goetsch for supplying some of the electron micrographs presented. We appreciate the communication of unpublished results by Liz Vallen and Mark Rose; Mike Snyder; Bessie Huang; Mike Rout and John Kilmartin; Mitsuhiro Yanagida; Loretta Goetsch, Mark Goebl, Steve Johnson, and Peter Baum. This work has been supported by NIH postdoctoral grant GM12701 to M.W. and NIH grant GM18541 to B.B.

REFERENCES

Baum, P., Furlong, C., and Byers, B. (1986). Yeast gene required for spindle pole body duplication: Homology of its product with Ca^{2+}-binding proteins. *Proc. Natl. Acad. Sci. U.S.A.* **83,** 5512–5516.

Baum, P., Yip, C., Goetsch, L., and Byers, B. (1988). A yeast gene essential for regulation of spindle pole duplication. *Mol. Cell. Biol.* **8,** 5386–5397.

Beach, D., Durkacz, B., and Nurse, P. (1982). Functionally homologous cell cycle control genes in budding and fission yeast. *Nature (London)* **300,** 706–709.

Berlin, V., Styles, C. A., and Fink, G. R. (1990). BIK1, a protein required for microtubule function during mating and mitosis in *Saccharomyces cerevisiae,* colocalizes with tubulin. *J. Cell Biol.* **111,** 2573–2586.

Byers, B. (1981a). Cytology of the yeast life cycle. *In* "Molecular Biology of the Yeast *Saccharomyces.* I. Life Cycle and Inheritance" (J. N. Strathern, E. W. Jones, and J. R. Broach, eds.), pp. 59–96. Cold Spring Harbor Laboratory Press, Cold Spring Harbor, New York.

Byers, B. (1981b). Multiple roles of the spindle pole bodies in the life cycle of *Saccharomyces cerevisiae. In* "Molecular Genetics in Yeast" (D. von Wettstein, A. Stenderup, M. Kielland-Brandt, and J. Friis, eds.), Alfred Benzon Symp. **16,** pp. 119–133. Munksgaard, Copenhagen.

Byers, B., and Goetsch, L. (1974). Duplication of spindle plaques and integration of the yeast cell cycle. *Cold Spring Harbor Symp. Quant. Biol.* **38,** 123–131.

Byers, B., and Goetsch, L. (1975a). Behavior of spindles and spindle plaques in the cell cycle and conjugation of *Saccharomyces cerevisiae. J. Bacteriol.* **124,** 511–523.

Byers, B., and Goetsch, L. (1975b). Electron microscopic observations in the meiotic karyotype of diploid and tetraploid *Saccharomyces cerevisiae. Proc. Natl. Acad. Sci. U.S.A.* **72,** 5056–5060.

Byers, B., Shriver, K., and Goetsch, L. (1978). The role of spindle pole bodies and modified microtubule ends in the initiation of microtubule assembly in *Saccharomyces cerevisiae. J. Cell Sci.* **30,** 331–352.

Conde, J., and Fink, G. R. (1976). A mutant of *Saccharomyces cerevisiae* defective for nuclear fusion. *Proc. Natl. Acad. Sci. U.S.A.* **73,** 3651–3655.

Cross, F., Hartwell, L. H., Jackson, C., and Konopka, J. B. (1988). Conjugation in *Saccharomyces cerevisiae. Ann. Rev. Cell Biol.* **4,** 429–457.

Davidow, L. S., Goetsch, L., and Byers, B. (1980). Preferential occurrence of nonsister spores in two-spored asci of *Saccharomyces cerevisiae:* Evidence for regulation of spore-wall formation by the spindle pole body. *Genetics* **94,** 581–595.

Delgado, M. A., and Conde, J. (1984). Benomyl prevents nuclear fusion in *Saccharomyces cerevisiae. Mol. Gen. Genet.* **193,** 188–189.

Dutcher, S. K., and Hartwell, L. H. (1982). The role of *S. cerevisiae* cell division cycle genes in nuclear fusion. *Genetics* **100,** 175–184.

Dutcher, S. K., and Hartwell, L. H. (1983). Genes that act before conjugation to prepare the *Saccharomyces cerevisiae* nucleus for caryogamy. *Cell* **33,** 203–210.

Fong, H. K. W., Hurley, J. B., Hopkins, R. S., Miake, R., Johnson, M. S., Doolittle, R. F., and Simon, M. I. (1986). Repetitive segmental structure of the transducin beta subunit: Homology with the *CDC4* gene and identification of related mRNAs. *Proc. Natl. Acad. Sci. U.S.A.* **83,** 2162–2166.

Goebl, M. G., Yochem, J., Jentsch, S., McGrath, J. P., Varshavsky, A., and Byers, B. (1988). The yeast cell cycle gene *CDC34* encodes a ubiquitin-conjugating enzyme. *Science* **241,** 1331–1335.

Hartwell, L. H., Mortimer, R. K., and Culotti, M. (1973). Genetic control of the cell division cycle in yeast. V. Genetic analysis of *cdc* mutants. *Genetics* **74,** 267–286.

Heidemann, S. R., and McIntosh, J. R. (1980). Visualization of the structural polarity of microtubules. *Nature (London)* **286,** 517–519.

Hirano, T., Funahashi, S., Uemura, T., and Yanagida, M. (1986). Isolation and characterization of *S. pombe cut* mutants that block nuclear division but not cytokinesis. *EMBO J.* **5,** 2973–2979.

Huang, B., Mengersen, A., and Lee, V. D. (1988). Molecular cloning of cDNA for caltractin, a basal body-associated Ca^{2+}-binding protein: Homology in its protein sequence with calmodulin and the yeast *CDC31* gene product. *J. Cell Biol.* **107,** 133–140.

Huffaker, T. C., Thomas, J. H., and Botstein, D. (1988). Diverse effects of β-tubulin mutations on microtubule formation and function. *J. Cell Biol.* **106,** 1997–2010.

Hymans, J. S., and Borisy, G. G. (1978). Nucleation of microtubules *in vitro* by isolated spindle pole bodies of the yeast *Saccharomyces cerevisiae. J. Cell Biol.* **78,** 401–414.

Kanbe, T., Kobayashi, I., and Tanaka, K. (1989). Dynamics of cytoplasmic organelles in the cell cycle of the fission yeast *Schizosaccharomyces pombe:* Three-dimensional reconstruction from serial sections. *J. Cell Sci.* **94,** 647–656.

Kilmartin, J. V., and Fogg, J. (1982). Partial purification of yeast spindle pole bodies. *In* "Microtubules in Microorganisms" (P. Cappuccinelli and N. R. Morris, eds.), pp. 157–170. Marcel Dekker, New York.

Lorincz, A. T., and Reed, S. I. (1984). Primary structure homology between the product of yeast cell division control gene *CDC28* and vertebrate oncogenes. *Nature (London)* **307,** 183–185.

Meluh, P. B., and Rose, M. D. (1990). *KAR3,* a kinesin-related gene required for yeast nuclear fusion. *Cell* **60,** 1029–1041.

Moens, P. B., and Rapport, E. (1971). Spindles, spindle plaques, and meiosis in yeast *Saccharomyces cerevisiae* (Hansen). *J. Cell Biol.* **50,** 344–361.

Nurse, P. (1990). Universal control mechanism regulating onset of M-phase. *Nature (London)* **344,** 503–508.

Peterson, J. B., Gray, R. H., and Ris, H. (1972). Meiotic spindle plaques in *Saccharomyces cerevisiae J. Cell Biol.* **53,** 837–841.

Peterson, J. B., and Ris, H. (1976). Electron-microscopic study of the spindle and chromosome movement in the yeast *Saccharomyces cerevisiae. J. Cell Sci.* **22,** 219–242.

Polaina, J., and Conde, J. (1982). Genes involved in the control of nuclear fusion during the sexual cycle of *Saccharomyces cerevisiae. Mol. Gen. Genet.* **186,** 253–258.

Pringle, J. R., and Hartwell, L.H. (1981). The *Saccharomyces cerevisiae* cell cycle. *In* "Molecular Biology of the Yeast *Saccharomyces.* I. Life Cycle and Inheritance" (J. N. Strathern, E. W. Jones, and J. R. Broach, eds.), pp. 97–142. Cold Spring Harbor Laboratory Press, Cold Spring Harbor, New York.

Robinow, C. F., and Marak, J. (1966). A fiber apparatus in the nucleus of the yeast cell. *J. Cell Biol.* **29,** 129–151.

Rose, M. D., Price, B. R., and Fink, G. R. (1986). *Saccharomyces cerevisiae* nuclear fusion requires prior activation by alpha factor. *Mol. Cell. Biol.* **6,** 3490–3497.

Rose, M. D., and Fink, G. R (1987). *KAR1,* a gene required for function of both intranuclear and extranuclear microtubules in yeast. *Cell* **48,** 1047–1060.

Rose, M. D., Misra, L. M., and Vogel, J. P. (1989). *KAR2,* a karyogamy gene, is the yeast homolog of mammalian BiP/GRP78 gene. *Cell* **57,** 1211–1221.

Rout, M. P., and Kilmartin, J. V. (1990). Components of the yeast spindle and spindle pole body. *J. Cell Biol.* **111,** 1913–1927.

Schild, D., and Byers, B. (1978). Meiotic effects of DNA-defective cell division cycle mutations of *Saccharomyces cerevisiae. Chromosoma (Berlin)* **70,** 109–130.

Schild, D., Ananthaswamy, H. N., and Mortimer, R. K. (1981). An endomitotic effect of a cell cycle mutation of *Saccharomyces cerevisiae*. *Genetics* **97,** 552–562.

Shuster, E. O., and Byers, B. (1989). Pachytene arrest and other meiotic effects of the start mutations in *Saccharomyces cerevisiae. Genetics* **123,** 29–43.

Snyder, M. (1989). The *SPA2* protein of yeast localizes to sites of cell growth. *J. Cell Biol.* **108,** 1419–1429.

Snyder, M., and Davis, R. W. (1988). *SPA1:* A gene important for chromosome segregation and other mitotic functions in *S. cerevisiae. Cell* **54,** 743–754.

Stearns, T., Evans, L., and Kirschner, M. (1991). γ-tubulin is a highly conserved component of the centrosome. *Cell* **65,** 825–836.

Tanaka, K., and Kanbe, T. (1986). Mitosis in the fission yeast *Schizosaccharomyces pombe* as revealed by freeze-substitution electron microscopy. *J. Cell Biol.* **80,** 253–268.

Thomas, J. H., and Botstein, D. (1986). A gene required for separation of chromosomes on the spindle apparatus in yeast. *Cell* **44,** 65–76.

Trueheart, J., Boeke, J. D., and Fink, G. R. (1987). Two genes required for cell fusion during yeast conjugation: Evidence for a pheromone-induced surface protein. *Mol. Cell. Biol.* **7,** 2316–2328.

Uzawa, S., Samejima, I., Hirano, T., Tanaka, K., and Yanagida, M. (1990). The fission yeast *cut1*$^+$ gene regulates spindle pole body duplication and has homology to the budding yeast *ESP1* gene. *Cell* **62,** 913–925.

Winey, M., Goetsch, L., Baum, P., and Byers, B. (1991). *MPS1* and *MPS2:* novel yeast genes defining distinct steps of spindle pole body duplication. *J. Cell Biol.* **114,** 745–754.

Yochem, J., and Byers, B. (1987). Structural comparison of the yeast cell division cycle gene *CDC4* and a related pseudogene. *J. Mol. Biol.* **195,** 233–245.

Zheng, Y., Jung, M. K., and Oakley, B. R. (1991). γ-tubulin is present in *Drosophila melanogaster* and *Homo sapiens* and is associated with the centrosome. *Cell* **65,** 817–823

Zickler, D., and Olson, L. W. (1975). The synaptonemal complex and the spindle plaque during meiosis in yeast. *Chromosoma* **50,** 1–23.

8

The Centrosome in Cell Division and Development of *Drosophila*

DAVID M. GLOVER

Cell Cycle Genetics Group
Cancer Research Campaign Laboratories
Department of Biochemistry
University of Dundee
Dundee DD1 4HN, Scotland

I. INTRODUCTION

Many of the issues that arose following the first description of the centrosome in 1887 by Van Beneden and Bovari remain unresolved. The suggested role of the centrosome in the initiation of embryonic development is poorly understood; the description of the role of the centrosome in the mitotic cycle has changed only in the level of its cytological resolution over the past century; and there remains a confusion of nomenclature that has persisted since the early investigators gave this organelle a large number of alternative names. In this chapter I will try at least to avoid this confusion by using the term centrosome to describe the entire organelle including both the centrioles—the cylindrical arrays of fused triplet microtubules—and the pericentriolar material that appears responsible for nucleating astral microtubules at interphase and spindle microtubules during mitosis.

Copyright © 1992 Academic Press, Inc.
All rights of reproduction in any form reserved.

II. MATERNAL CONTRIBUTION TO *DROSOPHILA* EGG

The process of oogenesis is critical in establishing the basic framework for the organization of the future embryo. It is the stage at which the shape of the egg is determined and in which positional cues are laid down that provide signals for subsequent zygotic development. Maternally active genes are required, for example, to establish correctly both anterior–posterior and dorsoventral polarity (see Anderson, 1989, for review). Oogenesis is a time during which many maternal factors are provided to the egg, not in the least the large dowry of proteins required for the rapid cleavage divisions of the newly fertilized embryo. The embryonic development of *Drosophila* is initiated by 13 rounds of rapid mitotic divisions, for which the mother must provide the embryo with all the necessary proteins. At least one centrosome-associated protein, the Bx63 antigen, is part of the maternal dowry (Frasch *et al.,* 1985; Whitfield *et al.,* 1988); it is likely that many other centrosomal components are provided in this way also. The Bx63 antigen is present throughout early embryonic development at an approximately constant level, as judged by Western blotting experiments. However, as development proceeds, it appears that gradually more and more of the antigen accumulates into functional centrosomes.

Yet the *Drosophila* egg is thought to begin its development with no centrosomes and, as with many other animal species, it has been generally accepted that the sperm provides centrosomes to the egg at fertilization. Oogenesis in *Drosophila* is initiated from stem cells that divide to produce daughter stem cells and cytoblasts. The cytoblast undergoes four synchronous mitoses with incomplete cytokinesis to produce a cyst of 16 connected cells. One of these cells will become the oocyte, whereas the others become the highly polyploid nurse cells. The nurse cells will provide the developing oocyte with its maternal dowry and then degenerate, their entire contents becoming part of the egg cytoplasm. As the nurse cells cease to divide, their centrioles are passed into the oocyte through a series of interconnecting cytoplasmic canals. The centrioles accumulate in a specific region of the oocyte and their subsequent fate is not clear (Mahowald and Strassheim, 1970; Carpenter, 1975). The incoming sperm is believed to contribute a centriole to the egg at fertilization. However, under certain conditions, functional centrosomes can be found in unfertilized eggs (see subsequent text). The precise nature of the maternal and paternal contributions of centrioles (or centrosomal material) therefore certainly requires detailed examination.

III. MITOSIS IN *DROSOPHILA* EMBRYO

Centrosome replication in *Drosophila* embryos is a process that normally occurs in synchrony with other mitotic events. As in the embryos of many other animal species, the mitotic divisions of newly fertilized *Drosophila* embryos occur in abbreviated cycles that are rapid successions of M and S phases. The embryo is a syncytium for the first 13 rounds of nuclear division which occur at approximately 10 min intervals. After the first 9 cycles of division, the majority of nuclei in the interior of the football-shaped embryo migrates to the cortex in a process that requires the integrity of microtubules (Zalokar and Erk, 1976). A small number of nuclei, the yolk nuclei, are left behind in the interior of the embryo. These drop out of the synchronous division cycles but continue replicating their DNA to become polyploid. The associated centrosomes are lost from the yolk nuclei at this time and their fate is not clear. At the surface, the majority of nuclei undergoes an additional 4 division cycles. Then, in interphase of cycle 14, individual cell membranes begin to grow down from the surface of the embryo around each of the nuclei (Zalokar and Erk, 1976; Foe and Alberts, 1983; Karr and Alberts, 1986; Warn *et al.,* 1987; Kellogg *et al.,* 1988). Thus in a period of about 2 hr the embryo becomes populated by about 6000 nuclei derived from the single zygotic nucleus formed by the fusion of the male and female pronuclei. Since all the proteins required to build several thousand sets of mitotic apparatus are maternally provided, it has been possible to isolate mutations that affect mitosis that show a maternal effect. Adult females homozygous for such mutations produce embryos that are unable to complete the cleavage divisions (see Glover, 1989, for a discussion of the isolation of such mutants).

The 14th mitotic cycle marks a significant change in the way mitosis is regulated and heralds the onset of embryonic morphogenesis. Gastrulation occurs immediately after cellularization; the newly formed cells of the embryo undergo precisely defined patterns of invaginations, movement, and migration. The posterior pole cells, which will become the germ line of the developing organism, are pushed dorsally and anteriorly by elongation of the embryo in its confined shell. The cell cycle lengthens at this phase of development and distinct G_2 periods appear. The mitotic divisions lose their synchrony and mitosis takes place only in a series of domains regulated both in time and position (Foe, 1989). The primary trigger for these divisions appears to be the gene *string,* the transcription of which anticipates the mitotic events (Edgar and O'Farrell, 1989; O'Farrell *et al.,* 1989). *string* is the homolog of the mitotic gene *cdc25* of the fission yeast, *Schizosaccharomyces pombe.* It is required for the activation of the $p34^{cdc2}$ kinase, an enzyme essential for the G2–M transition in all eukaryotes (Nurse, 1990). We shall see later in this chapter that the divisions that occur either late in embryogenesis or during larval development are subject to additional controls that permit the coordination of cell proliferation with development. This higher level of regulation extends to several aspects of the mitotic cycle including centrosome behavior.

IV. CENTROSOME DUPLICATION IN SYNCYTIAL EMBRYOS

The centrosome cycle can be conveniently followed in syncytial *Drosophila* embryos by immunofluorescence using antibodies to antigens such as Bx63 (Frasch *et al.,* 1985). The upper panels of Fig. 1, for example, show a field of synchronous prophase figures from a syncytial blastoderm embryo, illustrating the regular arrays of centrosomes throughout the cortex at the poles of the mitotic figures. For several decades, it has been apparent from studies on other organisms that the centrosomal replication cycle has some degree of autonomy. Several experiments carried out in the 1960s by Mazia indicated that a limited amount of centrosomal replication can occur in the absence of nuclear division. Perhaps some of the most dramatic examples of autonomous centrosome replication are now known in the *Drosophila* embryo. In 1986 we showed that the embryos produced by females homozygous for the mutation *gnu* (GNU embryos) provide a striking illustration of this phenomenon (Freeman *et al.,* 1986). In GNU embryos, DNA replication takes place in the absence of nuclear division to produce giant nuclei. The centrosomes in GNU embryos continue their replication cycles, however, and are functional to the degree that they nucleate asters of microtubules (see Fig. 1). However, these centrosomes never participate in directing the formation of spindle-like structures until later in the development of GNU embryos when the giant nuclei break down and liberate fragments of chromatin. Once this occurs, the occasional spindle-like structure can be seen in the embryo in association with chromatin fragments. It seems, therefore, that intimate contact is required between the condensed chromosomes and microtubules in order to initiate spindle formation per se. However, it is clear from the phenotype of the GNU embryo that the centrosome cycle can continue, independent of the nuclear cycle.

Is this independent replication of centrosomes in the GNU embryo a direct consequence of the *gnu* mutation? We believe that it is, instead, a secondary consequence of the effect of the mutation upon the control of DNA replication. It seems that the *gnu* mutation disrupts the control of DNA replication in the transition between the completion of female meiosis and the onset of zygotic development (Freeman and Glover, 1987). In the wild-type, egg development is arrested following the completion of female meiosis. Unless the egg is fertilized, the meiotic products, the three polar bodies and the female pronucleus, do not undergo any further DNA replication. In unfertilized GNU eggs, development does not arrest, DNA replication continues, and giant nuclei are formed. Moreover, free centrosomes are also seen in the cytoplasm of these unfertilized eggs just as in the fertilized ones. This indicates that noninseminated GNU eggs contain all the components required for centrosome formation and that these must have been provided maternally. Dirksen (1961) has seen the appearance of centrosomes also in artificially activated sea urchin eggs. Thus, under certain conditions in at least two species for which it is generally accepted that centrosomes are contributed

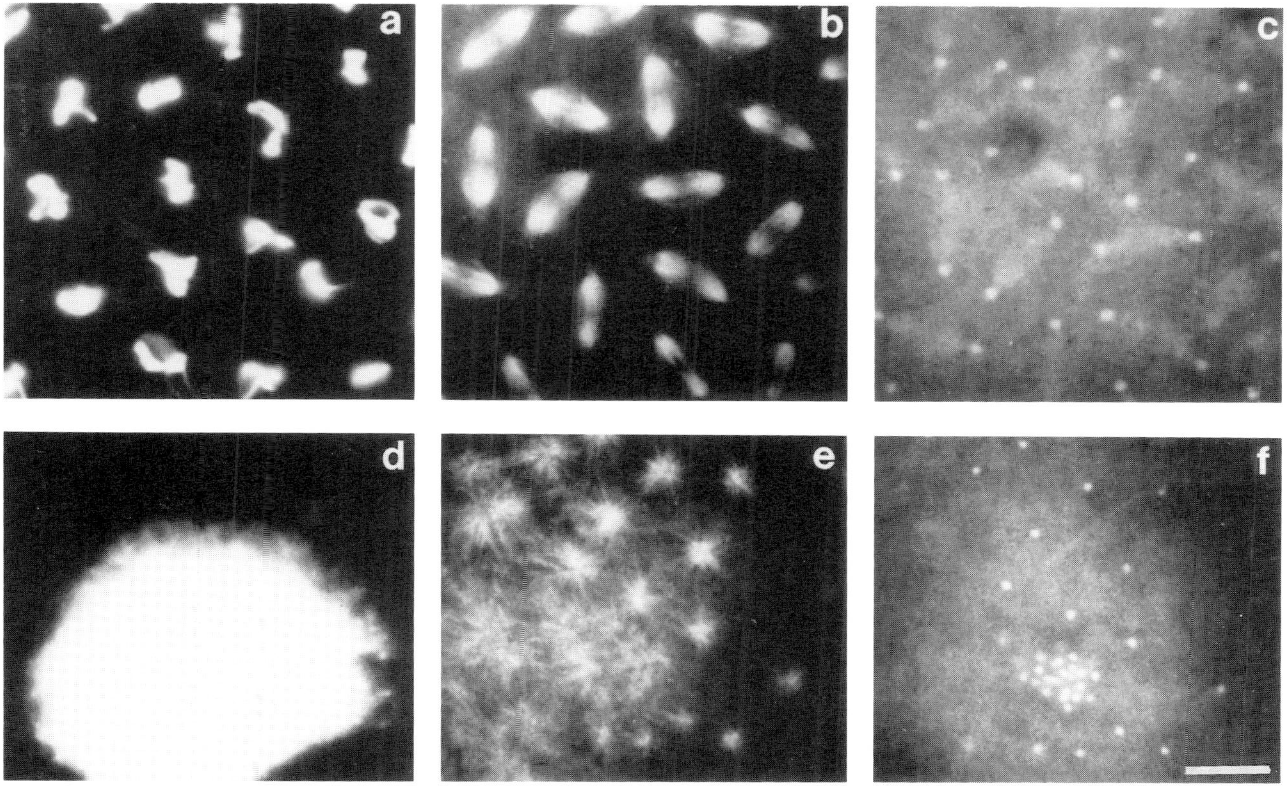

Fig. 1. Centrosome replication in wild-type and GNU embryos. (a–c) Field of prophase nuclei in the cortex of a wild-type embryo at blastoderm. (d–f) Distribution of centrosomes around a giant nucleus in a GNU embryo. Chromatin is stained with the DNA fluorochrome Hoechst (a,d); microtubules are revealed by immunostaining with antitubulin antibody YL1/2 (b,e); centrosomes are displayed by staining with monoclonal antibody Bx63 (c,f). Bar= 10μm.

paternally, these organelles can arise entirely from maternal products. Indeed, although *D. melanogaster* does not normally undergo parthenogenetic development, several other species of *Drosophila* can do so (Templeton, 1983). In these cases, centrosomes must be derived entirely from maternal components.

Although the phenotype of the GNU egg indicates that centrosome replication can be uncoupled from nuclear division, it does not rule out the possibility that it might be coupled to DNA replication. This possibility has been eliminated by experiments in which aphidicolin, an inhibitor of DNA polymerase, was injected into syncytial wild-type *Drosophila* embryos (Raff and Glover, 1988). This work showed that centrosomes could proceed through multiple rounds of division in the absence of DNA replication (Fig. 2). In addition, the continued cycling of many other aspects of mitosis was observed, including cycles of nuclear envelope breakdown and reformation, chromatin condensation and decondensation, and the cyclical budding events that occur at the cortex of the embryo in the syncytial blastoderm stages. Thus there seems to be no absolute requirement for the completion of S phase in order for both the nuclear and cytoplasmic events of M phase to continue to cycle. Perhaps a critical event in S phase continues to be triggered in a cyclical manner; even so, DNA synthesis is severely restricted and chromosome replication does not take place under these experimental conditions. This is in marked contrast to the well-studied cell cycles of the yeasts or mammalian cells in which mitosis cannot take place unless S phase has been satisfactorily completed.

This ability of centrosome replication and other aspects of mitosis to continue to cycle in the syncytial embryo independent of S phase is likely to be a specific property resulting from the abbreviated embryonic cycles. These rapid divisions populate the embryo with sufficient cells from which the organism can be built in subsequent programs of morphogenesis. The rapidity, uniformity, and relative simplicity of these rapid nuclear divisions in their common cytoplasm obviate the requirements for coordinating mitosis with either cell growth or morphogenesis, so a minimal number of control steps might be expected in these mitotic cycles.

V. ORGANIZATIONAL ROLE OF CENTROSOME IN EMBRYONIC DEVELOPMENT

Not all the cells in the *Drosophila* embryo form as late as the 14th cell cycle. During the 8th and 9th nuclear division cycles, the majority of nuclei migrates to

---

Fig. 2. Centrosome replication continues in embryos injected with aphidicolin. Embryos have been injected with aphidicolin at comparable stages of development (see Raff and Glover, 1988), allowed to develop for further intervals of 10, 45, and 90 min, and then fixed for immunostaining with anticentrosomal antibody Bx63 (*right*) and counterstained with Hoechst to reveal DNA (*left*). The increase in the number of centrosomes relative to nuclei can easily be observed. Bar: 20 μm.

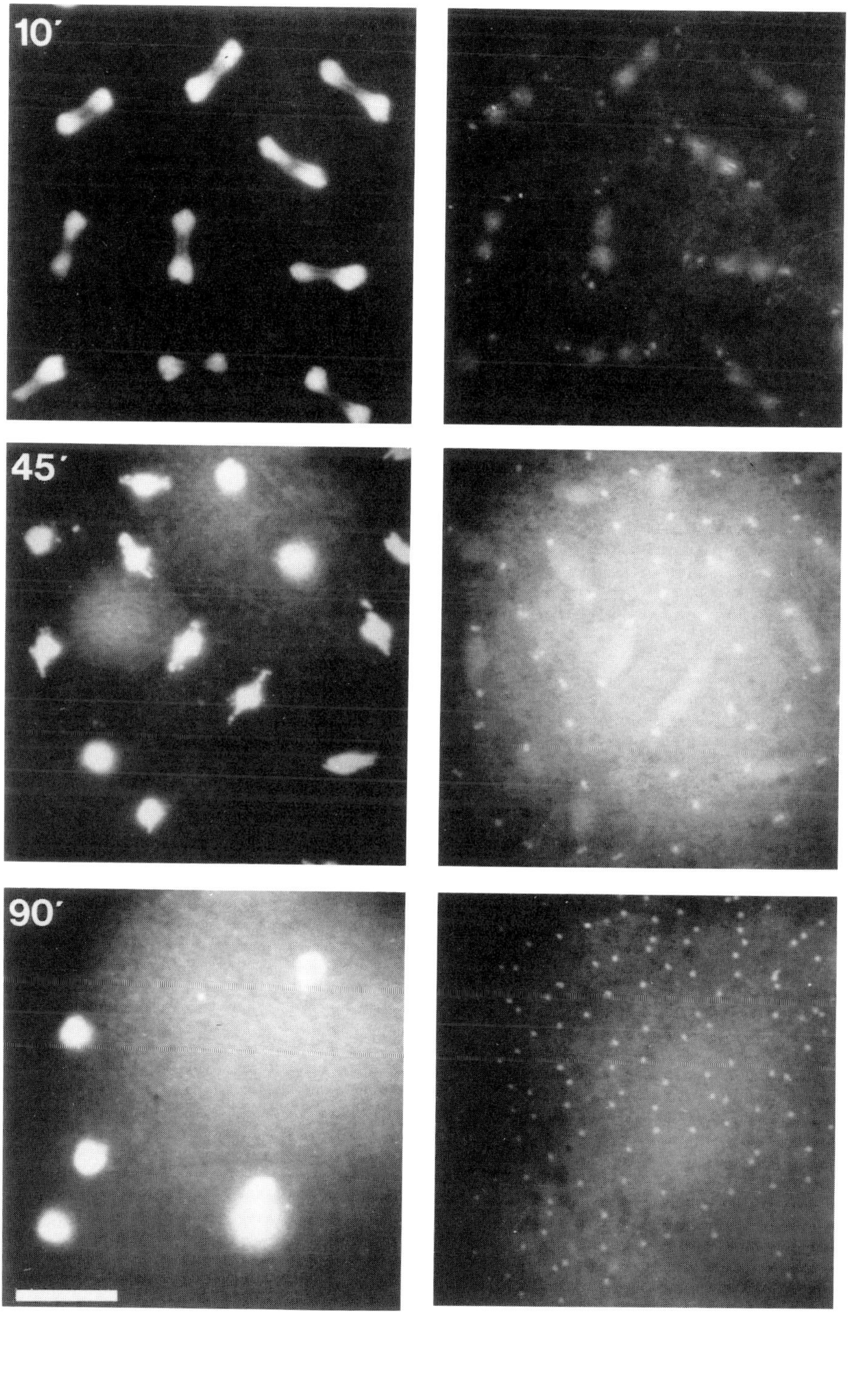
10′
45′
90′

the cortex of the developing embryo; once this has occurred, a small number of nuclei undergoes cellularization four rounds of mitosis before the rest do. These cells form at the posterior pole of the embryo and are destined to become the germ line of the developing organism. Pole cell formation requires maternally provided determinants that are localized in the posterior pole plasm (Illmensee and Mahowald, 1974, 1976; Okada *et al.,* 1974). These determinants are likely to be components of structures called polar granules that can be seen in the posterior pole plasm (Mahowald, 1962, 1968; Counce, 1963). The granules are composed of both RNA and protein and are absent in a number of mutants that fail to form pole cells. Several maternally active genes that are required for polar granule formation have been described (for review, see Anderson, 1989).

It is not an interaction between polar granules and nuclei that leads to the formation of pole cells, but one involving the polar granules and centrosomes or components intimately associated with them. This interaction was demonstrated by experiments in which aphidicolin was injected into syncytial embryos at a stage before nuclear migration had begun (Raff and Glover, 1989). Under these conditions, DNA synthesis is inhibited, and centrosomes dissociate from nuclei and migrate independently to the cortex of the embryo. Those that arrive at the posterior pole become incorporated into anucleate pole cells (Fig. 3). It seems likely that the centrosome and its associated microtubules are interacting in a highly specific way with one or more components of the polar granules to organize the formation of pole cells.

Observations made by Mahowald in the 1960s using electron microscopy indicated varying degrees of interactions between centrosomes and the polar granules in different species of *Drosophila* (Mahowald, 1962, 1968). More recently the distribution of components of the polar granules has been examined in the light microscope using molecular probes. Genes that encode several protein and RNA components of the polar granules have been cloned now. These include the gene *vasa,* which encodes a protein component of polar granules required for both pole cell formation and the correct posterior development of *Drosophila* (Lasko and Ashburner, 1988), and the gene for cyclin B, the transcript for which is present throughout the embryo but is concentrated at the posterior pole (Whitfield *et al.,* 1989). The distribution of *vasa* protein and cyclin B RNA is affected by mutations that prevent pole cell formation and is consistent with the idea that both are components of the polar granules (Lasko and Ashburner, 1990; Raff *et al.,* 1990). Once centrosomes arrive at the posterior pole, either associated with nuclei or independently following the microinjection of aphidicolin, cyclin B transcripts become closely associated with microtubules in proximity to the centrosomes. This response may be seen in Fig. 4, in which the chromosomes at the posterior pole are at various stages of anaphase. Cyclin B RNA, revealed by *in situ* hybridization, has a cuplike distribution over the spindle poles. It therefore seems that a number of interactions, direct or indirect, are taking place between centro-

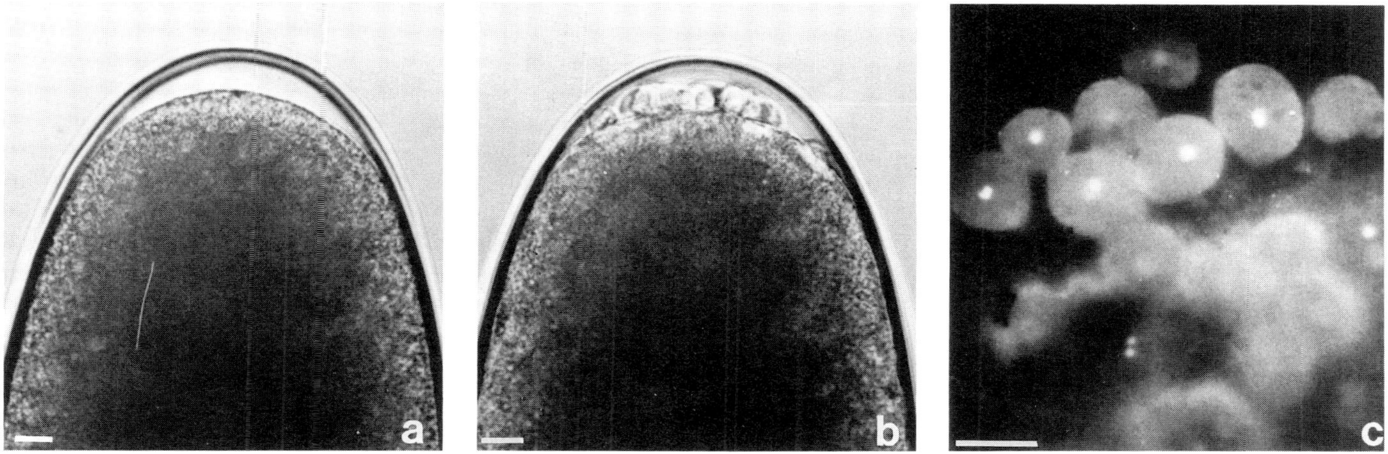

Fig. 3. Pole cells form around centrosomes in aphidicolin-injected embryos. (a) Phase contrast image of the posterior pole of an embryo injected with aphidicolin around nuclear cycle 7–8. (b) Pole cells have developed in the same embryo 40 min later. Bars: 25 μm. When such embryos are fixed and prepared for immunostaining, it is seen that nuclei remain in the interior and that centrosomes have migrated to the surface (see Raff and Glover, 1989). (c) Each pole cell has formed around a centrosome revealed by staining for the Bx63 antigen by immunofluorescence. Bar: 10 μm.

somes and polar granules, and that these interactions play an important role in cytoskeletal reorganization culminating in pole cell formation.

VI. CENTROSOME DUPLICATION IN MITOTIC MUTANTS

Mutations that affect mitosis in *Drosophila* may show considerable pleiotropy in their phenotypes. In this respect, *Drosophila* mitotic mutations differ from those of the yeasts, for example, which are largely conditional lethal (temperature-sensitive) mutations, selected so that the cells arrest at a specific stage of the cell

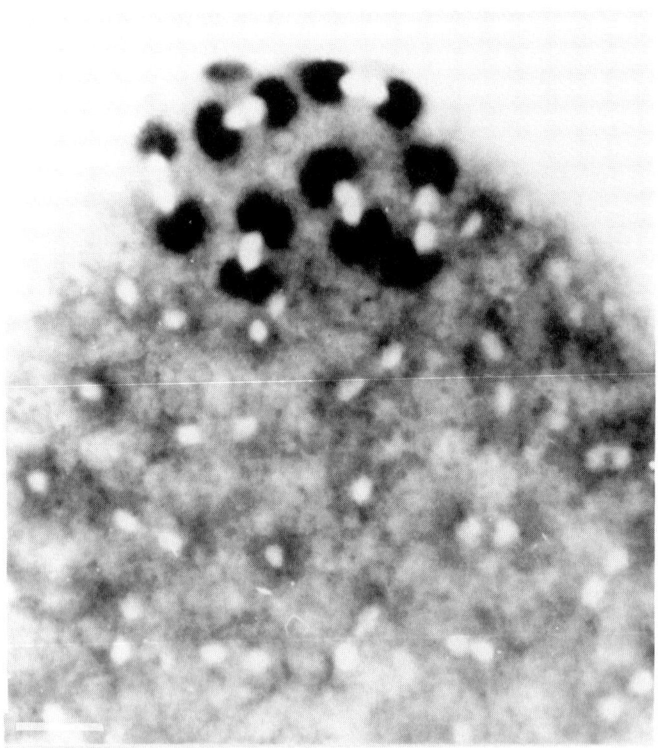

Fig. 4. Cyclin B transcripts associate around centrosomes at the posterior pole. The anaphase chromosomes in this embryo are stained with Hoechst and fluoresce brightly. The regions of dark staining show the cyclin B transcripts, detected by *in situ* hybridization with a digoxygenin DNA probe. Cyclin B transcripts appear to be components of the polar granules (see Raff *et al.*, 1990), which appear to become organized around centrosomes as they arrive at the posterior pole. Bar: 10 μm.

cycle following a shift from the nonpermissive to the permissive temperature (See Chapter 7). *Drosophila* mitotic mutants have been selected as recessive lethal mutations from their developmental phenotype. Consequently, mitotic aberrations often only begin to appear as essential wild-type proteins are depleted; therefore the stages of cell cycle arrest may not be precise. A large group of mitotic mutations is lethal in either the late larval or early pupal stages of development (Gatti *et al.,* 1983; Ripoll *et al.,* 1987; Gatti and Baker, 1989; Glover, 1989). Mitotic defects usually are not manifest until the late stages of development since the homozygous mutant embryo is supplied with sufficient wild-type gene product from its heterozygous mother to enable it to undergo embryogenesis. Most larval tissues then undergo growth and polyploidization, with the exception of imaginal tissues and cells in the central nervous system. These must proliferate during larval and pupal development to form the adult structures. Consequently, many mitotic mutants have small imaginal disks and show mitotic abnormalities in larval neuroblasts. The mitotic phenotypes can be examined in squashed preparations of larval brains in which the chromosomes are stained with orcein. In this way, mutations have been found that appear to affect several aspects of mitosis. Several mutations affect chromosome condensation, others lead to metaphase arrest, some show defects in chromosome separation at anaphase, and others appear to affect cytokinesis (see Glover, 1988, for review).

Mutations in *asp* lead to metaphase arrest; many cells also show polyploidy (Ripoll *et al.,* 1985). Studies on mutations at this locus exemplify how different effects on centrosome replication can be observed depending on the developmental stage at which the phenotype is studied. In addition to its effect on mitosis in the larval brain, mutations at this locus also affect cell division in other tissues. There is also, for example, an effect on meiotic divisions resulting in nondisjunction. *asp* mutations are thought to affect the stability of microtubules, a thesis that is supported by recent observations of mutant cells using indirect immunofluorescence. Immunostaining of the meiotic spindles in males indicates that the microtubules are abnormally long and wavy and that some spindles have only one pole and take the form of unopposed hemispindles. Additional alleles of *asp* have been isolated that, in heteroallelic combinations, permit the survival of females to adulthood. These females lay sufficient eggs to permit the analysis of the maternal effects of *asp* mutations on mitosis in the syncytial embryo (Gonzalez *et al.,* 1990). ASP embryos develop to varying degrees and show abnormalities of spindle microtubules similar to those seen in male meiosis. ASP embryos show asynchronous cycles during the syncytial stages, and can develop large areas that either are devoid of or have a reduced number of nuclei. These areas are often populated by free centrosomes. It seems likely that these centrosomes are undergoing independent cycles of replication such as occur in GNU or aphidicolin-injected embryos. Apparently a variety of mitotic abnormalities, ranging from alteration in the pattern of DNA synthesis to changes in the behavior of microtubules, can therefore result in the dissociation of centrosomes from nuclei followed by their autonomous replication. This embryonic phenotype contrasts with the effect of *asp* mutations on the behavior of centrosomes in larval neuroblasts. These metaphase-arrested cells often contain only a single centrosome and never more than two. Single centrosomes nucleate an unusually long array of microtubules that stretch across the whole of the cell, to which are attached the condensed chromosomes (Fig. 5). Cells having two centrosomes can be arrested in metaphase with either a replicated chromosome complement (4N), or even higher levels of ploidy. The effects of the mutation on centrosome behavior are therefore quite different depending on the developmental stage. In cells in which regulatory controls are imposed on mitosis, the centrosome cycle is more tightly coupled to the rest of the mitotic cycle than in the syncytial embryo.

Other mitotic mutants also display abnormal centrosome behavior. The mutation *mgr,* for example, results in polyploid larval neuroblasts in which the mitotic chromosomes are arranged in circles (Gonzalez *et al.,* 1988). It has been proposed that this arrangement represents a monopolar spindle, an interpretation that has gained support by applying scanning confocal laser microscopy to study the spatial organization of chromosomes, microtubules, and centrosomes (C. Gonzalez, unpublished observations). These monopolar structures differ from those seen in the *asp* mutation because the condensed chromosomes are arranged with their centromeres inward and arms pointing outward, as if they were under traction. A similar phenotype is seen in neuroblasts of larvae homozygous for strong mutant alleles of the mitotic locus *aurora* (M. H. Leibowitz and D. M. Glover, unpublished data). *aurora* is a gene originally identified by weak mutant alleles that show a maternal effect phenotype. Branched multipolar spindles appear in AURORA embryos at various stages of syncytial development. In the progression from anaphase to telophase, distinct centrosome staining disappears and the resulting telophase spindles appear to have microtubules nucleated from points along the nuclear envelope. Since centrosome behavior is affected at different developmental stages by mutations in *aurora,* the case that the gene is directly involved in centrosome behavior is quite strong. However, as with each of these genes, it is prudent to wait and see whether molecular analysis supports this idea.

VII. COORDINATING BIOCHEMICAL AND GENETIC APPROACHES TO UNDERSTANDING CENTROSOMAL FUNCTION

Further understanding of the biochemical functions of centrosomal components may indeed come from the molecular cloning and analysis of their corresponding genes. Such analysis may be achieved by two approaches. First, one might attempt to clone genes identified by mutations described earlier. Mutations can be mapped in *Drosophila,* both by meiotic recombination and, more importantly, with respect to chromosomal rearrangements. The latter approach usually necessitates placing

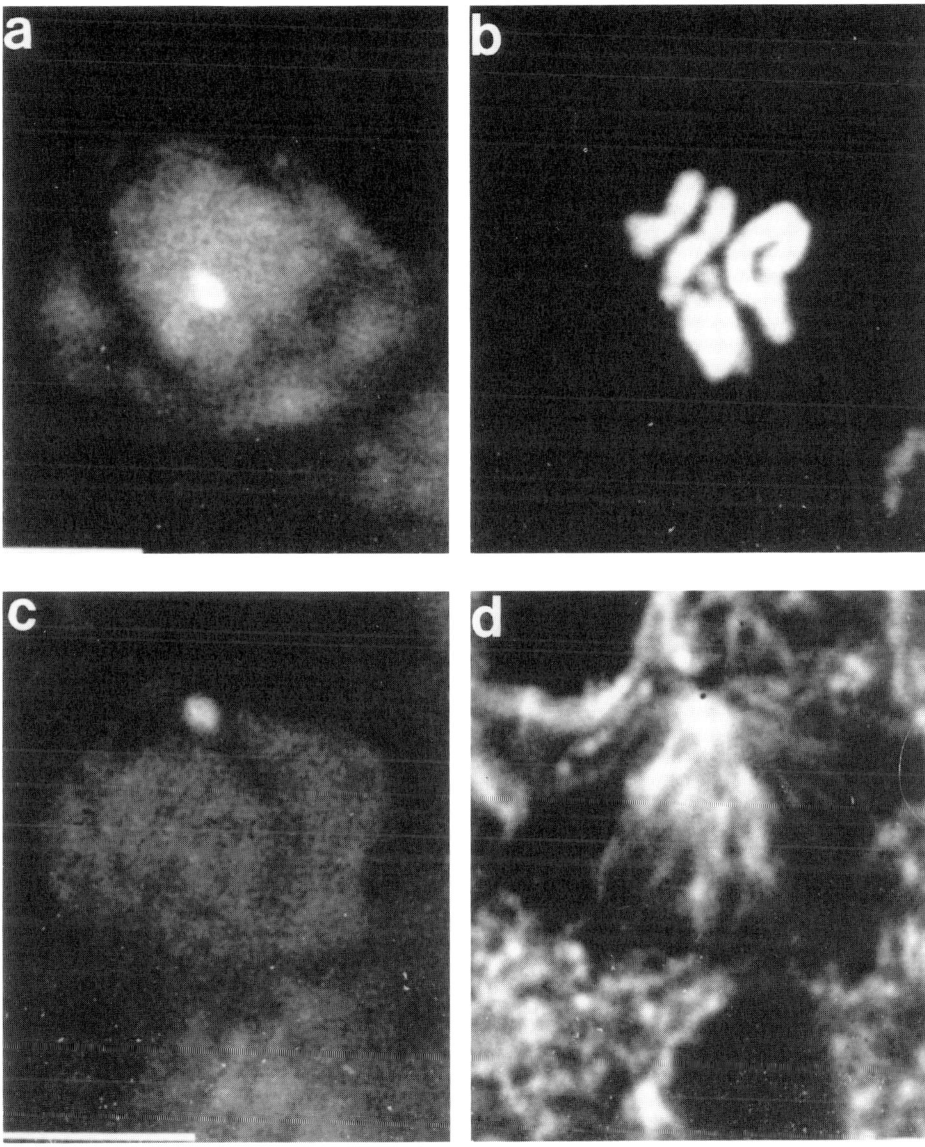

Fig. 5. Hemispindles in *asp* neuroblasts. Confocal micrographs of two neuroblasts from *asp* mutant larvae, showing that each cell contains a single centrosome as revealed by immunofluorescence staining of the Bx63 antigen (a,c). The upper cell is counterstained with propidium diiodide to reveal DNA (b); the lower cell is counterstained with an antitubulin antibody to reveal microtubules (d). Bar: 4 μm.

a chromosome carrying the mutation against its homologue from which regions are deleted and for which the extent of the deletion has been determined cytologically. In this way, a mutation can be localized with respect to the polytene chromosome map. Genes may then be cloned by chromosome walking, either from a gene at a known cytogenetic site or using DNA probes that have been microdissected from polytene chromosomes (Saunders, 1990). Alternatively, genes can be tagged with transposable elements. Mutations induced in this way have a built-in molecular marker, the transposon, that facilitates the molecular cloning of the gene. Once the gene has been cloned it may be sequenced and expressed, and antibodies can be raised against the protein, thereby providing biochemical access to the gene product.

A second approach is to identify centrosomal components by screening collections of antibodies. The Bx63 antibody was identified in this way by screening a collection of monoclonal antibodies raised against *Drosophila* embryonic and nuclear extracts (Frasch *et al.,* 1986). The gene encoding this antigen was then cloned by using the antibody to screen libraries of *Drosophila* DNA expressed in the bacteriophage λgt11 to generate polypeptides fused to β-galactosidase of *Escherichia coli* (Whitfield *et al.,* 1988). A similar approach should be possible using some of the antibodies described by Kellogg *et al.,* (1989). These workers have purified a number of proteins associated with microtubules from *Drosophila* embryos. The proteins have been fractionated by polyacrylamide gel electrophoresis and antisera have been raised against proteins present in individual bands. Several of these antibodies recognize centrosomal components, thus offering a route toward the cloning of their genes for further biochemical experiments. Once a *Drosophila* gene has been cloned, it may be localized by *in situ* hybridization to a precise position on the salivary gland chromosome map. It is then often possible to design experiments to introduce mutations into this chromosomal region in order ultimately to identify mutant alleles of the locus encoding the antigen. This will be particularly useful when applied to genes encoding known centrosome-associated antigens.

Antibodies against centrosomal components will, of course, provide biochemical tools that can be used to study the interactions of centrosomal proteins. Thus, over the next decade one can be confident that there will be a concerted analysis of the *Drosophila* centrosome using both biochemical and genetic approaches. Such studies will provide a greater possibility of gaining some molecular understanding of centrosome function in the not too distant future.

REFERENCES

Anderson, K. V. (1989). *Drosophila:* The maternal contribution. *In* "Genes and Embryos" (D. M. Glover, and B. D. Hames, eds.), pp. 1–38. IJRL Press, Oxford.

Carpenter, A. T. T. C. (1975). Electron microscopy of meiosis in *Drosophila melanogaster* females. *Chromosoma (Berlin)* **51,** 157–182.

Counce, S. J. (1963). Developmental biology of polar granules in *Drosophila* including observation on pole cell behavior and distribution during embryogenesis. *J. Morphol.* **112,** 129–145.

Dirksen, E. R. (1961). The presence of centrioles in artificially activated sea urchin eggs. *J. Biophys. Biochem. Cytol.* **11,** 244–247.

Edgar, B. A., and O'Farrell, P. H. (1989). Genetic control of cell division patterns in the *Drosophila embryo. Cell* **57,** 177–187.

Foe, V. E. (1989). Mitotic domains reveal early commitment of cells in *Drosophila* embryos. *Development* **107,** 1–22.

Foe, V. E., and Alberts, B. M. (1983). Studies of nuclear and cytoplasmic behavior during the five mitotic cycles that precede gastrulation in *Drosophila* embryogenesis. *J. Cell Sci.* **61,** 31–70.

Frasch, M., Glover, D. M., and Saumweber, H. (1986). Nuclear antigens follow different pathways into daughter nuclei during mitosis in *Drosophila* embryos. *J. Cell Sci.* **82,** 115–172.

Freeman, M., Nüsslein-Volhard, C., and Glover, D. M. (1986). The dissociation of nuclear and centrosomal division in *gnu,* a mutation causing giant nuclei in *Drosophila. Cell* **46,** 457–468.

Freeman, M., and Glover, D. M. (1987). The *gnu* mutation of *Drosophila* causes inappropriate DNA synthesis in unfertilised and fertilised eggs. *Genes Dev.* **1,** 924–930.

Gatti, M., Pimpinelli, S., Bove, C., Baker, B. S., Smith, D. A., Carpenter, A. T. C., and Ripoll, P. (1983). Genetic control of mitotic cell division in *Drosophila melanogaster. In* "Proceedings of the XV International Congress of Genetics" (V. L. Chopra, D. C. Joshi, R. P. Sharma, and H. C. Bansal, eds.), pp. 193–204. IBH Publishing, Oxford.

Gatti, M., and Baker, B. S. (1989). Genes controlling essential cell cycle functions in *Drosophila melanogaster. Genes Dev.* **3,** 438–452.

Glover, D. M. (1989). Mitosis in *Drosophila. J. Cell Sci.* **92,** 137–146.

Gonzalez, C., Casal, J., and Ripoll, P. (1988). Functional monopolar spindles caused by mutation in *mgr,* a cell division gene of *Drosophila melanogaster. J. Cell Sci.* **89,** 39–47.

Gonzalez, C., Saunders, R. D. C., Casal, J., Molina, I., Carmena, M., Ripoll, P., and Glover, D. M. (1990) Mutations as the *asp* locus of *Drosophila* lead to multiple free centrosomes in syncytial embryos, but restrict centrosome duplication in larval neuroblasts. *J. Cell Sci.* **96,** 605–616.

Illmensee, K., and Mahowald, A. P. (1974). Transplantation of posterior pole plasm: Induction of germ cells at the anterior pole of the egg. *Proc. Natl. Acad. Sci. U.S.A.* **7,** 1016–1020.

Illmensee, K., and Mahowald, A. P. (1976). The autonomous function of germ plasm in a somatic region of the *Drosophila* egg. *Exp. Cell Res.* **97,** 127–140.

Karr, T. L., and Alberts, B. M. (1986). Organization of the cytoskeleton in early *Drosophila* embryos. *J. Cell Biol.* **98,** 156–162.

Kellogg, D. R., Mitchison, T. J., and Alberts, B. M. (1988). Behavior of microtubules and actin filaments in living *Drosophila* embryos. *Development* **103,** 675–686.

Kellogg, D. R., Field, C. M., and Alberts, B. M. (1989). Identification of microtubules associated proteins in the centrosome, spindle, and kinetochore of the early *Drosophila* embryo. *J. Cell Biol.* **109,** 2977–2991.

Lasko, P. F., and Ashburner, M. (1988). The product of the *Drosophila* gene *vasa* is very similar to eukaryotic initiation factor-4A. *Nature (London)* **335,** 611–617.

Lasko, P. F., and Ashburner, M. (1990). Posterior localization of *vasa* protein correlates with, but is not sufficient for, pole cell development. *Genes Dev.* **4,** 905–921.

Mahowald, A. P. (1962). Fine structure of pole cells and polar granules in *Drosophila melanogaster. J. Exp. Zool.* **151,** 201–205.

Mahowald, A. P. (1968). Polar granules of *Drosophila.* II. Ultrastructural changes during early embryogenesis. *J. Exp. Zool.* **167,** 237–262.

Mahowald, A. P., and Strassheim, J. M. (1970). Intracellular migration of centrioles in the germarium of *Drosophila melanogaster*. *J. Cell Biol.* **45,** 306–320.

Nurse, P. (1990). Universal control mechanism regulating onset of M-phase. *Nature (London)* **344,** 503–508.

O'Farrell, P. H., Edgar, B. A., Lakich, D., and Lehner, C. (1989). Directing cell division during development. *Science* **246,** 635–640.

Okada, M., Kleinman, I. A., and Schneiderman, H. A. (1974). Restoration of fertility in sterilized *Drosophila* eggs by the transplantation of polar cytoplasm. *Dev. Biol.* **37,** 43–54.

Raff, J. W., and Glover, D. M. (1988). Nuclear and cytoplasmic mitotic cycles continue in *Drosophila* embryos in which DNA synthesis is inhibited with aphidicolin. *J. Cell Biol.* **107,** 2009–2019.

Raff, J. W., and Glover, D. M. (1989). Centrosomes, and not nuclei, initiate pole cell formation in *Drosophila* embryos. *Cell* **57,** 611–619.

Raff, J. W., Whitfield, W. G. F., and Glover, D. M. (1990). Two distinct mechanisms localize cyclin B transcripts in syncytial *Drosophila* embryos. *Development* **110,** 1249–1261.

Ripoll, P., Pimpinelli, S., Valdivia, M., and Avila, J. (1985). A cell division mutant of *Drosophila* with a functionally abnormal spindle. *Cell* **44,** 907–912.

Ripoll, P., Casal, J., and Gonzalez, C. (1987). Towards the genetic dissection of mitosis in *Drosophila*. *Bioessays* **7,** 204–210.

Saunders, R. D. C. (1990). Short cuts for chromosome walking: Chromosome microdissection and the polymerase chain reaction. *Bioessays,* **12,** 245–248.

Templeton, A. R. (1983). Natural and experimental parthenogenesis. *In* "Genetics and Biology of *Drosophila*" (M. Ashburner, Carson, and Thompson, eds.), Vol. 3C. Academic Press, London.

Warn, R. M., Flegg, L., and Warn, A. (1987). An investigation of microtubule organization and functions in living *Drosophila* embryos by injection of a fluorescently labeled antibody against tyrosinated α-tubulin. *J. Cell Biol.* **105,** 1721–1730.

Whitfield, W. G. F., Millar, S. E., Saumweber, H., Frasch, M., and Glover, D. M. (1988). Cloning of a gene encoding an antigen associated with the centrosome in *Drosophila*. *J. Cell Sci.* **89,** 467–480.

Whitfield, W. G. F., Gonzalez, C., Sanchez-Herrero, E., and Glover, D. M. (1989). Transcripts of one of two *Drosophila* cyclin genes become localized in pole cells during embryogenesis. *Nature (London)* **338,** 337–340.

Zalokar, M., and Erk, I. (1976). Division and migration of nuclei during early embryogenesis of *Drosophila melanogaster*. *J. Microbiol. Cell.* **25,** 97–106.

9

Control of Centrosome Inheritance in Echinoderm Development

GREENFIELD SLUDER

Worcester Foundation for Experimental Biology
Shrewsbury, Massachusetts 01545

I. INTRODUCTION

A. The Centrosome

Centrosomes in animal cells are the assembled structures that form the poles of the mitotic apparatus. They define the essential duality of the division process by

Copyright © 1992 Academic Press, Inc.
All rights of reproduction in any form reserved.

establishing the bipolar spindle axis (Mazia, 1961; Sluder and Reider, 1985b; Mazia 1987). Although their ultrastructure may vary from one organism to the next, these organelles have two well-established activities; they nucleate microtubules and they precisely double, or reproduce, before the cell enters mitosis. In higher animal cells, the radial array of microtubules nucleated by centrosomes forms the astral, continuous, and kinetochore fibers of the mitotic spindle (Nicklas, 1971; Gould and Borisy, 1977; Euteneuer and McIntosh, 1981; Wheatley, 1982). These microtubules are required for movement of chromosomes to the metphase plate (Rieder and Alexander, 1990), movement of chromosomes to the poles (Nicklas and Koch, 1972; Salmon, 1975; Rieder, 1982), spindle elongation (Masuda *et al.*, 1988), and formation of the cleavage furrow (Rappaport, 1986). Although the importance of centrosomes in cell division was recognized before the turn of the century (reviewed in Wilson, 1925; Fulton, 1971; Mazia, 1984), remarkably little is known about their composition or how they function. In part, this has been due to the difficulty of isolating enough of them in a form sufficiently pure for meaningful biochemical or molecular analyses.

At the ultrastructural level, the centrosomes of animal cells typically consist of a pair of orthogonally arranged centrioles surrounded by amorphous osmiophilic material, called the pericentriolar material, into which the microtubules of the aster and central spindle are inserted. (See Stubblefield and Brinkley, 1967; Brinkley and Stubblefield, 1970; Fulton, 1971; Wheatley, 1982; Bornens *et al.*, 1987; Vorobjev and Nadezhdina, 1987 for reviews of centrosome ultrastructure.) In some cells the pericentriolar material is distributed primarily around the older of the two centrioles; the younger centriole acquires pericentriolar material when it separates from its parent and starts to form its own daughter centriole (Rieder and Borisy, 1982; Vorobjev and Chentsov, 1982).

The pericentriolar material and the globular tip of the basal feet, which are attached to the wall of the centriole, act as the centrosomal microtubule organizing center (MTOC) during interphase and mitosis. When permeabilized cells are exposed to exogenous tubulin, microtubules are assembled from the pericentriolar material, not from the centrioles (Snyder and McIntosh, 1975; Kuriyama, 1984). In lysed cell preparations, isolated pieces of the pericentriolar material nucleate microtubules; naked centrioles in the same preparations show only elongation of their existing microtubules (Gould and Borisy, 1977). In many cells, denser aggregates of osmiophilic material, called pericentriolar satellites, are seen or in close proximity to the pericentriolar material. Although these satellites were once thought to be a condensed form of the pericentriolar material, recent work indicates that they represent concentrations of a centrin-like protein that do not nucleate microtubules (Baron and Salisbury, 1988). Although the pericentriolar material appears amorphous at the ultrastructural level, it has a fibrous substructure that is revealed by immunofluorescence straining of cells with various antibodies (Schatten *et al.*, 1987; Baron and Salisbury, 1988). These observations support the proposal that the centrosomal MTOC is formed by the condensation or folding of the linear fibers with microtubule nucleating sites along their length (Mazia, 1984, 1987).

The role of the centrioles in the centrosome is not well understood. A minimum function is to organize the axoneme of the primary cilium and motile cilium (Rieder and Borisy, 1982; Wheatley, 1982; Masuda and Sato, 1984). In this regard, the ability of centrioles to act as basal bodies in a wide variety of cells (reviewed in Fulton, 1971; Wheatley, 1982) and the ability of isolated basal bodies to induce asters when microinjected into *Xenopus* eggs (Heidemann and Kirschner, 1975; Heidemann *et al.,* 1977) has led to the belief that centrioles and basal bodies are functionally equivalent organelles.

The role of centrioles in the formation and reproduction of centrosomes remains controversial because a few types of animal cells and all higher plant cells propagate without recognizable centrioles (Fulton, 1971; Szollosi *et al., 1972; Debec al.,* 1982; Wheateley, 1982). Consequently, some investigators feel that the centrioles are dispensable (Dietz, 1964; Friedlander and Wahrman, 1970; Pickett-Heaps, 1971; Steffen *et al.,* 1986). Perhaps the centrioles are assembled by the centrosome and do not play a role in organizing the centrosome (Mazia, 1984). In contrast, other studies provide evidence that centrioles, or an entity spatially and mechanically associated with them, play a key role in determining the formation and reproduction of centrosomes (Heidemann and Kirschner, 1975; Maller *et al.,* 1976; Heidemann *et al.,* 1977; Sluder and Rieder, 1985a; Sluder *et al.,* 1989a). The controversy over the role of centrioles in the formation and reproduction of the centrosome is discussed by Sluder (1990).

B. The Reproduction of Centrosomes

The precise doubling, or reproduction, of the centrosome during interphase is an important event in the preparation of the cell for mitosis. The cell must not only precisely coordinate the steps of centrosome reproduction with nuclear events during the cell cycle but also must insure that the centrosome will divide into two, and only two daughters. If the centrosome fails to double or divides into more than two, the ensuing mitosis will inevitably be abnormal.

Centrosome reproduction starts late in the G_1 phase of the cell cycle, when the two centrioles lose their orthogonal arrangement (Kuriyama and Borisy, 1981). At the onset of DNA synthesis, short daughter centrioles are assembled at right angles to and separated slightly from the proximal end of each mature centriole Robbins *et al.*, 1968; Rattner and Phillips, 1973; Wheatley, 1982). These daughter centrioles, often called procentrioles, elongate during the S and G_2 phases, reaching their mature length in mitosis or the following G_1 phase (Kuriyama and Borisy, 1981). The centrosome as a whole typically splits at a variable time in G_2;

pairs of mother–daughter centrioles go to each daughter centrosome (Aubin *et al.*, 1980). However, one should bear in mind that the times in the cell cycle when these visible events of centrosome reproduction are first detectable probably occur near the end of the reproductive processes for these structures. The molecular doubling of centrosomal components and the assembly of the precursor structures must have occurred at an earlier time in the cell cycle.

In echinoderm zygotes, the experimental model used in the studies described later in this chapter, the centrioles separate and duplicate as the centrosome flattens during telophase (Mazia, 1984, 1987; Schatten *et al.*, 1986). These zygotes have no G_1 phase, and DNA synthesis starts in telophase (Hinegardner *et al.*, 1964). In G_2 the flattened centrosome splits and the daughter asters move to opposite sides of the nucleus by prophase (Paweletz *et al.*, 1984; Mazia, 1987).

Studies characterizing the functional properties of centrosome reproduction in echinoderm zygotes indicate the presence of a counting mechanism that insures that centrosomes will reproduce only in a one-to-two fashion (Mazia *et al.*, 1960; Sluder and Begg, 1985; Sluder and Rieder, 1985a; Sluder *et al.*, 1989a). Centrosome reproduction is not based on a simple fission mechanism in which an amorphous body subdivides into two roughly equal bodies. Instead, centrosomes behave as if they reproduce by a generative mechanism that depends on the replication and separation of an organizing entity we call the polar organizer. This polar organizer is the kernel around which the familiar structures of the centrosome are elaborated. Functional analyses indicate that each mitotic centrosome contains two polar organizers (Mazia *et al.*, 1960; Sluder and Begg, 1985). The reproduction of centrosomes is determined by the separation of these two polar organizers, their physical separation, and their replication to regenerate the duplex state of the centrosome. By functional criteria, we have shown that splitting and replication are distinct events that can be put out of phase with each other for several cell cycles (Sluder and Begg, 1985).

Presently, we do not know the composition of the polar organizers; they have an activity that we can manipulate experimentally. However, further work with sea urchin zygotes has shown that the splitting and duplication of polar organizers is coincident with the duplication and distribution of centrioles (Sluder and Rieder, 1985a; Sluder *et al.*, 1989a). Centrosomes containing two centrioles have full reproductive capacity, that is, they reproduce in a one-to-two fashion before each mitosis. Centrosomes with one centriole have half the normal reproductive capacity; they do not double before the next mitosis and then double prior to the following mitosis. Centrosomes with no centrioles have no reproductive capacity; they do not double between mitoses. Thus, polar organizers behave as if they are each associated with a centriole. We cannot say that the centriole is the polar organizer because some cells, such as those of higher plants, propagate without centrioles. Presumably such acentriolar cells have polar organizers without nine triplet microtubule structures to signal their location (see Sluder, 1990, for a discussion of this issue).

II. CENTROSOME INHERITANCE IN DEVELOPMENT

A. Our Approach

In this chapter we review our studies on how centrosome inheritance is controlled, using sea urchin and starfish zygotes as model systems. The gametes of these echinoderms have a number of practical advantages over other experimental systems. Chief among these advantages are the availability of large numbers of gametes, the rapid cell cycle of the zygotes, and the simple culture conditions used (seawater). Also, the optical clarity of these zygotes allows for direct *in vivo* observations of centrosome behavior without the need for fixation and sectioning. By characterizing the functional response of living zygotes to specific experimental perturbations, we have been able to test possible ways in which the fertilized egg could control centrosome inheritance. This approach has allowed us to provide information that could not be obtained from conventional biochemical, molecular, or immunological methods.

B. Terminology

We will use the terms oocyte, egg, and zygote in the following ways. An oocyte is a female gamete prior to the point in its development at which it can be naturally fertilized. For sea urchins and starfish, this is prophase of meiosis I, before germinal vesicle breakdown. An egg is a female gamete that is capable of being fertilized. A zygote is an egg that has been fertilized. In very early development, a zygote is sometimes called a fertilized egg; later in development, it is often called an embryo.

C. The Problem

In the development of fertilized animal eggs, specific mechanisms must exist at the gamete or zygotes levels to control centrosome inheritance. Both the sperm and egg inherit a centrosome from the meiosis II spindle at the end of gametogenesis. Thus, both gametes could contribute a centrosome at fertilization that could be used in establishing the first and subsequent mitotic spindles. Such biparental inheritance of centrosomes was implicit in Fol's "quadrille of centers" (discussed in Wilson, 1925) and was also shown in drawings of the early development of *Crepidula* (Conklin, 1901). Although such biparental inheritance of centrosomes may seem logical, it would pose awkward control problems for the zygote. Prior to first mitosis, the zygote must duplicate its chromosomes; however, centrosomes normally start to reproduce at or near the time of DNA synthesis. Thus, to use both male and female centrosomes in development, the zygote would have to prevent centrosome reproduction during the first interphase and then immediately return to the normal regulatory pathways in which centrosome duplication is temporally linked to the onset of DNA synthesis. If such a lapse in the normal pattern of centrosome reproduction did not occur in first interphase, the centrosomes from both parents would double; consequently the zygote would enter first mitosis with four centrosomes. Aneuploidy induced by the ensuing four-way division would severely compromise the viability of the embryo. Thus, control of centrosomal inheritance is of fundamental importance to all animal systems regardless of which gamete actually contributes the centrosome used in development.

For all organisms examined carefully, the zygote exhibits uniparental inheritance of the centrosome used in development. Examples of maternal and paternal inheritance of centrosomes have been reported. In those animals that naturally undergo parthenogenetic development, for example, the honeybee drone and certain populations of whip-tailed lizards, centrosome inheritance must be solely maternal (reviewed in Beatty, 1967). Other animal forms, such as the mouse (Szollosi *et al.*, 1972; Schatten *et al.*, 1988) and perhaps the domestic pig (Szollosi and Hunter, 1973), also appear to use the maternal or egg centrosome in development, although the sperm could, in principle, contribute a functional centrosome to the zygote. However, other mammals, such as the domestic sheep, may use the paternal centrosome in development (Le Guen and Crozet, 1989). Other examples of paternal inheritance of centrosomes can be found in the echinoderms, as will be shown later in this chapter.

Regardless of which parental centrosome an organism uses in development, the important biological problem to be addressed is how the centrosome of one gamete is predictably suppressed before or after fertilization to insure the uniparental inheritance of centrosomes. This issue, first raised early in this century (Wilson, 1925), has not been investigated in a systematic manner.

III. CENTROSOME INHERITANCE IN SEA URCHIN EGGS

A. Postfertilization Events

Sea urchin eggs complete their meiotic reduction divisions and form a female pronucleus while they are in the ovary. They are fertilized at this stage of gametogenesis after they are shed into seawater (Fig. 1a). Just after fertilization, the sperm head rotates in the cortical cytoplasm of the egg (Fig. 1b). A small radial array of microtubules, called the sperm aster, is assembled at the junction between the sperm nucleus and mitochondrion. As the sperm aster grows larger, it moves away from the cortex into deeper cytoplasm in response to pulling forces generated by microtubules directed toward the cell center (Hamaguchi and Hiramoto, 1986). Concurrently, the compact sperm nucleus is reorganized into a male pronucleus (Fig. 1c). At this time, there is no visible aster of maternal origin at the female pronucleus or elsewhere in the zygote. When the microtubules growing from the sperm aster reach the female pronucleus, the pronucleus migrates rapidly toward the astral focus where it comes into close apposition with the male pronucleus (Fig. 1d). Just prior to pronuclear fusion, or syngamy, the sperm aster appears to double (Fig. 1e) and the first round of DNA synthesis begins (Hinegardener *et al.,* 1964). The doubling of the sperm aster involves the duplication of the two sperm centrioles and the distribution of mother–daughter centriole pairs to the two daughter centrosomes (Paweletz *et al.,* 1987a,b). In early prophase, these centrosomes move to opposite sides of the zygote nucleus and later form the poles of the first mitotic spindle (Fig, 1f,g).

B. Origin of the Centrosome Used in Development

From descriptive studies of postfertilization events, early cytologists postulated that the centrosomes used in the development of the sea urchin zygote come solely from the sperm. Their best experimental evidence for the paternal inheritance of centrosomes came from observations that polyspermic fertilization led to multipolar divisions and the finding that enucleate egg fragments, when fertilized,

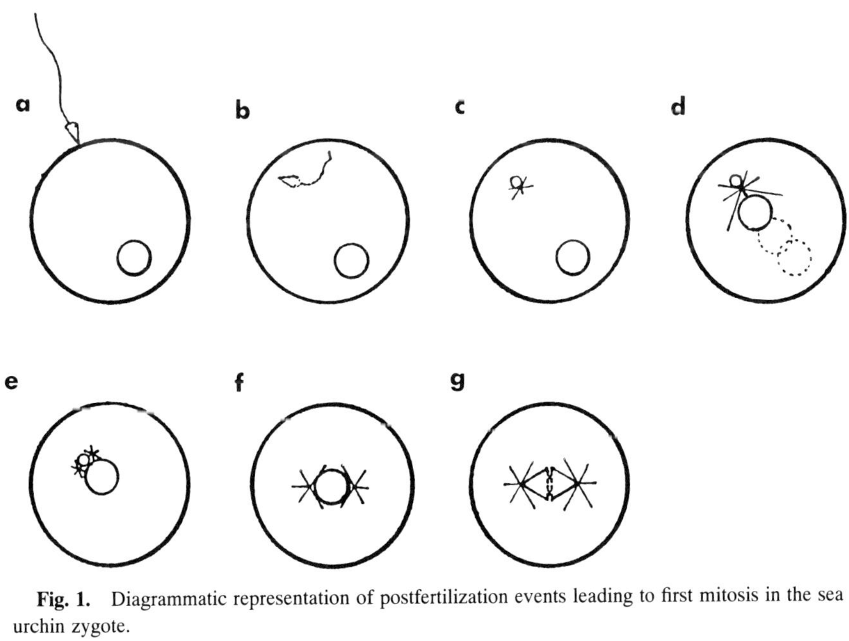

Fig. 1. Diagrammatic representation of postfertilization events leading to first mitosis in the sea urchin zygote.

divided in a normal bipolar fashion (reviewed in Wilson, 1925). Although the paternal inheritance of centrosomes in sea urchin eggs was widely accepted at the time, the appealing simplicity of the story was clouded by Loeb's demonstration of artificial parthenogenesis. He showed that unfertilized eggs could be induced to undergo normal cleavage divisions and develop into plutei using solely maternal centrosomes (reviewed in Loeb, 1913). How the egg can assemble or express one or more centrosomes of maternal origin after parthenogenetic activation, yet suppress the maternal centrosome in normal development, has been a mystery.

Our work with sea urchin zygotes has provided a direct demonstration of the paternal inheritance of centrosomes in normal development and has given us some insight into the fate of the maternal centrosome (Sluder and Rieder, 1985b). The male and female pronuclei are kept separate experimentally after fertilization in order to determine which pronucleus is associated with the mitotic centrosomes at first mitosis. To this end, unfertilized eggs are briefly treated with Colcemid to prevent future microtubule assembly (method described in Sluder, 1976). Drug-treated eggs fertilize in a normal fashion and form a male pronucleus at the periphery of the zygote. However, the pronuclei remain separate, because no sperm aster is assembled. Zygotes in which the pronuclei are separated by almost the diameter of the cell are selected for further observation. At the onset of first mitosis, both pronuclei break down, leaving small irregular hyaline areas about the same size as the original nuclei. Shortly after nuclear envelope breakdown, individual zygotes are irradiated on the microscope with 366-nm light for 15 sec. to photochemically inactivate the Colcemid (Aronson and Inoué, 1970; Borisy, 1971; Sluder, 1976). Immediately after irradiation, a functional bipolar spindle of normal appearance forms in association with the male chromosomes. The female pronucleus, on the other hand, develops in one of two fashions. In most cases, it remains as a small hyaline area throughout mitosis; there is no indication of spindle assembly and the chromosomes appear to be randomly distributed in the nuclear area. In a few cases, a small radial array of microtubules, called a monaster, is formed in the female pronuclear area and persists until the eggs enter telophase; there is never any indication of bipolar spindle assembly. In all cases, the mitotic centrosomes are associated with the sperm components.

To characterize the ultrastructure of the male and female pronuclear areas, the development of individual zygotes with well-separated pronuclei is followed under the polarizing microscope until the male spindle is in metaphase, at which time the zygotes are circled on the coverslip with a diamond scribe, fixed, and embedded. Each zygote previously observed *in vivo* is relocated, excised, and serially sectioned (0.25 μm) for observation with a high voltage electron microscope. Reconstructions of the male pronuclear region reveal a typical spindle morphology with kinetochore and nonkinetochore microtubules converging on two well-defined centrosomes, each containing a pair of centrioles. The female pronuclear hyaline areas, in contrast, contain well-condensed chromosomes with prominent kinetochores surrounded by the convoluted fenestrated remnants of the nuclear envelope. No microtubules or centrioles are associated with the female chromosomes or nuclear envelope.

We also completely reconstructed zygotes that assembled monasters at the site of the female pronucleus. The focus of the monaster consists of a hollow spherical array of small patches of pericentriolar-like material into which the numerous radially oriented microtubules terminate. Centrioles are not found in or near the monaster. The maternal chromosomes are distributed around the monaster with one kinetochore oriented toward the astral center and attached to microtubules.

C. Fate of the Maternal Centrosome

These results provide clear functional and structural evidence that the centrosome used in the development of the sea urchin zygote is contributed by the sperm alone. This raises the important issue of what happens to the maternal centrosome inherited by the egg at the end of meiosis II. Is it disassembled, or is it intact but suppressed? The female centriole must, at least, either lose its association with the female pronucleus or be degraded. The occasional formation of a monaster at the site of the female pronucleus, however, suggests that a remnant of the maternal centrosomal MTOC can persist into first mitosis. If so, why is this MTOC sometimes active at mitosis when the male and female pronuclei are kept separate but never active when these nuclei come together in normal development?

Insight into these issues was provided by additional experiments in which eggs are fertilized and shortly thereafter, fragmented with a nylon screen (Sluder *et al.*, 1989a). Since this fragmentation occurs before syngamy, four classes of fragments are obtained: those with no nuclei, those with both male and female pronuclei, those with just a male pronucleus, and those with just a female pronucleus. Since female pronuclei are always clearly larger than male pronuclei early in the first cell cycle, those fragments containing only a female pronucleus can be unambiguously identified for further observation.

At first mitosis, 88% of the zygote fragments with only a female pronucleus assemble a single monaster after nuclear envelope breakdown. Although similar in appearance to sperm-derived asters, these monasters often have a less tightly focused center and can be more variable in their overall appearance. Serial section reconstruction (0.25 μm) of these monasters reveals that the monastral focus consists of numerous patches of pericentriolar-like material into which the radially oriented microtubules are inserted. These patches are arranged as a hollow sphere surrounding a tightly packed mass of membranous vesicles. As expected, these monasters do not contain any centrioles.

During mitosis, the size and birefringence (a measure of the quantity of oriented microtubules per unit volume) of the monaster increases. When the cell enters telophase, the monaster is disassembled and nuclear envelopes reform around individual chromosomes, giving a number of micronuclei that often fuse to form a single nucleus. At the second and subsequent mitoses, only a single monaster is assembled in each egg fragment. In many cases, a progressive loss of the radial symmetry of the monastral microtubules from one mitosis to the next is observed. However, there is never any indication of bipolarity or multipolarity in their organization. This lack of bipolarity is confirmed by the reformation of a single nucleus (or a single aggregate of micronuclei) at each telophase.

The inability of monasters to regularly double between mitoses is not an artifact of the fragmentation procedure. Examination of the development of zygote fragments containing only a male pronucleus reveals that they always form bipolar spindles and divide in a normal fashion at each mitosis.

In the remaining 12% of the zygote fragments studied, the female pronucleus breaks down at mitosis leaving an irregular hyaline area. At each telophase, nuclear envelopes reform around individual chromosomes or groups of chromosomes. Sometimes these fragments form hyaline areas at the first two mitoses and then form a monaster at third mitosis. Culturing the zygote fragments at slightly higher temperatures increases the percentage of cells that form monasters rather than hyaline areas at first mitosis. Since higher temperatures promote spindle microtubule assembly in sea urchin zygotes (Inoué and Sato, 1967; Inoué *et al.*, 1975), we conclude that the MTOC activity associated with the female pronucleus is not as efficient as that associated with the paternal centrosomes. This may explain why monasters form in only a minority of the whole zygotes in which the pronuclei are kept apart (Sluder and Rieder, 1985b). The maternal MTOC is less likely to nucleate microtubules when competing for tubulin subunits with two paternal centrosomes than when it is by itself.

These results indicate that the maternal centrosome is only partially degraded after meiosis II. The female centriole is either disassembled or dissociated from the female pronucleus, and the reproductive capacity of the female centrosome, as defined by the ability to regularly double between mitoses, is lost. However, the MTOC activity of the maternal centrosome persists, as evidenced by the formation of a monaster when the female pronucleus is kept separate from the male pronucleus. In normal development, therefore, this active remnant of the maternal MTOC must be either disassembled or incorporated into the paternal centrosomes at syngamy.

IV. CENTROSOME INHERITANCE IN STARFISH

A. Experimental System

To gain a better understanding of how the echinoderm zygote selectively suppresses the activity of the maternal centrosome, we examined the pattern of centrosome inheritance in starfish zygotes (Sluder *et al.*, 1989b). Unlike sea urchins, starfish shed the female gametes as oocytes in prophase of meiosis I (Fig. 2a). Under the influence of 1-methyladenine, which induces oocyte shedding (Kanatani, 1973), the germinal vesicle breaks down (Fig. 2b). Shortly thereafter, a meiosis I spindle is assembled with one pole attached to the egg surface (Fig. 2c). The egg is fertilized at this stage and the reduction divisions are completed. Shortly after fertilization, a small sperm aster is assembled at the junction of the sperm head and tail (Fig. 2d). In telophase of meiosis I, the first polar body is extruded (Fig. 2e). The retained female centrosome splits to form the poles of the meiosis II spindle (Fig. 2e,f). In telophase of meiosis II, the second polar body is extruded (Fig. 2g) and the female pronucleus forms near the zygote cortex (Fig. 2h). Shortly after the completion of meiosis II, the female pronucleus starts to move into the sperm aster (Fig 2i). At this time, during interphase, the microtubule nucleating activity of the retained maternal centrosome is diminished to such a degree that the position of this centrosome can no longer be determined. Shortly after syngamy, the zygote nucleus breaks down and the first mitotic spindle is assembled (Fig. 2j). Mitotic spindles are always noticeably larger than either of the meiotic spindles. At some time prior to first mitosis, one parental centrosome must be suppressed. Since this event occurs after the oocyte is shed, the process is accessible to observation and experimental intervention *in vivo.*

B. Origin of the Centrosome Used in Development

At the completion of meiosis, the starfish zygote inherits the single maternal centrosome that was the deeper or cytoplasmic pole of the meiosis II spindle. To

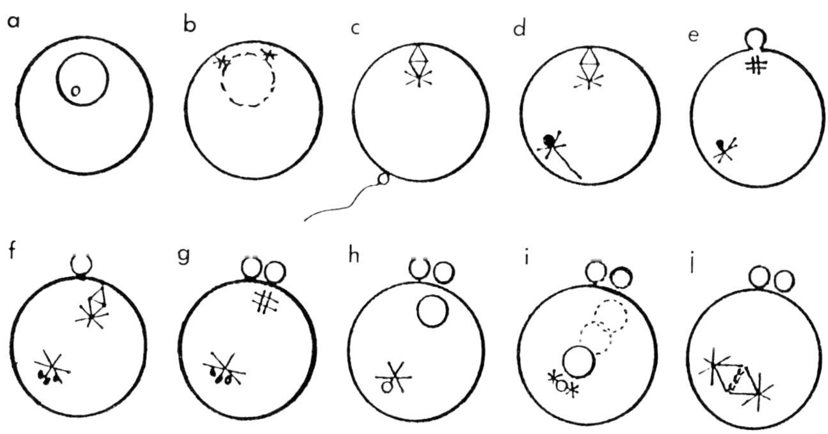

Fig. 2. Diagrammatic representation of postfertilization events leading to first mitosis in the starfish zygote.

determine which parental centrosome is used during development, we performed several simple experimental manipulations on fertilized eggs. First, we follow the development of slightly flattened zygotes in which the site of sperm entry is distant from the meiosis I spindle. In such cases, the female pronucleus does not move into the sperm aster but remains near the polar bodies. When the male and female pronuclei break down, a bipolar spindle is assembled at the site of the sperm pronucleus. By contrast, a monaster is assembled at the site of the female pronucleus. As the zygote proceeds from one mitosis to the next, the sperm centrosomes double with each cell cycle. However, the monaster does not double between mitoses and a single maternal nucleus is reformed at each telophase. Next, we induce polyspermy and follow the development of the sperm components. Polyspermic zygotes develop a number of sperm asters during meiosis I and II. After meiosis II, the female pronucleus moves into one of the sperm asters. All sperm asters double before the first mitosis and bipolar spindles are assembled at the sites of the sperm pronuclei. Finally, we follow cases in which the slight flattening of the zygote prevents both the extrusion of polar bodies and syngamy. In such cases, between two and four female pronuclei form at telophase of meiosis II. At first mitosis, the female pronuclei break down leaving small hyaline areas or monasters. During this time, the sperm aster doubles and organizes a bipolar spindle at first mitosis. Between subsequent mitoses, the sperm centrosomes double while the monaster(s) of maternal origin do not.

These observations indicate that the sperm alone contributes the centrosome used in the development of the starfish zygote. Thus, the zygote just after fertilization contains two maternal centrosomes that are destined to lose their independent existence and a paternal centrosome that will survive to be used in development. Clearly the organism can differentially control the stability of male and female centrosomes, even when they initially reside in a common cytoplasm. In the experiments described in the next sections, we test several possible ways in which the fertilized egg could exert such control over centrosomal inheritance.

C. Activity of the Sperm Centrosome during Meiosis

In principle, the starfish zygote could control centrosome inheritance by "masking" the paternal centrosome from putative cytoplasmic conditions that would destabilize all active centrosomes during the meiotic divisions. Such a mechanism was suggested by reports that the sperm centrosome in the fertilized egg of the surf clam *Spisula* does not initiate microtubule assembly until the zygote has completed meiosis II (Longo and Anderson, 1970; Kuriyama *et al.*, 1986).

How soon after sperm entry will the paternal centrosome in the starfish zygote start to assemble microtubules? A tiny sperm aster becomes apparent at the junction of the globular sperm head and the sperm tail within a few minutes of fertilization. This aster persists throughout both meiotic divisions, becoming noticeably larger during meiosis II. The sperm aster doubles at the time of syngamy and the daughter asters become more birefringent as the zygote approaches mitosis. The fact that the sperm centrosome is active in microtubule nucleation from the time of fertilization indicates that the capacity of the sperm centrosome to survive the meiotic divisions is not due to "masking" or compartmentalization to protect it from destabilizing cytoplasmic conditions during meiosis.

D. Artificially Activated Eggs

To determine if the fertilizing sperm releases factors that cause the eventual destabilization of the maternal centrosomes, we artificially activate unfertilized eggs with the Ca^{2+} ionophore A23187. A 10–15 min ionophone treatment triggers the elevation of the fertilization envelope and completion of the meiotic sequence. After meiosis II, each egg reforms a single female pronucleus. At the first, second, and third mitosis only, a single monaster is assembled in each egg. The monaster may be less radially symmetrical at the second and third mitoses. In no case did bipolar or multipolar spindles form at the second or third mitosis. As in sea urchin eggs, only the MTOC activity of the maternal centrosome is expressed at mitosis; the ability of this centrosome to double between mitoses is lost. Since these artificially activated eggs do not contain sperm components, we conclude that the reproductive capacity of the maternal centrosome in normal development is not inhibited by factors released from the sperm. The fate of the maternal centrosome in normal development must be determined by a change in cytoplasmic conditions of the egg alone.

E. Ultrastructure of Maternal Centrosomes during Meiosis

Our previous demonstration that the reproductive capacity of centrosomes in sea urchin zygotes is linked to the number of centrioles they contain (Sluder and Rieder, 1985a, Sluder *et al.,* 1989a) suggests that a loss of the maternal centrioles from starfish zygotes could determine the functional disappearance of the maternal centrosome before first mitosis. This possibility is consistent with random thin section studies showing that unfertilized sea urchin eggs do not contain centrioles (Verkey and Moyer, 1967; Anderson, 1968; Long and Anderson, 1968, 1969; Moy *et al.,* 1977; Paweletz and Mazia, 1979). The centrioles of the starfish zygote could be degraded before meiosis I or they could all be segregated to the polar bodies during the reduction divisions.

To address this issue, we systematically reconstructed from 0.25-μm serial sections the centrosomes of both meiosis I and meiosis II spindles. We were particularly interested in determining whether or not both poles of the first meiotic spindle contained pairs of centrioles and, if so, how these centrioles were distributed during the meiotic process. In addition, we reconstructed first and second polar bodies in order to unambiguously count the number of centrioles segregated from the oocyte.

Results show that each centrosome of the meiosis I spindle contains exactly two centrioles. In telophase of meiosis I, a pair of centrioles is segregated to the first polar body and the retained centrosome contains a pair of centrioles. At meiosis II each spindle pole contains only one centriole; thus, the pair of centrioles retained by the zygote at the end of meiosis I must separate without duplicating. In telophase of meiosis II, one centriole is segregated to the second polar body. As a consequence, the maternal centrosome retained by the starfish zygote at the end of meiosis II contains a single centriole. This finding indicates that the loss of the reproductive capacity of the maternal centrosome cannot be due solely to the segregation of all maternal centrioles to the polar bodies during meiosis I and II. We have yet to determine the fate of the single retained maternal centriole at the ultrastructural level; it may either be degraded or lose its association with the female pronucleus. Even if this centriole remains associated with the female pronucleus, the maternal centrosome has lost the ability to organize an independent aster in close proximity to the male centrosome as well as the capacity to double between mitoses.

F. Role of Regional Cytoplasmic Conditions

Although the paternal and maternal centrosomes reside in the same cytoplasm during meiosis, their different fates might be determined by their locations in different regions of the zygote. Unlike the maternal centrosomes, which remain at the cell cortex during the reduction divisions, the paternal centrosome moves into deeper cytoplasm shortly after fertilization. Possible differences in cytoplasmic conditions at the cell cortex and in the cell interior might lead to the selective destabilization of the maternal centrosomes.

To test this possibility, we use a microneedle to pull the meiosis I spindle from its cortical location and move it to deeper cytoplasm shortly after fertilization. By quantitatively focusing through the cell after the manipulation, we determine the position of the detached meiosis I spindle relative to the upper and lower cell surfaces. The zygotes studied are those in which the meiotic spindle was moved an average of 29 μm from the nearest cell surface. The example shown in Fig. 3 illustrates the results of one such experiment. Before detachment, the meiosis I spindle is seen in bright contrast at the upper cortex of the just-fertilized zygote (Fig. 3a). After manipulation, the spindle is left 41 μm from the nearest cell surface and is rotated (Fig. 3b). This spindle is functional and has a normal appearance in the deeper cytoplasm. During anaphase of meiosis I, the sperm aster can be seen below and to the left of the meiotic spindle (Fig. 3c). In this example, only one of the poles of the meiotic spindle splits into two asters at meiosis II (Fig. 3d.) However, in other trials, both poles of the meiosis I spindle split to form a total of four asters. In telophase of meiosis II, both maternal and paternal asters become large and weakly birefringent (Fig. 3e). At the end of meiosis II, female pronuclei form and move into the doubling sperm aster. The first mitosis, a single bipolar spindle is assembled from the paternal centrosomes (Fig. 3f). Since a multipolar spindle does not form, the maternal centrosomes must be lost; they no longer exist as independent entities.

From these results we conclude that maternal and paternal centrosomes have different intrinsic stabilities at the end of meiosis II, although they reside in the same cytoplasmic environment. The cortical location of the maternal centrosomes does not determine the loss of their reproductive capacity. We note that this functional difference between paternal and maternal centrosomes is also evident

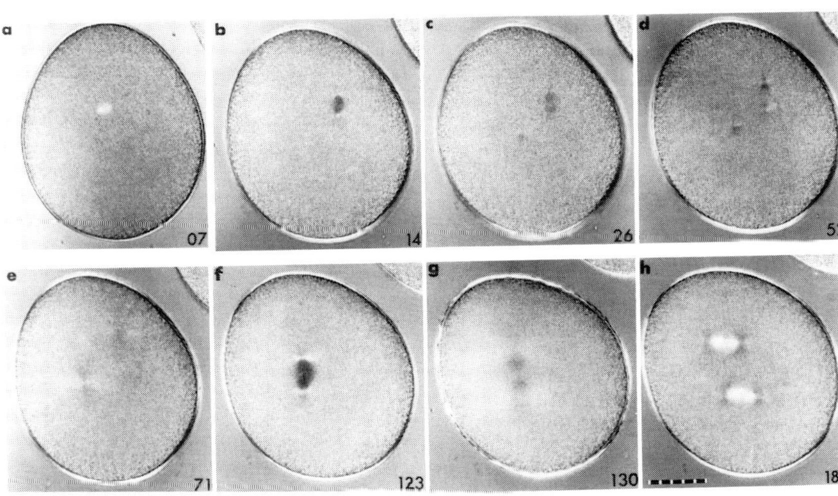

Fig. 3. Detachment of the meiosis I spindle from the cortex of a fertilized *Pisaster* zygote. (a) Before detachment the spindle is seen in bright contrast at the egg surface. (b) After detachment this spindle is 41 μm from the nearest surface. It appears in dark contrast because it was rotated during the micromanipulation. (c) Anaphase of meiosis I. The sperm aster is visible below and to the left of the spindle. (d) Meiosis II. The lower pole of the meiotic spindle has split. (e) Telophase of meiosis II. (f) First mitosis. Only the two paternal centrosomes are functional. (g) Telophase of first mitosis. (h) Second mitosis. The paternal centrosomes have reproduced and two spindles are assembled. Slight flattening of the zygote has caused the cleavage furrow to recede. Minutes after fertilization are shown in the lower corner of each frame. Polarization microscopy. 10 μm per scale division. [Reproduced from Sluder *et al.* (1989b) by copyright permission of Academic Press.]

in normal development. At the end of meiosis I, the retained maternal centrosome subdivides to form the two poles of the meiosis II spindle. However, the sperm centrosome does not split at this time; it does so only in late telophase of meiosis II. In concluding that maternal and paternal centrosomes are functionally different, we note that every female starfish starts life with her father's centrosomes. Therefore, the properties of her original paternal centrosome must be changed in some way when she forms oocytes.

The results of these micromanipulation experiments also indicate that all the maternal centrosomes are functionally equivalent. When the meiosis I spindle is moved into deeper cytoplasm, polar bodies are not formed and the zygote retains all maternal centrosomes. Since the functional independence of all maternal centrosomes is lost after meiosis II, the zygote does not control centrosome inheritance by specifically segregating potentially stable maternal centrosomes to the polar bodies. These results are in contrast to the results of studies on the parthenogenetic activation of some starfish eggs and eggs from other marine forms (Obata and Nemoto, 1984, and references cited therein). Successful parthenogenetic development was achieved only if first and/or second polar body extrusion was inhibited by the activating agent, suggesting that some female centrosomes are stable and could survive if retained by the zygote.

G. Spindle Transfers

When do the female centrosomes lose their reproductive capacity. Obviously this could occur early in gametogenesis, during the meiotic divisions, or just after meiosis II. If the reproductive capacity of the maternal centrosomes is eliminated during gametogenesis, the centrosomes that organize the meiosis I spindle could function as spindle poles during meiosis but would not be stable or able to reproduce later in development. If so, the doubling of the female centrosome between meiosis I and meiosis II would represent simply a subdivision of the centrosome, a possibility consistent with our observation that the meiosis II centrosomes have only a single centriole each. On the other hand, if the reproductive capacity of the maternal centrosomes is lost during or after meiosis, then the centrosomes of the meiosis I spindle should be stable and reproduce under appropriate cytoplasmic conditions.

Our approach to this problem was to transfer meiosis I spindles from newly fertilized zygotes into zygotes that had already reached first mitosis. We reasoned that the centrosomes of the meiosis I spindle should not reproduce between the early mitotic divisions if they have no reproductive capacity. If they do have reproductive capacity, they might reproduce between mitoses, assuming that they are stable under the cytoplasmic conditions of the early cleavage stages.

We start by aspirating a small volume of cytoplasm containing a meiosis I spindle from a newly fertilized zygote into a microinjection pipette (Fig. 4a). The pipette is quickly withdrawn from the zygote and capped with inert oil (Fig. 4b). Then a second microinjection preparation containing previously fertilized zygotes that have reached prophase of first mitosis is put on the microscope, and a zygote is microinjected with the cytoplasm containing the meiosis I spindle (Fig. 4c). Generally, the meiotic spindle retains its bipolar configuration throughout the transfer operation and first mitosis inside the recipient zygote. We follow the development of the injected spindle during the next two or three mitoses to determine if the meiotic centrosomes continue to be stable in a mitotic cytoplasm and to ascertain how they double between mitoses, if at all. The transferred spindle is easily recognized by its smaller size and its location at the site of the injection.

The mitotic spindle of the recipient zygote always develops in a normal fashion after the transfer operation; each centrosome of this spindle reproduces in the expected one-to-two-to-four fashion between successive mitoses. The meiotic centrosomes, however, show four patterns of behavior after first mitosis; these are diagrammed in Fig. 5. Rarely, the meiotic centrosomes do not appear to be stable and do not reproduce; only a disorganized monaster is formed at the second and subsequent mitoses (Fig. 5A). More commonly, the injected material forms a monopolar spindle with a single tightly focused aster at the second mitosis. This aster doubles before the third and subsequent mitoses, forming bipolar spindles of normal appearance (Fig. 5B). In other cases, a bipolar spindle is formed at the site of the injected material at the second mitosis; these centrosomes reproduce to form two bipolar spindles at the third mitosis (Fig. 5C). Finally, we sometimes observed four asters at the second mitosis at the site of the injected material (Fig. 5D).

These results, albeit variable, demonstrate that the centrosomes of the meiosis

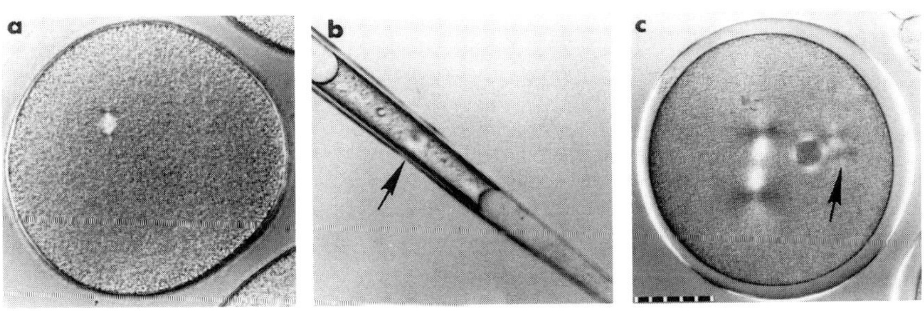

Fig. 4. Transfer of a meiosis I spindle from a newly fertilized egg into a zygote in first mitosis. (a) The meiotic spindle is seen in bright contrast at the upper surface of the egg. (b) This spindle has been aspirated into a micropipette and capped with oil. The spindle is indicated by the arrow. (c) The meiotic spindle after it has been microinjected into a zygote; for this particular trial, the recipient is a sea urchin zygote. The larger mitotic spindle is to the left of the transferred spindle (arrow). The spherical inclusions are oil droplets expelled from the micropipette. Polarization microscopy. 10 μm per scale division.

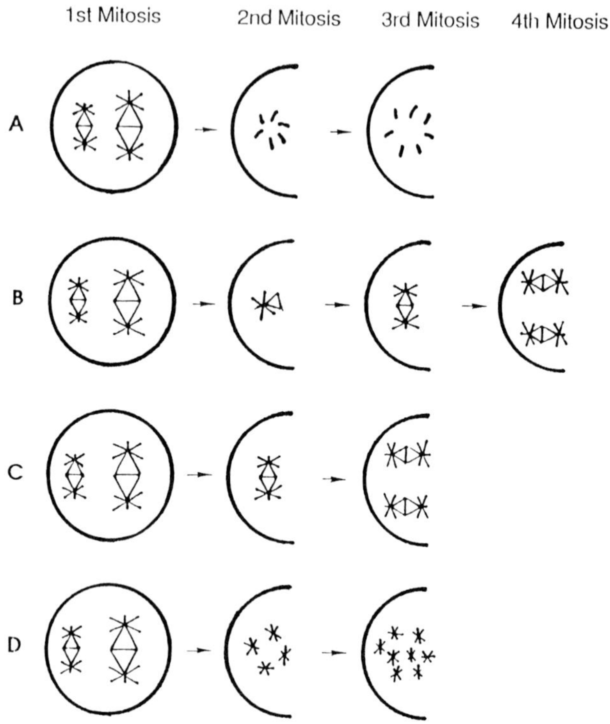

Fig. 5. Diagrammatic representation of the development of meiosis I spindles after transfer into fertilized starfish zygotes in first mitosis. For clarity of presentation only the development of the transferred meiotic spindles is shown; the mitotic spindles of the recipient zygotes develop in a normal fashion. Meiotic spindles are smaller than mitotic spindles during the first mitosis. The four patterns of development we observed are shown in rows A–D.

I spindle can have reproductive capacity. The variability in the number of centrosomes expressed at the second mitosis may result from damage to the centrosomes during the transfer operation, perhaps by proteolysis or mechanical disruption, or to the possibility that some portion of a meiosis I aster remains bound to the donor zygote cortex. In any case, the centrosomes surviving to second mitosis are capable of reproducing in a normal fashion.

In contrast to these results, transfers of meiosis II spindles into zygotes in first mitosis reveal that these centrosomes have a noticeably lower incidence of stability and reproductive capacity. Relatively few meiosis II centrosomes are stable or retain any reproductive capacity beyond first mitosis. In most cases, the transferred spindle gives rise to an irregular monaster at second and subsequent mitoses. Thus, the reproductive capacity of the meiosis II centrosomes appears to be significantly diminished on average. This finding suggests that the functional loss of the maternal centrosome's reproductive capacity in normal development must be due to a degradative process that starts during the reduction divisions. However, once zygote development has proceeded to first mitosis, this degradative activity no longer exists.

V. ORIGIN OF CENTROSOMES IN PARTHENOGENETICALLY ACTIVATED EGGS

Given the strictly paternal inheritance of centrosomes in sea urchin and starfish zygotes, the origin of the centrosomes used in the parthenogenetic development of unfertilized sea urchin eggs (Loeb, 1913; Harvey, 1956; VonLedebur-Villiger, 1972; Miki-Nourmura, 1977) and starfish eggs (Delage and Goldsmith, 1913; Picard and Doree, 1983; Obata and Nemoto, 1984) is a mystery. Parthenogenesis in sea urchin eggs is typically induced by activating the cell cycle of unfertilized eggs with agents such as ammoniated seawater or butyric acid-containing seawater and then treating the eggs with hypertonic seawater to express maternal centrosomes. Short treatments with hypertonic seawater fail to elicit supernumerary asters; only a monaster is formed at first mitosis. Slightly longer treatments (the exact duration of which depends on the particular batch of eggs used, the time in the spawning season, and the seawater temperature) lead to the formation of one or two asters that reproduce in a normal fashion and support the regular cleavages of early development. However, longer treatments with hypertonic seawater generally lead to the assembly of many supernumerary asters, often called cytasters. These cytasters are of strictly maternal origin, have centrioles (Dirksen, 1961; Kuriyama and Borisy, 1983; Kallenbach, 1985), and can reproduce in a normal fashion between mitoses (G. Sluder, unpublished observations).

If the original maternal centrosome was suppressed after meiosis II, where do these many centrosomes come from? Some investigators have proposed that parthenogenetic activation leads to the re-expression of the retained maternal centrosome (Briggs and King, 1959; Sachs and Anderson, 1970; Obata and Nemoto, 1984; Mazia, 1978, 1988). However, we suggest that this is unlikely since prolonged treatment with hypertonic seawater leads to the formation of 30 or more cytasters by the time of first mitosis (Miki-Nourmura, 1977; Mazia, 1978; Kuriyama and Borisy, 1983; Kallenbach, 1985). Unless unusual mechanisms are activated there is not enough time before first mitosis for that many centrosomes to form by the pathway conventionally used to form new centrosomes from an existing centrosome or centrosome precursor (in this case the putative maternal centrosome) at the time of DNA synthesis. Furthermore, treatment with hypertonic seawater will induce numerous cytasters in fertilized eggs by the time of first mitosis. In such cases, the sperm centrosome doubles only once, not several times, prior to first mitosis; the many cytasters appear throughout the cytoplasm as the zygote enters mitosis (G. Sluder, unpublished observations). Finally, the putative re-expression and multiple reproduction of the maternal centrosome retained from meiosis II does not easily explain the induction of odd numbers of cytasters (G. Sluder, unpublished observations) or the formation of cytasters in enucleated egg fragments (Yatsu, 1905; Wilson, 1925; Harvey, 1956; Kato and Sugiyama, 1971).

These observations suggest that parthenogenetic activation, at least in sea urchin eggs, does not re-express the original maternal centrosome retained from meiosis II but that it leads to the assembly of a variable number of new centrosomes from pre-existing cytoplasmic stores of centrosomal subunits. Indeed, we have shown that sea urchin eggs contain, even before fertilization, complete pools of all the subunits necessary to assemble at least 16 centrosomes. Neither transcription, translation, nor nuclear DNA synthesis are required for the repeated reproduction of the sperm centrosome (Sluder *et al.,* 1986, 1990; Sluder and Lewis, 1987). Perhaps parthenogenetic activation causes the self-assembly of a variable number of precursor structures for centrosomes (perhaps polar organizers) directly from the pre-existing cytoplasmic pools of subunits. These precursor structures then organize complete centrosomes by the time of first mitosis. If the duration of the hypertonic treatment is just right, only one or two precursor structures are formed; the centrosomes they organize can then support quasi-normal development.

VI. CONCLUDING REMARKS

The significance of our investigations into the inheritance of centrosomes in echinoderm zygotes extends beyond the demonstration of which gamete contributes the centrosome used in the development of the echinoderm zygote. We are principally interested in the fundamental biological problem of how an organism can predictably suppress the centrosome of one gamete and not the other. This issue was brought into sharp focus for us by the starfish zygote in which the centrosomes of both parents are active in the same cytoplasm during the meiotic divisions, yet the maternal centrosome ceases to function as an independent entity by the time of first mitosis.

This work leaves us with at least three intriguing questions for further study. What are the changes in cytoplasmic conditions during meiosis that alter the female centrosomes so that they can no longer reproduce? What is the target in these centrosomes that is degraded? Finally, what differences between female and male centrosomes make only one of them susceptible to these changed cytoplasmic conditions?

ACKNOWLEDGMENTS

I am pleased to express my gratitude to my collaborators C. L. Rieder, F. J. Miller, K. Lewis, and E. Davison for the substantial contributions they have made to the work described here. I thank F. J. Miller for the preparation of the drawings and photographic plates. I also thank T. Moss, A. Moss, and M. Melan for helpful suggestions on the manuscript. The efforts of Rebecca Bell and Carol Tuttle in the typing of the manuscript are most appreciated. The work on starfish oocytes was started in the laboratory of Daniel Mazia; I thank him for his financial support, use of his facilities, and his advice. Most of the work described in this article was supported by NIH GM 30758 to G. Sluder, NCI PO 3012708 to the Worcester Foundation for Experimental Biology, NIH GM-40198 to C. L. Rieder, and Biotechnological Resource grant RR 01219 to support the Wadsworth Center's high-voltage electron microscope.

REFERENCES

Anderson, E. (1968). Oocyte differentiation in the sea urchin *Arbacia punctulata,* with particular reference to the origin of cortical granules and their participation in the cortical reaction. *J. Cell Biol.* **37,** 514–539.

Aronson, J., and Inoué, S. (1970). Reversal by light of the action of *N*-methyl *N*-desacetyl colchicine on mitosis. *J. Cell Biol.* **45,** 470–477.

Aubin, J. E., Osborn M., and Weber, K. (1980). Variations in the distribution and migration of centriole duplexes in mitotic PtK_2 cells studied by immunofluorescence microscopy. *J. Cell Sci.* **43,** 177–194.

Baron, A. T., and Salisbury, J. L. (1988). Identification and localization of a novel, cytoskeletal, centrosome-associated protein in PtK_2 cells. *J. Cell Biol* **107,** 2669–2678.

Beatty, R. A. (1967). Parthenogenesis in vertebrates. *In* "Fertilization" (C. B. Metz and A. Monroy, eds.), pp. 413–440, Academic Press, New York.

Borisy, G. G. (1971). Mechanism of U.V. reversal of colchicine inhibition of microtubule aggregation *in vitro. Proc. 11th Ann. Meeting Am. Soc. Cell Biol* **27,** 35a.

Bornens, M., Paintrand, M., Berges, J. Marty, M. C., and Karsenti, E. (1987). Structural and chemical characterization of isolated centrosomes. *Cell Motil. Cytosk.* **8,** 238–249.

Briggs, R., and King, T. J. (1959). Nucleocytoplasmic Interactions in Eggs and Embryos. *In* "The Cell" (J. Brachet and A. Mirsky, eds.), pp. 537–617, Academic Press, New York.

Brinkely, B. R., and Stubblefield, E. (1970). Ultrastructure and interaction of the kinetochore and centriole in mitosis and meiosis. *In* "Advances in Cell Biology" (D. M. Prescott, L. Goldstein, and E. McConkey, eds.) pp. 119–180. Appleton-Century Crofts, Amsterdam.

Conklin, E. G. (1901). Centrosome and centrosphere in the maturation, fertilization, and cleavage of *Crepidula. Anat. Anz.* **19,** 280–287

Debec, A., Szollozi, A., and Szollozi, D., (1982). A *Drosophila melanogaster* cell line lacking centriole. *Biol Cell* **44,** 133–138.

Delage, Y., and Goldsmith M. (1913). La Parthenogenese Naturelle et Experimentale. Flammarion, Paris.

Dietz, R. (1964). The dispensability of the centrioles in the spermatocyte division of *Pales ferruginea* (Nematocera). *Proc. Oxford Chromosome Conf. 1st* **1,** 161–166.

Dirksen, E. R. (1961). The presence of centrioles in artificially activated sea urchin eggs. *J. Biophys. Biochem. Cytol.* **11,** 244–247.

Euteneuer, U., and McIntosh, J. R. (1981). Structural polarity of kinetochore microtubules in PtK_1 cells. *J. Cell. Biol.* **89,** 338–345.

Friedlander, M., and Wahrman, J. (1970). Giant centrioles in neuropteran meiosis. *J. Cell Sci.* **1,** 129–144.

Fulton, C. (1971). Centrioles. *In* "Origin and Continuity of Cell Organelles" (W. Beerman, J. Reinert, and H. Ursprung, eds.), pp. 170–221. Springer-Verlag, New York.

Gould, R. R., and Borisy, G. G. (1977). the pericentiolar material in Chinese hamster ovary cells nucleates microtubule formation. *J. Cell Biol.* **73,** 601–615.

Hamaguchi, M. S., and Hiramoto, Y. (1986). Analysis of the role of astral rays in pronuclear migration in sand-dollar *Clypeaster japonicus* eggs by the Colcemid–UV method. *Devel. Growth Diff.* **28,** 143–156.

Harvey, E. B. (1956). The American *Arbacia* and Other Sea Urchins. Princeton University Press, Princeton, New Jersey.

Heidemann, S. R., and Kirschner, M. W. (1975). Aster formation in eggs of *Xenopus laevis.* Induction by isolated basal bodies. *J. Cell Biol.* **67,** 105–117.

Heidemann, S. R., Sander, G., and Kirschner, M. W. (1977). Evidence for a functional role of RNA in centrioles. *Cell* **10,** 337–350.

Hinegardner, R. T., Rao, B., and Feldman, D. E. (1964). The DNA synthetic period during early development of the sea urchin egg. *Exp. Cell Res.* **36,** 53–61.

Inoué, S., and Sato H. (1967). Cell motility by labile association of molecules: The nature of mitotic spindle fibers and their role in chromosome movement. *J. Gen. Physiol.* **50,** 259–292.

Inoué, S., Fuseler, J., Salmon, E. D., and Ellis, G. W. (1975). Functional organization of mitotic microtubules; physical chemistry of the *in vivo* equilibrium system. *Biophys. J.* **15,** 725–744.

Kallenbach, R. J. (1985). Ultrastructural analysis of the initiation and development of cytasters in sea-urchin eggs. *J. Cell Sci.* **73,** 261–278.

Kanatani, H. (1973). Maturation-inducing substances in starfish. *Int. Rev. Cytol.* **35,** 253–298.

Kato, K. H., and Sugiyama, M. (1971). On the *de novo* formation of the centriole in the activated sea urchin egg. *Dev. Growth Diff.* **13,** 359–366.

Kuriyama, R. (1984). Activity and stability of centrosomes in Chinese hamster ovary cells in nucleation of microtubules *in vitro. J. Cell. Sci.* **66,** 277–296.

Kuriyama, R., and Borisy, G. G. (1981). Centriole cycle in Chinese hamster ovary cells as determined by whole mount electron microscopy. *J. Cell. Biol.* **91,** 814–821.

Kuriyama, R., and Borisy, G. G. (1983). Cytasters induced within unfertilized sea-urchin eggs. *J. Cell. Sci.* **61,** 175–190.

Kuriyama, R., Borisy, G. G., and Masui, Y. (1986). Microtubule cycles in oocytes of the surf clam, *Spisula solidissima:* An immunofluorescence study. *Devel. Biol.* **114,** 151–160.

Le Guen, P., and Crozet, N. (1989). Microtubule and centrosome distribution during sheep fertilization. *Eur. J. Cell Biol.* **48,** 239–249.

Loeb, J. (1913). Artificial Parthenogenesis and Fertilization. University of Chicago Press, Chicago.

Longo, F. J., and Anderson E. (1968). The fine structure of pronuclear development and fusion in the sea urchin. *Arbacia punctulata. J. Cell Biol.* **39,** 339–368.

Longo, F. J., and Anderson, E. (1969). Cytological aspects of fertilization in the lamellibranch, *Mytilus edulis.* I. Polar body formation and development of the female pronucleus. *J. Exp. Zool.* **172,** 69–96.

Longo, F. J., and Anderson E. (1970). An untrastructural analysis of fertilization in the surf calm, *Spisula solidissima. J. Vet. Res.* **33,** 495–514

Maller, J. Poccia, D., Nishioka, D., Kidd, P., Gerhart, J. and Hartman H. (1976). Spindle formation and cleavage in *Xenopus* eggs injected with centriole-containing fractions from sperm. *Exp. Cell Res.* **99,** 285–294.

Masuda, H., McDonald, K. L., and Cande, W. Z. (1988). The mechanism of anaphase spindle elongation: Uncoupling of tubulin incorporation and microtubule sliding during *in vitro* spindle reactivation. *J. Cell Biol.* **107,** 623–633.

Masuda, M., and Sato, H. (1984). Reversible resorption of cila and the centriole cycle in dividing cells of sea urchin blastulae. *Zool. Sci.* **1,** 445–462.

Mazia, D. (1961). Mitosis and physiology of cell division. *In* "The Cell, Biochemistry, Physiology, Morphology" (J. Brachet and A. Mirsky, eds.). Vol. 3, pp. 77–412. Academic Press, New York.

Mazia, D. (1978). Origin of twoness in cell reproduction. *ICN-UCLA Symp. Mol. Cell Biol.* **12,** 1–14.

Mazia, D. (1984). Centrosomes and mitotic poles. *Exp. Cell Res.* **153,** 1–15.

Mazia, D. (1987). The chromosome cycle and the centrosome cycle in the mitotic cycle. *Int. Rev. Cyt.* **100,** 49–92.

Mazia D. (1988). Mitotic poles in artificial parthenogenesis: A letter to Katsuma Dan. *Zool. Sci.* **5,** 519–527.

Mazia, D., Harris, P. J., and Bibring, T. (1960). The multiplicity of mitotic centers and the time-course of their duplication and separation. *J. Biophys. Biochem. Cytol.* **7,** 1–10.

Miki-Nourmura, T. (1977). Studies on the *de novo* formation of centrioles: Aster formation in the activated eggs of sea urchins. *J. Cell Sci.* **24,** 203–216.

Moy, G. W., Brandriff, B., and Vacquier, V. D. (1977). Cytasters from sea urchin eggs parthenogenetically activated by procaine. *J. Cell Biol.* **73,** 788-793.

Nicklas, R. B. (1971). Mitosis. *In* "Advances in Cell Biology" (D.M. Prescott, L. Goldstein, and J. E. McConkey, eds.), Vol. 2, pp. 225–297. Appleton-Century-Crofts, New York.

Nicklas, R. B., and Koch, C. A. (1972). Chromosome micromanipulation. Part 4. Polarized motions within the spindle and models for mitosis. *Chromosoma (Berlin)* **39,** 1–26.

Obata, L., and Nemoto, S. (1984). Artificial parthenogenesis in starfish eggs; Production of parthenogenetic development through suppression of polar body formation by methylxanthine. *Biol. Bull.* **166,** 525–536.

Paweletz, N., and Mazia, D. (1979). Fine structure of the mitotic cycle of unfertilized sea urchin eggs activated by ammoniacal sea water. *Eur. J. Cell Biol.* **20,** 37–44.

Paweletz, N., Mazia D., and Finze, E. M. (1984). The centrosome cycle in the mitotic cycle of sea urchin eggs. *Exp. Cell Res.* **152,** 47–65.

Paweletz, N., Mazia, D., and Finze, E. M. (1987a). Fine structural studies of the bipolarization of the mitotic apparatus in the fertilized sea urchin egg. I. The structure and behavior of centrosomes before fusion of the pronuclei. *Eur. J. Cell Biol.* **44,** 195–204.

Paweletz, N., Mazia D., and Finze, E. M. (1987b). Fine structural studies of the bipolarization of the mitotic apparatus in the fertilized sea urchin egg. II. Bipolarization before first mitosis. *Eur. J. Cell Biol.* **44,** 205–213.

Picard, A. and Doree, M. (1983). Hormone-induced parthenogenetic activation of mature starfish oocytes. *Exp. Cell Res.* **145,** 315–323.

Pickett Heaps, J. D (1971). The autonomy of the centriole: Fact or fallacy. Cytobios, **3,** 205–214.

Rappaport, R. (1986). Establishment of the mechanism of cytokinesis in animal cells. *Int. Rev. Cyt.* **105,** 245–281.

Rattner, J. B., and Phillips, S. G. (1973). Independence of centriole formation and DNA synthesis. *J. Cell Biol.* **57,** 359–372.

Rieder, C. L. (1982). The formation, structure, and composition of the mammalian kinetochore and kinetochore fiber. *Int. Rev. Cytol.* **79,** 1–58.

Rieder, C. L., and Borisy, G. G. (1982). The centrosome cycle in PtK_2 cells: Asymmetric distribution and structural changes in the pericentriolar material. *Biol. Cell* **44,** 117–132.

Rieder, C. L., and Alexander, S. P. (1990). Kinetochores are transported poleward along a single astral microtubule during chromosome attachment to the spindle in newt lung cells. *J. Cell Biol.* **110,** 81–95.

Robbins, E., Jentzsch, G., and Micali, A. (1968). The centriole cycles in synchronized HeLa cells. *J. Cell Biol.* **36,** 329–339.

Sachs, M. I., and Anderson E. (1970). A cytological study of artificial parthenogenesis in the sea urchin, *Arbacia punctulata. J. Cell Biol.* **47,** 140–158.

Salmon, E. D. (1975). Spindle microtubule thermodynamics of *in vivo* assembly and role in chromosome movement. *Ann, N.Y. Acad. Sci* **253,** 383–400.

Schatten, H., Schatten G., Mazia D., Balczon, R., and Simerly C. (1986). Behavior of centrosomes during fertilization and cell division in mouse oocytes and in sea urchin eggs. *Proc. Natl. Acad. Sci. U.S.A.* **83,** 105–109.

Schatten, H., Walter, M., Mazia, D., Biessmann, H., Paweletz, N., Coffe, G., and Schatten, G. (1987). Centrosome detection in sea urchin eggs with a monoclonal antibody against *Drosophila* intermediate filament proteins: Characterization of stages of the division cycle of centrosomes. *Proc. Natl. Acad. Sci. U.S.A.* **84,** 8488–8492.

Schatten, H., Howard, C., Coffe, G., Simerly, C., and Schatten, G. (1988). Centrosomes, centrioles and post-translationally modified microtubules during fertilization. *Zool. Sci.* **5,** 585–601.

Sluder, G. (1976). Experimental manipulation of the amount of tubulin available for assembly into the spindle of dividing sea urchin eggs. *J. Cell Biol.* **70,** 75–85.

Sluder, G. (1990). Experimental analysis of centrosome reproduction in echinoderm eggs. *Adv. Cell Biol.* **3,** 221–250.

Sluder, G., and Begg, D. A. (1985). Experimental analysis of the reproduction of spindle poles. *J. Cell Sci.* **76,** 35–51.

Sluder, G., and Rieder, C. L. (1985a). Centriole number and the reproductive capacity of spindle poles. *J. Cell Biol.* **100,** 887–896.

Sluder, G., and Reider, C. L. (1985b). Experimental separation of pronuclei in fertilized sea-urchin eggs; Chromosomes do not organize a spindle in the absence of centrosomes. *J. Cell. Biol.* **100,** 897–903.

Sluder, G., Miller, F. J., and Rieder, C. L. (1986). The reproduction of centrosomes: Nuclear versus cytoplasmic controls. *J. Cell. Biol.* **103,** 1873–1882.

Sluder, G., and Lewis, K. (1987). Long term effects of Aphidicolin on the cell division cycle of sea urchin eggs. *J. Exp. Zool.* **244,** 89–100.

Sluder, G., Miller, F. J., and Rieder, C. L. (1989a). Reproductive capacity of sea urchin centrosomes without centrioles. *Cell Motil. Cytosk.* **13,** 264–273.

Sluder, G., Miller, F. J., Lewis, K., Davison, E. D., and Rieder, C. L. (1989b). Centrosome inheritance in starfish zygotes: Selective loss of the maternal centrosomes after fertilization. *Dev. Biol.* **131,** 567–579.

Sluder, G., Miller, F. J., Cole, R., and Rieder, C. L. (1990). Protein synthesis and the cell cycle: Centrosome reproduction in sea urchin eggs is not under translational control. *J. Cell Biol.* **110,** 2025–2032.

Snyder, J. A., and McIntosh, J. R. (1975). Initiation and growth of microtubules from mitotic centers in lysed mammalian cells. *J. Cell Biol.* **67,** 744–760.

Steffen, W., Fuge, H., Dietz, P., Bastemeyer, M., and Muller, G. (1986). Aster-free spindle poles in insect spermatocytes: Evidence for chromosome induced spindle formation? *J. Cell Biol.* **102,** 1679-1687.

Stubblefield, E., and Brinkley, B. R. (1967). Architecture and function of the mammalian centriole. *Symp. Int. Soc. Cell Biol.* **6,** 175–218.

Szollosi, D., and Hunter, R. H. F. (1973). Ultrastructural aspects of fertilization in the domestic pig: Sperm penetration and pronucleus formation. *J. Anat.* **116,** 181-206.

Szollosi, D., Calarco, P., and Donahue, P. (1972). Absence of centrioles in the first and second meiotic spindles of mouse oocytes. *J. Cell Sci.* **11,** 521–541.

Verkey, C. A., and Moyer, F. H. (1967). Fine structural changes during sea urchin oogenesis. *J. Exp. Zool.* **164** 195–226.

Von Ledebur-Villiger, M. (1972). Cytology and nucleic acid synthesis of parthenogenetically activated sea urchin eggs. *Exp. Cell Res.* **72,** 285–308.

Vorobjev, I. A., and Chentsov, Y. S. (1982). Centrioles in the cell cycle. I. Epithelial cells. *J. Cell Biol.* **98,** 938–949.

Vorobjev, I. A., and Nadezhdina, E. S. (1987). The centrosome and its role in the organization of microtubules. *Int. Rev. Cyt.* **106,** 227–293.

Wheatley, D. N. (1982). The Centriole: A Central Enigma of Cell Biology. Elsevier Biomedical Press, New York.

Wilson, E. B. (1925). The Cell in Development and Heredity. MacMillan, New York.

Yatsu, N. (1905). The formation of centrosomes in enucleated egg-fragments. *J. Exp. Zool.* **2,** 287–312.

10

The Lymphocyte Centrosome

DAVID L. BROWN, J. DAVID KNOX, AND MICHELINE PAULIN-LEVASSEUR

Department of Biology
University of Ottawa
Ottawa K1N 6N5 Canada

I. INTRODUCTION

In mature unactivated B and T cells, DNA is tightly packed into large heterochromatin masses and exhibits very little transcriptional activity; the nucleus is surrounded by only a thin layer of cytoplasm. The centrosome is situated in a nuclear cleft and a small number of microtubules extend from it and form a submembranous radial array. When a foreign antigen is presented to an unactivated lymphocyte possessing a corresponding receptor, a specific immune response—such as the generation of cytotoxic, helper, suppressor, or other effector cells—may be triggered. In all such cases, the specific immune response is preceded by the proliferation and expansion of clones capable of recognizing and

Copyright © 1992 Academic Press, Inc.
All rights of reproduction in any form reserved.

responding to that particular antigen. The proliferative response involves large increases in transcription and translation, in the volume of the nucleus and cytoplasm, and in the number of microtubules assembled from the centrosome.

The centrosome and associated microtubule system have been implicated in a variety of lymphocyte functions. In this chapter we will discuss changes in structure and organization of the centrosome–microtubule system and the possible roles of these changes in activation, proliferation, and effector functions. We will emphasize those features that are unique to the lymphocyte and its usefulness as a model system for the analysis of centrosome structure and function.

II. THE UNACTIVATED LYMPHOCYTE

A. Structure and Metabolic Activity

The small lymphocytes present in peripheral blood and the lymphatic system continuously circulate throughout the body as quiescent cells. These cells, each committed to a single antigenic specificity, represent the millions of different lymphocyte clones that are capable of responding to an almost infinite number of foreign antigens. Until they respond to the appropriate stimulus (Section III) these cells have only very low levels of gene transcription and protein synthesis (see Crabtree, 1989, for review).

The small unactivated lymphocytes are nondividing G_0 cells. Typically they have diameters of 6–8 μm and few organelles are present in the thin layer of dense granular cytoplasm. The nucleus, which occupies most of the cell volume, has highly aggregated peripheral chromatin, only sparse interchromatinic material, and a small undifferentiated nucleolus (Fig. 1a). The nuclei of unactivated lymphocytes have been analyzed morphometrically and biochemically and have been shown to have several features that distinguish them from active nuclei (Setterfield *et al.*, 1983, 1985; Bladon *et al.*, 1988). The inactive appearance was the most likely reason that the functions of small lymphocytes were not recognized until relatively recently (Gowans *et al.*, 1962).

B. The Centrosome

The single centrosome of the unactivated lymphocyte is usually located adjacent to a nuclear cleft and, invariably, the Golgi apparatus is associated with it (Fig. 1a, b). Immunofluorescence staining with antitubulin reveals a sparse microtubule system that forms a radial array extending from the centrosome and around the periphery of the cell near the plasma membrane (Fig. 1c). Frequently, the vimentin intermediate filaments parallel the distribution of microtubules and also appear to extend from the centrosome (Fig. 1d, Paulin-Levasseur and Brown, 1987b).

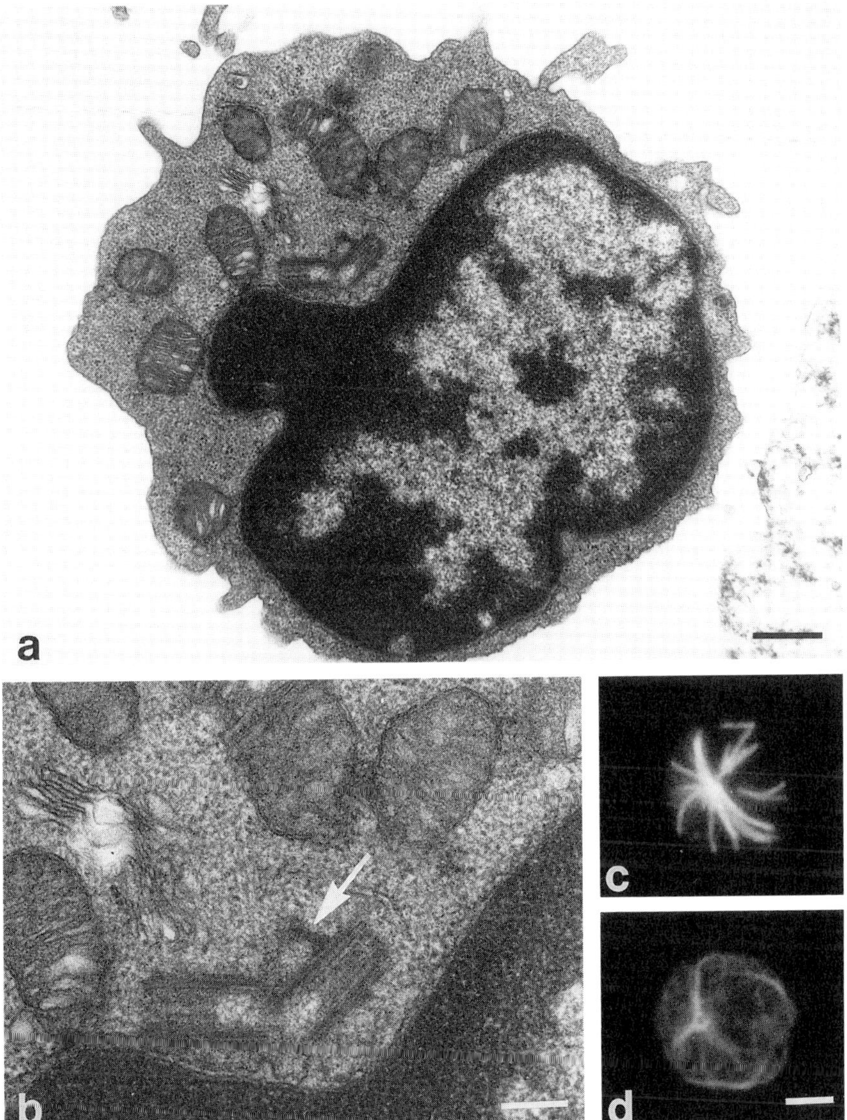

Fig. 1. Unactivated mouse splenic lymphocytes. (a) Electron micrograph showing the highly aggregated peripheral chromatin of the inactive nucleus, a thin layer of dense granular cytoplasm, and, in the nuclear cleft, the centrosome and associated Golgi apparatus. Bar: 0.50 μm. (b) Higher magnification view of the centrosomal region of (a) showing the electron-dense distal core of the centrioles and a single satellite body (arrow) associated with one of the centrioles. Bar: 0.25 μm. (c,d) Double immunofluorescence staining of a cell with antibodies to tubulin (c) and vimentin (d), demonstrating the partially coincident radial distribution of these two filamentous systems. Bar: 2.0 μm. (Fig. 1c,d reproduced from Paulin-Levasseur and Brown, 1987b, with permission).

The organization of the centrosome in unactivated mouse splenic lymphocytes has been examined by electron microscopy of serial sections (Schweitzer and Brown, 1984). The overall shape of the centrosome is roughly ovoid and the configuration of the two centrioles is approximately orthogonal (Fig. 1b). An electron-dense core is evident at the distal end of both centrioles. The centriole pair is surrounded by amorphous pericentriolar material that includes denser aggregates, the satellite bodies (SBs). In unactivated lymphocytes, approximately nine SBs are detected; these are exclusively associated with one of the centrioles (Figs. 1b, 4a). A similar centrosome organization has been reported for calf thymocytes (Komesli *et al.,* 1989; see Chapter 1).

The preferential, or exclusive, association of the SBs with one centriole of the pair in inactivated splenocytes and thymocytes is consistent with the observations in other cell types by Vorobjev and Chentsov (1982) and Rieder and Borisy (1982). They examined the centriole cycles in pig kidney embryo cells and PtK_2 cells, respectively, and reported that pericentriolar structures and the capacity to assemble microtubules were associated only with the parent centriole of the centrosome. Similarly, in the unactivated splenocyte the 40–45 microtubules detected by serial sectioning terminated at one centriole of the pair and most (an average of 32) could be traced to the SBs (Schweitzer and Brown, 1984).

C. Microtubule Functions

The initial interest in the microtubule system of lymphocytes was stimulated by the pioneering studies of Edelman and co-workers, which implicated the submembranous array of microtubules as a likely candidate for a role in signal transmission (reviewed in Edelman, 1976). A part of the evidence supporting this role was derived from studies of the effects of antimicrotubule drugs, such as Colcemid, on a process of cell-surface receptor redistribution termed capping (for review, see Bourguignon and Bourguignon, 1984).

Capping involves the cross-linking of lymphocyte cell-surface receptors by multivalent ligands, such as the plant lectin concanavalin A (con A) or anti-immunoglobulin (anti-Ig). This induces the temperature-dependent redistribution of receptor–ligand complexes to form patches that are subsequently transported in an energy-dependent manner to one pole of the cell. Ligand-mediated receptor patching and capping has been suggested to be important in a number of immune responses (Bourguignon and Bourguignon, 1984; Kupfer and Singer, 1989a).

Good evidence suggests that the actin microfilament system is directly involved in lymphocyte surface receptor redistribution: (1) the inhibitor of microfilament function, cytochalasin B, effectively inhibits capping induced by anti-Ig; (2) immunofluorescence staining has shown that actin and a variety of actin-binding proteins become localized to the region below surface receptor caps; and (3) ligand-bound receptors become associated with the detergent-insoluble cytoskeletal fraction (Bourguignon and Bourguignon, 1984; Kupfer and Singer, 1989b).

Although it has been clear for some time that the microtubule system is involved in the redistribution of lymphocyte surface receptors (Edelman, 1976), the precise function of microtubules and intermediate filaments in the process is unclear. Figure 2 summarizes the results of several of our studies investigating cytoskeletal reorganizations during the capping of surface immunoglobulin (sIg) on mouse B lymphocytes induced by anti-Ig.

As diagrammed in Fig. 2a, the sIg–anti-Ig cap is normally formed over the region of the cytoplasm containing the centrosome. A colchicine or Colcemid treatment that results in disassembly of the microtubules does not inhibit cap formation, but the cap is now formed randomly on the cell surface with respect to the centrosome (Fig. 2b). These results led us to suggest (Rogers *et al.,* 1981) that the submembranous microtubule system provides the spatial control to direct the microfilament-dependent translocation of sIg–anti-Ig complexes to form a cap over a specified region of the cytoplasm. Our subsequent studies (Paatero and Brown, 1982), examining the effects of the microtubule-assembly-promoting drug taxol on the organization of microtubules and on the capping of sIg are consistent with this suggestion.

In lymphocytes, taxol causes an extensive reorganization of the microtubule system and the formation of one to several centrosome-associated bundles of microtubules. Concomitantly, the centrosome with its associated Golgi apparatus is displaced from its normal position in the nuclear cleft to a position near the plasma membrane. The microtubules of the bundles in taxol-treated cells no longer form a submembranous array, but most frequently pass through the cytoplasm near the surface of the nucleus (Brown *et al.,* 1985; Section III,C). This reorganization of microtubules by taxol has the same effect as the removal of the microtubules by Colcemid, leading to random cap position with respect to the centrosome (Fig. 2c; Paatero and Brown, 1982). This indicates that it is not just the presence of microtubules, but their particular organization, that is required to direct the site of cap formation over the region of the cell containing the centrosome and Golgi apparatus where endocytosis occurs.

Contrary to some reports (e.g., Yahara and Kakimoto-Sameshima, 1978), we have observed that the radial pattern of microtubules is maintained throughout patch and cap formation. In contrast, coincident with cap formation, vimentin is relocalized as a diffuse accumulation under the cap and discrete filament arrays are no longer detected by immunofluorescence (Fig. 2d; Paulin-Levasseur and Brown, 1987b). A similar relocalization of vimentin staining underneath the cap has been observed by others (Bourguignon and Bourguignon, 1981; Dellagi and Brouet, 1982) and has led to the suggestion that this intermediate filament reorganization might function in the capping process.

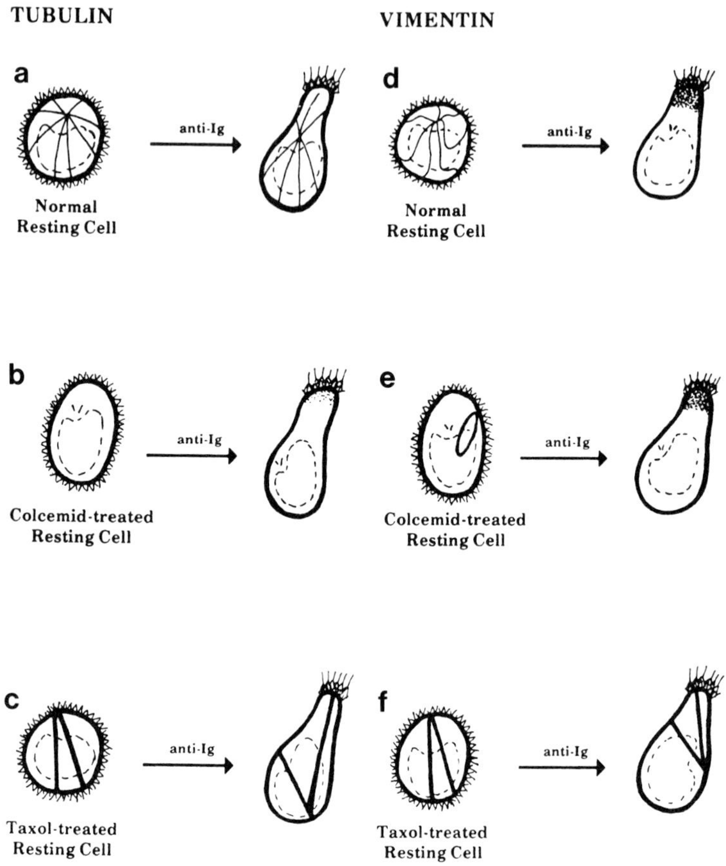

Fig. 2. Organization of tubulin and vimentin during the capping of surface immunoglobulin in normal and drug-treated mouse splenic lymphocytes. Refer to text for description.

As diagrammed in Fig. 2e, studies have shown that the disassembly of microtubules leads to the formation of a ring-like structure of vimentin filaments that shows no spatial relationship to the position of the centrosome. It is interesting that, coincident with cap formation, which under these conditions also is not related to the position of the centrosome, this ring-like structure disappears and vimentin forms a diffuse aggregate under the cap that is indistinguishable from the one seen in control cells. Clearly the formation of the diffuse aggregate and its localization to the cap can occur independent of the microtubule system.

In taxol-treated lymphocytes, vimentin filaments colocalize with the microtubule bundles and remain colocalized with the stable microtubule bundles during cap formation (Fig. 2f). This result shows that neither the formation of the vimentin aggregate nor its relocalization to the cap is a requirement for cap formation (Fig. 2f). These results, however, do not rule out a role for the intermediate filament system in capping; it is possible (as discussed in Paulin-Levasseur and Brown, 1987b) that this system contributes to the directionality of sIg redistribution by transducing spatial information between microtubules and microfilaments.

As has been observed in a number of cell types including lymphocytes (see Thyberg and Moskalewski, 1985, for review), the disassembly of microtubules by Colcemid or colchicine produces a disorganization of the Golgi apparatus and a dissociation from the centrosome. Recent studies have shown that microtubule disassembly is required, but is not sufficient for Golgi apparatus fragmentation and dispersal (Turner and Tartakoff, 1989), whereas microtubules are absolutely required for the reassembly of the Golgi apparatus (Ho *et al.*, 1989). Also, as described earlier (Fig. 2e) and observed in other cell types (Bloom *et al.*, 1985), microtubule disassembly leads to a collapse of the intermediate filament system, suggesting that the microtubule system functions as a scaffold for its maintenance. The minimal microtubule array in the unactivated lymphocyte, therefore, appears to carry out some of the same functions of maintaining cell organization and polarity as do the more elaborate microtubule arrays in cycling cells.

D. Functions of the Unactivated Lymphocyte

We have discussed previously the analogy that can be made between the unactivated lymphocyte and the spermatozoon (Kaplan *et al.*, 1987). During spermatogenesis, the substitution of sperm-specific basic proteins, such as the protamines, for histones is correlated with the condensation and dense packaging of DNA that results in a nucleus that is inactive in transcription and replication (e.g., Roca and Mezquita, 1989). Similarly, during the differentiation of lymphocytes into mature unactivated B and T cells, their chromatin becomes highly aggregated (Fig. 1a) and exhibits little transcriptional activity.

In the case of the lymphocyte, this differentiation results in a cell that is superbly designed for transporting membrane receptors throughout the body. In this mode, and as a consequence of its ability to bind nonself molecules in the appropriate context, the largely inactive lymphocyte genome is activated. The activated lymphocyte responds to soluble growth factors known as lymphokines, progresses through the cell cycle, and proliferates. The cells resulting from this proliferation either become effector cells and immediately function in a specific immune response (Section IV) or revert to an inactive state, becoming memory cells awaiting a subsequent encounter with the nonself molecule that engendered their differentiation. Changes in the organization and potential functions of the centrosome and microtubule system during lymphocyte activation and proliferation are discussed in the following section.

III. THE CENTROSOME AND LYMPHOCYTE ACTIVATION

A. The Proliferative Response

As discussed in Section II, D, the initial function of the small unactivated lymphocyte is to recognize and bind a specific ligand (i.e., antigen in an appropriate context) and to proliferate. For reviews of B and T lymphocyte activation see DeFranco (1987) and Gardner (1989). Usually, specific ligand binding results in the proliferation of a limited number of lymphocyte clones capable of recognizing that ligand. Even before the role of this proliferative response was known, however, it had been shown that small peripheral blood lymphocytes in culture could be induced to divide by plant mitogens (Hungerford *et al.,* 1959).

Although the activation by plant mitogens, such as con A and phytohemagglutinin, is less specific than that by an antigen, it initiates the same series of events and results in lymphocyte proliferation and differentiation. The advantage of mitogenic activation as an experimental tool is that, in comparison with antigen, a much larger number of clones is triggered to enter the cell cycle. This has greatly facilitated the analysis of the molecular and structural changes characteristic of the proliferative response.

Following the earliest signal transduction events of activation, a complex series of gene activations occurs as the cells progress through the cell cycle and divide. Some genes, such as the proto-oncogenes c-*myc* and c-*fos,* are transcribed within minutes of activation and independent of protein synthesis (Crabtree, 1989). During the next few hours, a large number of genes is activated and total RNA and protein synthesis is elevated severalfold as the cells proceed through G_1. The response of the lymphocyte population is not synchronous and the duration of G_1 varies with lymphocytes isolated from different species. In mouse splenic lymphocytes activated by con A, for example, DNA replication is first detected after 20–24 hr and peaks at about 48 hr; in human peripheral blood lymphocytes also activated by con A, both the onset and peak of DNA synthesis occur later.

The increased transcriptional and translational activities detected in G_1 correlate with a large increase in cell size, a phenomenon referred to as blastogenesis. During blastogenesis, nuclear structure is extensively remodeled, nuclear volume increases by an order of magnitude due primarily to an amplification of interchromatinic material, aggregated chromatin is dispersed into small clumps, and a prominent nucleolus is formed (Fig. 3a; Setterfield *et al.,* 1983, 1985; Bladon *et al.,* 1988; Brown *et al.,* 1989). All these changes precede and appear to be required for the entry of cells into S phase. Extensive changes also occur in the volume and content of the cytoplasm. Generally, the cytoplasm appears less dense than that of the unactivated cell and contains more organelles (Fig. 3a; Rudd *et al.,* 1979; Brown *et al.,* 1985).

B. Centrosome Restructuring

Coincident with the changes in cytoplasmic and nuclear morphologies, major alterations in the extent and organization of cytoskeletal proteins occur during blastogenesis. Of particular interest are changes in the structure and microtubule assembly capacity of the centrosome.

1. Structure and Microtubule Assembly Activity in Vivo

The diagrammatic representations, derived from electron microscopy of serial sections (Schweitzer and Brown, 1984), in Fig. 4a,b show the major structural changes of the centrosome during cell activation. There is an overall increase in centrosome size that appears to be due to both an increase in the amount of pericentriolar material and the separation of the two centrioles. The SBs, which were exclusively associated with one centriole of the pair in the unactivated cell, now occur primarily in the region between the proximal ends of the two centrioles and are more numerous and larger (Fig. 3b). In other cell types it has been shown that the daughter centrioles acquire pericentriolar material and function as microtubule organizing centers only when they serve as parent centrioles in the next centriole cycle (Rieder and Borisy, 1982; Vorobjev and Chentsov, 1982). We have suggested that the relocalization of SBs observed in activated lymphocytes may represent a stage in the partitioning of microtubule organizing material between the old and new centrioles.

The restructuring of the centrosome during activation correlates with a 4- to 5-fold increase in its microtubule assembly capacity. Many microtubules are observed extending from the SBs situated between the two centrioles (Fig. 3b) as well as from the amorphous pericentriolar material (Schweitzer and Brown, 1984). By serial section electron microscopy we estimated that a minimum of 8 microtubules ended on each of the 12 SBs of the activated cell centrosome for a total of 96; about the same number ended in the amorphous pericentriolar material.

2. Structure and Assembly Activity in Vitro

The increase in microtubule assembly capacity of the centrosome during lymphocyte activation was confirmed by *in vitro* studies (Schweitzer and Brown, 1984). In these experiments, unactivated and activated cell populations were treated with low temperature or colchicine to disassemble existing microtubules. They were then premeabilized with low concentrations of Triton X-100 and incubated at 37°C with phosphocellulose-purified bovine brain tubulin under conditions that did not permit spontaneous assembly. As assayed by electron

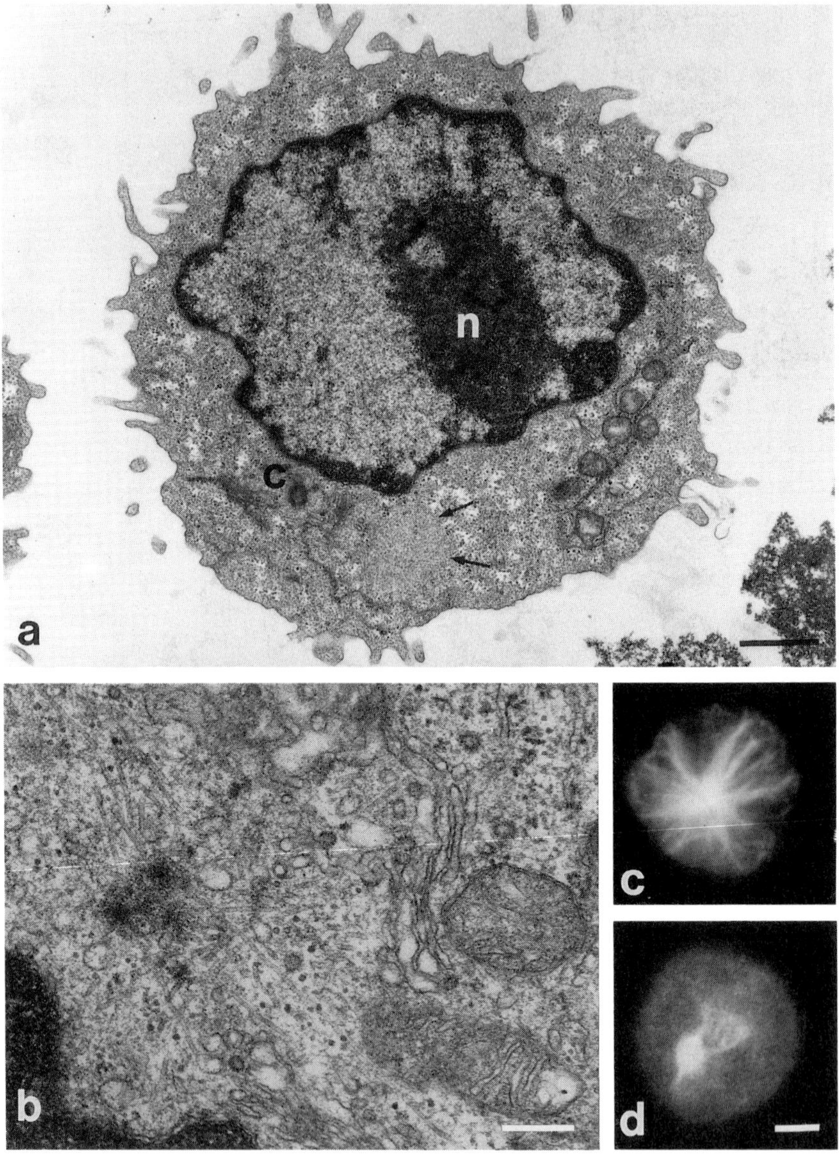

Fig. 3. Activated lymphocytes cultured for 48 hr in the presence of 4 μg/ml concanavalin A. (a) Electron micrograph showing the activated nucleus with disaggregated chromatin and a prominent nucleolus (n). Note the fibrogranular mass (arrows) located in the vicinity of the centrosome (c). Bar: 1.0 μm. (b) Electron micrograph of a section between the proximal ends of the two centrioles showing microtubules ending on satellite bodies. Bar: 0.25 μm (c,d) Double immunofluorescence staining of a cell with antibodies to tubulin (c) and vimentin (d), showing the increased number and length of the microtubules ending on the centrosome and the vimentin aggregate located in this region. Bar: 2.5 μm. (Fig. 3c and d reproduced from Paulin-Levasseur and Brown, 1987a, with permission).

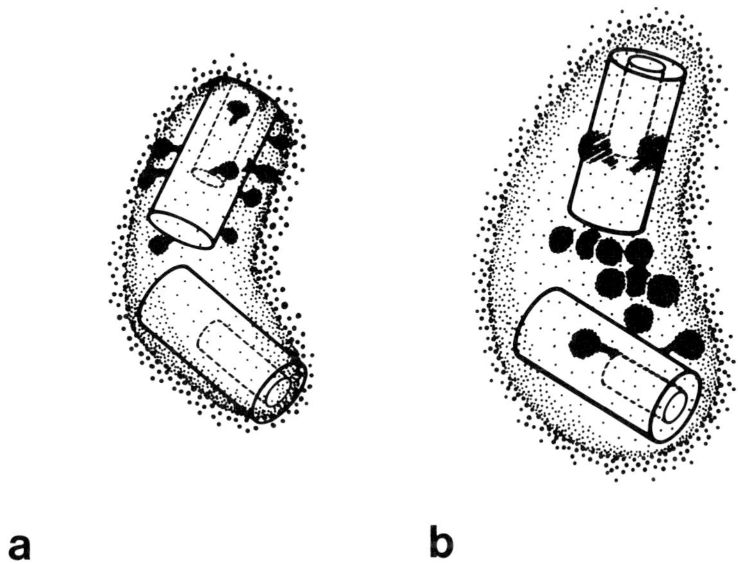

Fig. 4. Diagrammatic representations of the unactivated (a) and activated (b) lymphocyte centrosomes. The large stipple represents ribosomes which appear to be excluded from the pericentriolar material, shown in fine stipple. The dashed cylinders in the centrioles represent the dense core material present at their distal ends. Refer to text for additional description. (Reproduced from Schweitzer and Brown, 1984, with permission).

microscopy, the centrosomes of activated lymphocytes consistently initiated the assembly of larger numbers of microtubules than those of unactivated lymphocytes. In addition, the distribution of initiation sites *in vitro* closely resembled that observed *in vivo.* In both situations, microtubules assembled primarily from the SBs of unactivated cell centrosomes, whereas both SBs and amorphous pericentriolar material initiated assembly in the activated cell centrosomes.

An interesting observation in these studies was that microtubule disassembly by low temperature also resulted in the disaggregation of the SBs. This disaggregation was not simply a result of microtubule disassembly, since SBs were well preserved in cells in which microtubules had been disassembled by colchicine. The disaggregation of the SBs by low temperature was reversible *in vivo* and *in vitro;* at least *in vitro* it appeared to be dependent on microtubule reassembly, since SBs did not reform in control samples warmed in the absence of tubulin.

A second unexpected observation from these assembly experiments was that the centrosomes of unactivated and activated lymphocytes were able to initiate more microtubules *in vitro* (Schweitzer and Brown, 1984) than *in vivo* (Waterhouse *et al.,* 1983) as the exogenous tubulin concentration was increased. These results contrast those obtained for interphase centrosomes in permeabilized 3T3 cells

(Brinkley *et al.,* 1981) and mitotic centrosomes isolated from CHO cells (Kuriyama, 1984), in which the number of microtubules initiated *in vitro* becomes saturated at low tubulin concentrations. Our interpretation of this phenomenon is that the lymphocyte centrosome has assembly sites that are not occupied and that some component necessary for microtubule assembly, such as tubulin, is limiting. The observation that taxol treatment of lymphocytes promotes increased assembly from the centrosome (see Section III,C,1) is consistent with this interpretation. Our *in vitro* results with splenic lymphocyte centrosomes are also consistent with the recent study of Bré and Karsenti (1990) who analyzed the *in vitro* assembly of brain microtubule proteins by centrosomes isolated from a human lymphoblastoid cell line. They show that the extent of microtubule nucleation from the centrosome can be modulated by altering the ratio of tau:tubulin and suggest that the number of nucleated microtubules may be determined by factors that are not necessarily irreversibly bound to centrosomes.

3. *Calmodulin Association with the Centrosome*

What regulates the increased microtubule assembly from the centrosome is not known. Calmodulin has been suggested to play a role in microtubule stabilization, particularly in the assembly of the kinetochore microtubules of the spindle apparatus (Sweet *et al.,* 1989, and references therein). We have examined the possibility that calmodulin might be involved in promoting increased microtubule assembly during lymphocyte activation (Campos-Gonzalez, R. and D. L. Brown, unpublished observations). In unactivated lymphocytes, calmodulin is not detected in association with the centrosome or microtubules (Fig. 5a,b). However, coincident with the increases in microtubule assembly during lymphocyte activation, calmodulin localizes to the centrosome and to the proximal ends of the microtubules terminating on it (Fig. 5c,d). This localization is disrupted by low-temperature-induced disassembly of microtubules; relocalization of calmodulin to the centrosome occurs coincident with microtubule reassembly when the cells are rewarmed (Fig. 5e,f). This association of calmodulin with the centrosome and the observation that the reassembly of microtubules after cold disassembly is inhibited by calmodulin antagonists (data not shown) implicate calmodulin in the temporal regulation of microtubule assembly during lymphocyte activation. Calmodulin may be functioning directly to regulate centrosome activity in some way, or it may be promoting increased assembly by stabilizing microtubules in the manner suggested by Sweet *et al.* (1989).

4. *Tubulin Content of Activated Lymphocytes*

The large increase in microtubule assembly during lymphocyte activation also appears to be dependent on a similar large increase in cellular tubulin content (Waterhouse *et al.,* 1983). Determinations of tubulin contents, made by measurement of peptides specific to tubulin, showed a differential increase in tubulin as

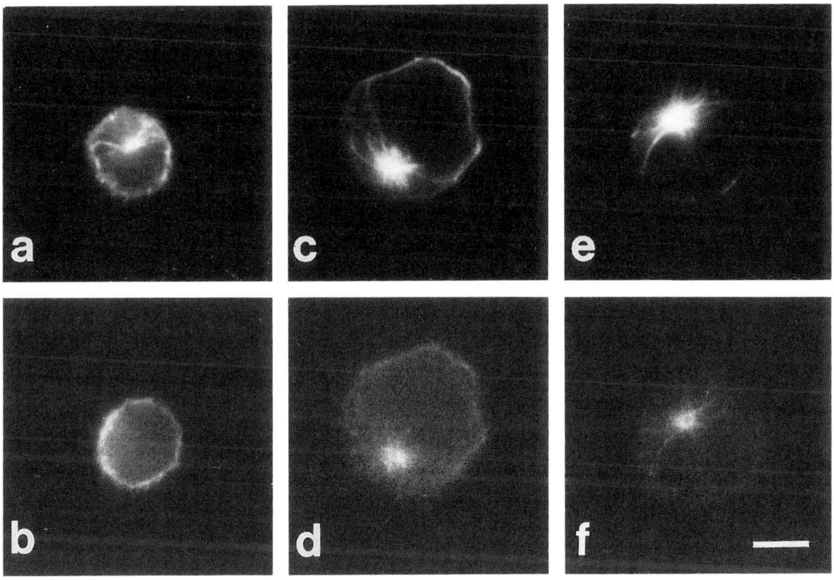

Fig. 5. Double immunofluorescence staining of tubulin (a,c,e) and calmodulin (b,d,f). (a,b) No calmodulin appears associated with the centrosome or microtubules in an unactivated lymphocyte. (c,d) Calmodulin is colocalized with the centrosome and the proximal ends of microtubules in an activated lymphocyte that has been cultured for 48 hr in the presence of 4 μg/ml concanavalin A. (e,f) Calmodulin relocalizes to the centrosomal region during microtubule reassembly in an activated cell that was warmed to 37°C for 5 min after a 30 min incubation at 4°C. Bar: 5.0 μm.

a percentage of total cell protein. Tubulin increased by 50% 24 hr after con A activation and remained at almost the same level after 48 hr, whereas the major increase in total cell protein occurred between 24 and 48 hr. This results in the steady increase in tubulin content per cell over the activation period to a value 2.5-fold higher than that seen in the unactivated population.

In addition to the increase in the number of assembled microtubules during lymphocyte activation, individual microtubules show an increase in length concomitant with the enlargement of the cell. Throughout activation, the growing microtubule array maintains the same radial organization, extending from the centrosome to the cell periphery just under the plasma membrane (Fig. 3c). In contrast, the vimentin intermediate filament network undergoes extensive reorganization during lymphocyte activation and mitosis (Paulin-Levasseur and Brown, 1987a, 1989).

5. *Reorganization of Vimentin in Activated Lymphocytes*

As described in Section II, C, the vimentin filaments in unactivated lymphocytes are partially coincident with the radial pattern of microtubules and appear to extend from the centrosome region. On the other hand, in the large lymphocytes fully activated by con A, a filamentous vimentin network is not detected by immunofluorescence staining. Rather, vimentin is observed as a nonfilamentous aggregate located near the centrosome (Fig. 3d). By electron microscopy, a fibrogranular mass, which may correspond to the aggregate observed by immunofluorescence, was observed in this vicinity (Fig. 3a). This reorganization occurs as a late event in the activation sequence, after the major increase in microtubule assembly has occurred, and may be associated with the entry of cells into S phase.

The vimentin network also reorganizes extensively during mitosis and cytokinesis (Paulin-Levasseur and Brown, 1987a, 1989). When the cells enter mitosis, the vimentin aggregate formed during activation is not detected and a vimentin filament network extends from the developing spindle poles and forms a cage-like structure that encloses the spindle apparatus. During cytokinesis, the vimentin filaments disappear; in each early G_1 daughter cell, a vimentin aggregate is associated with the daughter centrosome that has relocalized to a position near the midbody.

Recently it has become clear that intermediate filaments in several cell types are highly dynamic components of the cytoskeleton (for reviews, see Steinert and Liem, 1990; Stewart, 1990). Our results indicate that the vimentin intermediate filaments of lymphocytes also are highly dynamic, both during the interphase activation sequence and during cell division. The functions of such dynamic reorganizations are not known for lymphocytes or for any other cell type. Our studies with drugs that affect microtubule assembly, taxol in particular (see Paulin-Levasseur and Brown, 1987a; Section III,C,1), have suggested that vimentin dynamics may be important in the progression of lymphocytes through the cell cycle.

C. Requirements for Microtubule Organization in Activation

1. Mitogenic Activation

The treatment of lymphocytes with drugs, such as colchicine and Colcemid, that result in microtubule disassembly has been used extensively as an approach to indicate which lymphocyte functions are dependent on the presence of microtubules. Using this approach, we (Rudd *et al.*, 1979) and others (Thyberg, 1984, for review) have shown that mitogenic stimulation of lymphocytes is strongly inhibited or delayed by these drugs.

A problem with these types of studies has been to determine if the effect on mitogenic activation is directly caused by the removal of the microtubules. It is known that in lymphocytes (Thyberg *et al.*, 1977; Rogers *et al.*, 1981), as in other cell types (Thyberg and Moskalewski, 1985; Ho *et al.*, 1989; Turner and Tartakoff, 1989), disassembly of the microtubule system leads to an extensive disorganization of other cytoplasmic components, including the Golgi apparatus. As an alternative to the use of microtubule-disrupting drugs, we have carried out a series of studies examining the effects of the microtubule-assembly-promoting drug taxol on mitogenic activation.

Taxol has been shown to promote assembly of and to stabilize microtubules *in vivo* (Manfredi *et al.*, 1982) and, in some types of mammalian cells, to cause random microtubule assembly independent of the centrosome (e.g., De Brabander *et al.*, 1981). Such random assembly can have effects similar to those of microtubule removal by drugs such as colchicine, including a loss of cell polarity (De Brabander *et al.*, 1981) and a disruption of the Golgi apparatus (Sandoval *et al.*, 1984). In contrast, in unactivated and activated lymphocytes, a 10 μM taxol treatment for 4 hr promotes the assembly of microtubules only from the centrosome. As a result, one to several microtubule bundles are formed (Fig. 6). These bundles extend through the cytoplasm near the surface of the nucleus, distant from the submembranous position of the microtubules in untreated cells. In taxol-treated cells, the Golgi apparatus is not disrupted and it maintains its association with the centrosome, which may be displaced toward the plasma membrane (Fig. 6; Brown *et al.*, 1985).

To determine the effects of this microtubular reorganization on mitogenic activation, lymphocytes were preincubated with taxol for 4 hr and then activated by con A for 48 hr in the continued presence of taxol. Our results show that many of the major events of mitogenic activation can proceed normally in the presence of an extensively reorganized microtubule system. These events include the increase in cell size, the structural changes of the nucleus and cytoplasm (Fig. 6) that occur in G_1, and the replication of DNA in S phase that ensues (Fig. 7). These observations do not support earlier suggestions (Edelman *et al.*, 1973; Yahara and Edelman, 1973) that the normal submembranous microtubule array is required for transmembrane communication of the mitogenic signal. Our observations lead us to suggest that the role of the microtubule system in the mitogenic activation of lymphocytes is in maintaining cell polarity and the organization of the Golgi apparatus. This role apparently can be carried out by either the normal radial microtubule array or by the taxol-induced microtubule bundles, at least up to mitosis. When cells enter mitosis, the bundles are disassembled, multiple asters are formed, and the cells are blocked from further progression.

The microtubules of the taxol-induced bundles appear to be highly stable, even when the taxol is removed from the culture medium (Brown *et al.*, 1985); they are relatively resistant to disassembly by colchicine, low temperature, and methyl mercury (unpublished results). The ability of taxol-treated cells to proceed through mitogenic activation suggests that the dynamic behavior of microtubules, proposed to normally occur in interphase and mitotic cells (Kirschner and Mitchison,

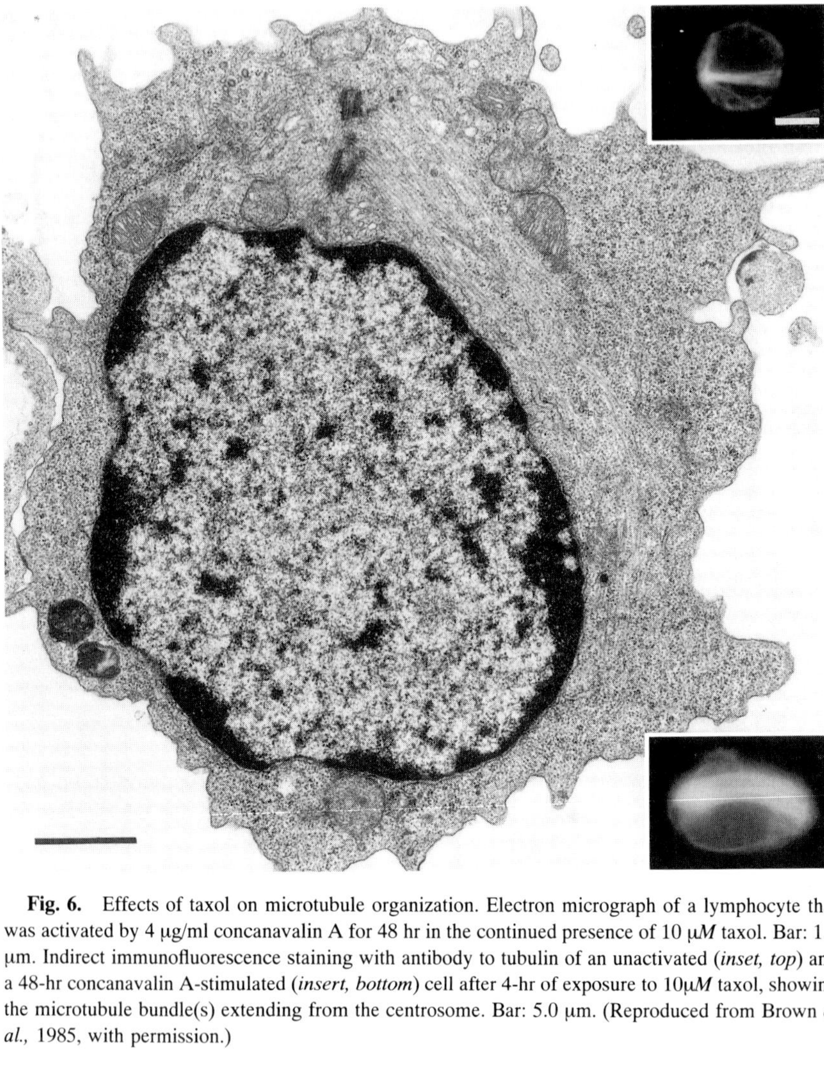

Fig. 6. Effects of taxol on microtubule organization. Electron micrograph of a lymphocyte that was activated by 4 μg/ml concanavalin A for 48 hr in the continued presence of 10 μ*M* taxol. Bar: 1.0 μm. Indirect immunofluorescence staining with antibody to tubulin of an unactivated (*inset, top*) and a 48-hr concanavalin A-stimulated (*insert, bottom*) cell after 4-hr of exposure to 10μ*M* taxol, showing the microtubule bundle(s) extending from the centrosome. Bar: 5.0 μm. (Reproduced from Brown *et al.*, 1985, with permission.)

1986), is not required in this process. The fact that the stable microtubules are disassembled at mitosis, however, indicates that the mechanism of disassembly normally triggered when cells enter mitosis is still functional. Verde *et al.* (1990) provide a recent discussion of the role of *cdc2* kinase in controlling the interphase–metaphase transition of microtubules.

Although dynamic microtubules do not appear to be required for mitogenic activation, the dynamic reorganization of vimentin filaments described in Section

III,B,5 may be. Taxol treatments of unactivated lymphocytes lead to a colocalization of filamentous vimentin with the microtubule bundles that persists over 48 hr in culture. In contrast, when such cells are activated by con A, even in the continued presence of taxol, vimentin dissociates from the microtubule bundles and undergoes a sequence of rearrangements that resembles that seen in control cells (Paulin-Levasseur and Brown, 1987a). What function this rearrangement plays in not known, but it is clear that the mechanisms responsible for the vimentin rearrangement are operating normally during the mitogenic activation of taxol-treated cells.

2. *Activation in the Mixed Lymphocyte Reaction*

The mixed lymphocyte reaction (MLR) is the *in vitro* counterpart of the *in vivo* allogeneic response involved in tumor and transplant rejection (Bain *et al.,* 1964). Cell proliferation and the generation of cytotoxic lymphocytes in the MLR result from cellular interactions between two populations of leukocytes differing in their major histocompatibility complex loci. Studies using the microtubule-disrupting agent colchicine have implicated microtubules in the cell–cell interactions that are necessary for the subsequent lymphocyte proliferation in the MLR (Ranney and Pincus, 1976; Stenzel *et al.,* 1978; Suthanthiran *et al.,* 1980). We have described previously the problems inherent in such microtubule-disassembly treatments and

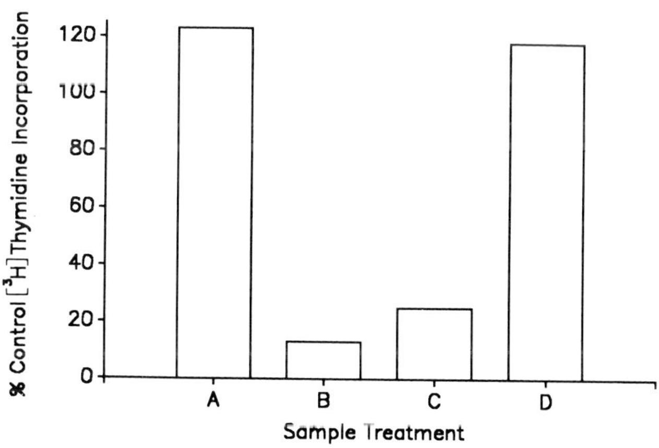

Fig. 7. Effect of 10 μM taxol on DNA synthesis as a percentage of control measured at 48 hr. (A) Concanavalin A activation of BALB/c splenocytes in the presence of taxol. (B) MLR activation with BALB/c responders and C57Bl/6 mitomycin C-treated stimulators in the presence of taxol. (C) MLR activation with BALB/c responders pretreated 4 hr with taxol and C57Bl/6 mitomycin C-treated stimulators. (D) MLR activation with BALB/c responders and C57Bl/6 mitomycin C-treated stimulators pretreated with taxol for 4 hr.

suggested that taxol treatments would provide a more direct test of the requirements for microtubule organization in such lymphocyte functions.

To generate an MLR, the splenic lymphocytes from two strains of mice with different cell-surface alloantigens are cocultured. This results in the activation and proliferation of responding cells in both populations, a response known as a two-way MLR. We have examined taxol effects on the functions of both the stimulator and responder populations of lymphocytes in one-way MLR assays (Roy *et al.,* 1988) in which one of the lymphocyte populations is treated with mitomycin C, an inhibitor of DNA synthesis. The treated population is capable of stimulating the proliferation of the untreated allogeneic lymphocytes, but is unable to replicate DNA and proliferate itself.

As summarized in Fig. 7 and discussed earlier, taxol has no effect on proliferation of splenic lymphocyte populations induced by con A, as assayed by [^{3}H]thymidine incorporation at 48 hr. In contrast, proliferation in the MLR in the continued presence of taxol is effectively inhibited. To determine whether the inhibition was due to an effect on only one of the populations, either the responders or the stimulators were pretreated with 10 μM taxol for 4 hr prior to the initiation of the MLR. As described, this treatment results in a reorganization of the microtubule system that persists for a least 2 days after removal of the drug. These studies demonstrated that taxol does not affect the ability of lymphocytes to stimulate an MLR, but the capacity of taxol-treated lymphocytes to respond to allogenic cells is inhibited (Fig. 7).

Microscopic analysis of nuclear morphologies in these populations and autoradiography of [^{3}H]thymidine incorporation indicate that the taxol inhibition occurs early in alloactivation, prior to DNA replication. The responding ability of taxol-treated lymphocytes is not restored by the addition of interleukin 2, indicating that the inhibition occurs very early in G_1. This is not like the effect of taxol on mitogenic activation, in which cells complete DNA replication and are then blocked in mitosis. Based on these results, we have suggested that the presence of the normal submembranous microtubule system is required for the cell–cell interactions that induce the MLR (Roy *et al.,* 1988). For example, the microtubule system may be involved in modulating the distribution of membrane receptors in the interaction of dendritic antigen-presenting cells and T lymphocytes. This interaction initiates the proliferation of the responding cells in the MLR. For a detailed discussion of the various cell–cell interactions in the MLR, see Metlay *et al.* (1989).

IV. THE CENTROSOME AND EFFECTOR FUNCTIONS

During their development, both T and B lymphocytes acquire receptors for antigens that commit them to a single antigenic specificity for the rest of their lifespan. When these cells encounter their specific antigen, in the appropriate context, they are activated and proliferate, giving rise to either memory cells or to terminally differentiated effector cells. The two major types of effector T cells are T helper (T_H) cells, that induce the proliferation and differentiation of specific antigen-presenting cells (APCs), and cytotoxic T lymphocytes (CTLs), that lyse specific target cells. T cells only recognize antigenic determinants in conjunction with major histocompatibility complex. This restriction insures that a T cell will be triggered only when it interacts with a cell bearing both the appropriate cell-surface antigen and the antigenic determinant for which the T cell is specific. A successful interaction appears to require a reorganization of the centrosome–microtubule array in the effector T cell.

A. Centrosome Orientation

By immunofluorescence staining with antibodies to tubulin, Geiger *et al.* (1982) examined the microtubule organization in CTLs bound to their targets. They reported that the centrosome of the CTL, but not that of the target cell, was found in the region of cell–cell contact and suggested that the CTL plasma membrane proximal to the centrosome was particularly active in forming intercellular contacts. However, a random orientation of the centrosome was observed in conjugates formed between CTLs and lysis-resistant target cells, providing evidence that cell binding does not occur at a polarized site on the plasma membrane of CTLs (Kupfer and Dennert, 1984). In addition, when conjugates were formed in the absence of Ca^{2+}, the centrosome of the CTL was randomly located with respect to the target cell (Kupfer *et al.*, 1985). Orientation of the centrosome was restored within minutes by the addition of Ca^{2+}, demonstrating that orientation can occur subsequent to target cell binding. More recent work by Gray *et al.* (1987) has shown a Ca^{2+} influx in CTLs in response to target cell binding, suggesting that centrosome orientation would normally be effected after cell binding.

A similar Ca^{2+}-dependent orientation of the centrosome has been observed in $T_H$cells interacting with APCs (Kupfer *et al.*, 1986, 1987). The rapid orientation of the centrosome in effector T cells to face bound target cells has therefore provided suitable model systems for studying the role of microtubule organization in the establishment of cell polarity and its function in cell–cell interactions.

Kirschner and Mitchinson (1986) have proposed a model for the development of a polarized microtubule system based on the selective stabilization of a subset of microtubules in a dynamic array. Using immunofluorescence straining with an antibody to detyrosinated α-tubulin as a marker for stable microtubules, Gundersen and Bulinski (1988) have tested this model in fibroblasts orienting toward an experimental wound. They showed that the formation of a polarized array of stable microtubules preceded the onset of cell migration into the wound and closely paralleled the orientation of the centrosome toward the wound edge. They suggested that the selective stabilization of microtubules may be involved in repositioning the centrosome.

Our results indicate that selective stabilization of microtubules is not involved in the orientation of the CTL centrosome toward the target cell. In these studies, EL4 specific CTLs were isolated from peritoneal exudates 10–12 days after the injection of EL4 leukemia cells into the peritoneal cavity of allogeneic *B*alb/c mice. These CTLs were mixed in a 1:1 ratio with cultured target EL4 cells and then prepared for immunofluorescence staining with antibodies to detyrosinated α-tubulin or acetylated α-tubulin (another marker for stable microtubules; Piperno and Fuller, 1985). Previously we have demonstrated that murine T cells are able to generate detyrosinated and acetylated α-tubulin (J. D. Knox and D. L. Brown, unpublished results). As shown in Fig. 8, no subset of stable microtubules was detected in CTLs conjugated to target cells. This indicated that, if a more stable subset of microtubules was formed during the orientation of the centrosome toward the target cell, it did not persist long enough for posttranslational modifications to occur.

We have further tested the involvement of selective stabilization of microtubules in the repositioning of the centrosome with the microtubule-assembly-promoting drug taxol. As detailed in Section III,C,1, the treatment of lymphocytes with taxol promotes the assembly of microtubules from the centrosome to form one to several highly stable microtubule bundles. To determine the effect of the stabilization of the entire microtubule array on the orientation of the centrosome, CTLs were preincubated with taxol for 4 hr before being mixed with target cells and prepared for immunofluorescence staining. These studies showed that 1 μM taxol, a dose that results in extensive microtubule bundling, had no effect on the orientation of the centrosome toward the target cell. To determine the effects of microtubular reorganization on effector functions, taxol-treated CTLs were mixed with ^{51}Cr-labeled target cells and a cytolytic activity assay was performed. This experiment showed that taxol treatment had no effect on lytic ability (J. D. Knox and D. L. Brown, unpublished results). We conclude that the selective stabilization of a subset of microtubules is required neither for movement of the centrosome toward the target cell nor for the effective CTL-mediated lysis.

B. Role of the Microtubule System in Effector Functions

As discussed earlier, in lymphocytes the Golgi apparatus and centrosome are closely associated (see Section II,C). This has led to the suggestion that the orientation of the centrosome could be involved in the targeting of Golgi-derived vesicles to the area of cell contact with the target cell. Vesicular granules containing one or more cytotoxic components such as perforins (pore-forming pro-

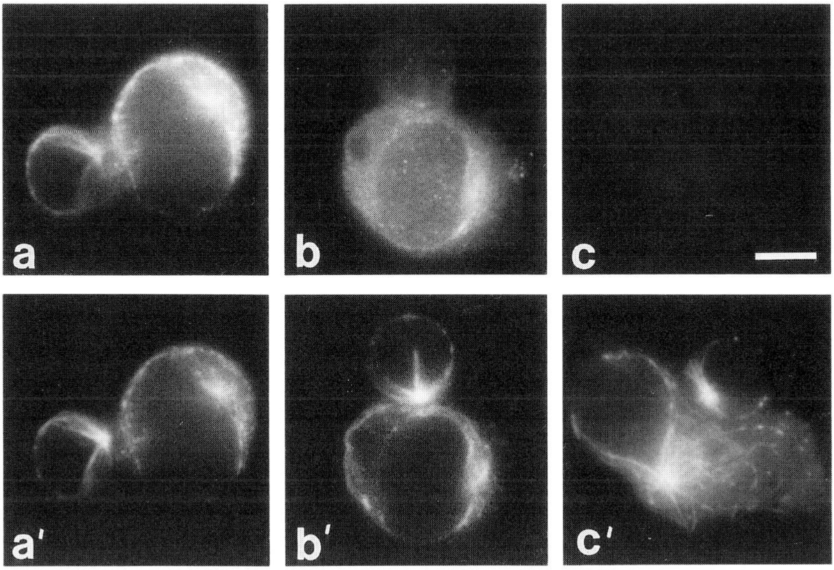

Fig. 8. Double immunofluorescence of conjugates formed between CTLs isolated from BALB/c mice and cultured EL4 cells as targets. (a–c) Micrographs of cells stained with antibodies specific for tyrosinated (a), detyrosinated (b), and acetylated (c) α-tubulin, respectively. (a'–c') The corresponding conjugates in the second row have been stained to show all microtubules (a'–b') or tyrosinated microtubules (c'). Bar: 5.0 μm.

teins), serine proteases, and tumor necrosis factors have been isolated from a wide range of CTL cell lines (for reviews, see Brunet *et al.,* 1988; Clark *et al.,* 1988; Young *et al.,* 1988).

Experiments with CTLs in which the centrosome was reversibly inhibited from orienting toward the target cell, by removal of Ca^{2+} from the medium or by treatment of the cells with microtubule-disassembling drugs, indicated that the orientation of the centrosome is a prerequisite for the effective lysis of the target cell and supported the directed secretion model (Kupfer *et al.,* 1983, 1985). However, there is now evidence that CTL lysis of certain target cells can occur in the absence of Ca^{2+} and without any appreciable degranulation. This has led to the suggestion that there are two, and possibly three, different lytic pathways used by CTLs in target cell killing (Ostergaard and Clark, 1989). Regardless of the pathway used, target cell lysis in the absence of centrosome orientation has never been observed (Ostergaard and Clark, 1987). Therefore, although the function of centrosome orientation remains uncertain, it appears to be required for a successful interaction between the CTL and the target cell.

Our recent analysis of the effects of hyperthermia on murine CTLs illustrates the importance of centrosome–microtubule organization for cytolytic killing of target cells (Knox *et al.,* 1991). These studies showed, by immunofluorescence and confocal microscopy using a monoclonal antitubulin and a rabbit autoimmune serum that binds a centriole-associated protein, that the organization of the centrosome–microtubule array was disrupted by hyperthermia and that cytolytic activity also was severely impaired. Electron microscopy indicated that the disorganization of the microtubules may have resulted from an aggregation of the pericentriolar material of the centrosome. The recovery of cytolytic activity following the hyperthermia treatment was coincident with the reorganization of the microtubules about the centrosome. These findings lead us to suggest that the initial inhibitory effect of hyperthermia on CTL function resulted from the disruption of centrosome and microtubule organization.

For T_H cells, a number of T-cell-derived lymphokines can serve as APC growth and differentiation factors *in vitro* (Coffman *et al.,* 1988). The action of these lymphokines and growth factors is generally nonspecific, so they can stimulate any APC carrying the appropriate receptors *in vitro.* The orientation of the centrosome inside an antigen-specific T_H cell to direct the secretion of these growth and differentiation factors toward a bound APC would clearly enhance the specificity of the immune response. Consistent with the results of studies of CTL–target cell conjugates, there is a very strong correlation between a specific effective cell interaction and orientation of the centrosome inside the T_H cell toward its bound APC (Kupfer and Singer, 1989b). The evidence, though indirect, suggests that in both cases the function of this orientation is to direct secretion into the tight intercellular space between the T effector cell and its bound target.

Another similarity between the two systems is the localization of the cytoskeletal protein talin under the effector T cell membrane where it is in contact with its specific target (Kupfer and Singer, 1989a). Talin, or one of its isoforms, is present in a wide range of different cell types. It is often localized to sites of intercellular adhesion along with a transmembrane glycoprotein of the integrin family, which includes the lymphocyte-function-associated antigen 1 (LFA-1). There is strong evidence that LFA-1 molecules are involved in cell–cell adhesion (Springer *et al.*, 1987; Springer, 1990). In murine T cells, LFA-1 and talin are not normally associated with one another but become directly or indirectly linked when an appropriate signal is received by the cell (Burn *et al.,* 1988). This has led to the suggestion that talin may function in the recruitment of LFA-1 to the contact region to further stabilize specific cell–cell adhesion (Kupfer and Singer, 1989a). Interestingly, the redistribution of talin was not dependent on the presence of Ca^{2+} in the medium and was therefore not tightly coupled to the orientation of the centrosome. Nevertheless, the formation of a talin cap in the cortical region of the cell could be involved in directing the orientation of the centrosome.

ACKNOWLEDGMENTS

We are grateful to G. Piperno (The Rockfeller University, New York) and G. Gundersen (Department of Anatomy and Cell Biology, Columbia University, New York), for providing the antibodies to acetylated and detyrosinated α-tubulin. We also thank the National Cancer Institute for the generous provision of taxol. This work was funded by grants from the Natural Sciences and Engineering Research Council of Canada to David L. Brown.

REFERENCES

Bain, B., Vas M., and Lowenstein, L. (1964). The development of large immature mononuclear cells in mixed leukocyte cultures. *Blood* **23,** 108–116.

Bladon, T., Brasch K., Brown, D. L., and Setterfield, G. (1988). Changes in structure and protein composition of bovine lymphocyte nuclear matrix during concanavalin A-induced mitogenesis. *Biochem. Cell Biol.* **66,** 40–53.

Bloom, G. S., Luca, F. C., and Vallee, R. B. (1985). Cross-linking of intermediate filaments to microtubules by microtubule associated protein 2. *Ann. N.Y. Acad. Sci.* **455,** 18–31.

Bourguignon, L. Y. W., and Bourguignon, G. J. (1981). Immunocytochemical localization of intermediate filament proteins during lymphocyte capping. *Cell Biol. Int. Rep.* **5,** 783–789.

Bourguignon, L. Y. W., and Bourguignon, G. J. (1984). Capping and the cytoskeleton. *Int. Rev. Cytol.* **87,** 195–224.

Bré, M. H., and Karsenti, E. (1990). Effects of brain microtubule-associated proteins on microtubule dynamics and the nucleating activity of centrosomes. *Cell Mot. Cytoskel.* **15,** 88–98.

Brinkley, B. R., Cox, S. M., Pepper, D. A., Wible, L., Brenner, S. L., and Pardue, R. L. (1981). Tubulin assembly sites and the organization of cytoplasmic microtubules in cultured mammalian cells. *J. Cell Biol.* **90,** 554–562.

Brown, D. L., Little, J. E., Chaly, N., Schweitzer, I., and Paulin-Levasseur, M. (1985). Effects of taxol on microtubule organization in mouse splenic lymphocytes and on response to mitogenic stimulation. *Eur. J. Cell Biol.* **37,** 130–139.

Brown, D. L., Chaly, N., and Kaplan, J. G. (1989). Assembly of the lymphocyte nucleus for transcription and replication. *In* "Cellular Basis of Immune Modulation" (J. G. Kaplan, B. R. Green, and R. C. Bleackley, eds.), pp. 67–77. Liss, New York.

Brunet, J.-F., Denizot, F., and Goldstein, P. (1988). A differential molecular biology search for genes preferentially expressed in functional T lymphocytes: The CTLA genes. *Immunol. Rev.* **103,** 21–36.

Burn, P., Kupfer, A., and Singer, S. J. (1988). Dynamic membrane–cytoskeletal interactions: Specific association of integrin and talin arises *in vivo* after phorbol ester treatment of peripheral blood lymphocytes. *Proc. Natl. Acad. Sci. U.S.A.* **85,** 497–501.

Clark, W., Ostergaard, H., Gorman, K., and Torbett, B. (1988). Molecular mechanisms of CTL-mediated lysis: A cellular perspective. *Immunol. Rev.* **103,** 38–51.

Coffman, R. L., Seymour, B. W. P., Lebman, D. A., Hiraki, D. D., Christiansen, J. A., Shrader, B., Cherwinski, H. M., Savelkoul, H. F. J., Finkelman, F. D., Bond, M. W., and Mosmann, T. R. (1988). The role of helper T cell products in mouse B cell differentiation and isotype regulation. *Immunol. Rev.* **102,** 5–28.

Crabtree, G. R. (1989). Contingent genetic regulatory events in T lymphocyte activation. *Science* **243,** 355–361.

De Brabander, M., Geuens, G., Nuydens, R., Willebrords, R., and De Mey, J. (1981). Taxol induces the assembly of free microtubules in living cells and blocks the organizing capacity of the centrosome and kinetochores. *Proc. Natl. Acad. Sci. U.S.A.* **78,** 5608–5612.

DeFranco, A. L. (1987). Molecular aspects of B-lymphocyte activation. *Ann. Rev. Cell Biol.* **3,** 143–178.

Dellagi, K., and Brouet, J.-C. (1982). Redistribution of intermediate filaments during capping of lymphocyte surface molecules. *Nature (London),* **298,** 284–286.

Edelman, G. M. (1976). Surface modulation in cell recognition and growth. *Science* **192,** 218–226.

Edelman, G. M., Yahara, I., and Wang, J. L. (1973). Receptor mobility and the receptor-cytoplasmic interaction in lymphocytes. *Proc. Natl. Acad. Sci. U.S.A.,* **70,** 1442–1445.

Gardner, P. (1989). Calcium and T lymphocyte activation. *Cell* **59,** 15–20.

Geiger, B., Rosen, D., and Berke, G. (1982). Spatial relationships of microtubule-organizing centers and the contact area of cytotoxic T lymphocytes and target cells. *J. Cell Biol.* **95,** 137–143.

Gowans, J. L., McGregor, D. D., Cowen, D. M. and Ford, C. E. (1962). Initiation of immune responses by small lymphocytes. *Nature (London)* **196,** 651–655.

Gray, L., Gnarra, J., and Engelhard, V. (1987). Demonstration of a calcium influx in cytolytic T lymphocytes in response to target cell binding. *J. Immunol.* **138,** 63–69.

Gundersen, G. G., and Bulinski, J. C. (1988). Selective stabilization of microtubules oriented toward the direction of cell migration. *Proc. Natl. Acad. Sci. U.S.A.* **85,** 5946–5950.

Ho, W. C., Allan, V. J., van Meer, G., Berger, E. G., and Kreis, T. E. (1989). Reclustering of scattered Golgi elements occurs along microtubules. *Eur. J. Cell Biol.* **48,** 250–263.

Hungerford, D. A., Donnelly, A. J., Nowell, P. C., and Beck, S. (1959). The chromosome constitution of a human phenotypic intersex. *Am. J. Hum. Genet.* **11,** 215–236.

Kaplan, J. G., Brown, D. L., Chaly, N., Greer, W. L., Prasad, K. V. S., Severini, A., and Sahai, B. M. (1987). Structural and evolutionary implications of the packaging of DNA for differentiation and proliferation in the lymphocyte. *J. Mol. Evol.* **26,** 173–179.

Kirschner, M., and Mitchison, T. (1986). Beyond self–assembly: From microtubules to morphogenesis. *Cell* **45,** 329–342.

Komesli, S., Tournier, F., Paintrand, M., Margolis, R. L., Job, D., and Bornens, M. (1989). Mass isolation of calf thymus centrosomes: Identification of a specific configuration. *J. Cell Biol.* **109,** 2869–2878.

Knox, J. D., Mitchel, R. E. J., and Brown, D. L. (1991). Effects of hyperthermia on microtubule organization and cytolytic activity of murine cytotoxic T lymphocytes. *Exp. Cell Res.* **194,** 275–283.

Kupfer, A., Dennert, G., and Singer, S. J. (1983). Polarization of the Golgi apparatus and the microtubule-organizing center within cloned natural killer cells bound to their targets. *Proc. Natl. Acad. Sci. U.S.A.* **80,** 7224–7228.

Kupfer, A., and Dennert, G. (1984). Reorientation of the microtubule organizing center and the Golgi apparatus in cloned cytotoxic lymphocytes triggered by binding to lysable target cells. *J. Immunol.* **133,** 2762–2766.

Kupfer, A., Dennert, G., and Singer, S. J. (1985). The reorientation of the Golgi apparatus and the microtubule-organizing center in the cytotoxic effector cell is a prerequisite in the lysis of bound target cells. *J. Mol. Cell Immunol.* **2,** 37–49.

Kupfer, A., Swain, S. L., Janeway, C. A., Jr., and Singer, S. J. (1986). The specific direct interaction of helper T cells and antigen-presenting B cells. *Proc. Natl. Acad. Sci. U.S.A.* **83,** 6080–6083.

Kupfer, A., Swain, S. L., and Singer, S. J. (1987). The specific direct interaction of helper T cells and antigen-presenting B cells. II. Reorientation of the microtubule organizing center and reorganization of the membrane-associated cytoskeleton inside the bound helper T cells. *J. Exp. Med.* **165,** 1565–1580.

Kupfer, A., and Singer, S. J. (1989a). Cell biology of cytotoxic and helper T cell functions:

Immunofluorescence microscopic studies of single cells and cell couples. *Ann. Rev. Immunol.* **7,** 309–337.

Kupfer, A., and Singer, S. J. (1989b). The specific interaction of helper T cells and antigen-presenting B cells. IV. Membrane and cytoskeletal reorganizations in the bound T cell as a function of antigen dose. *J. Exp. Med.* **170,** 1697–1713.

Kuriyama, R. (1984). Activity and stability of centrosomes in Chinese hamster ovary cells in nucleation of microtubules *in vitro. J. Cell Sci.* **66,** 277–295.

Manfredi, J. J., Parness, J., and Horwitz, S. B. (1982). Taxol binds to cellular microtubules. *J. Cell Biol.* **94,** 688–696.

Metlay, J. P., Pure, E., and Steinman, R. M. (1989). Control of the immune response at the level of antigen-presenting cells: A comparison of the function of dendritic cells and B lymphocytes. *Adv. Immunol.* **47,** 45–116.

Ostergaard, H. L., and Clark, W. R. (1987). The role of Ca^{2+} in activation of mature cytotoxic T lymphocytes for lysis. *J. Immunol.* **139,** 3573–3579.

Ostergaard, H. L., and Clark, W. R. (1989). Evidence for multiple lytic pathways used by cytotoxic T lymphocytes. *J. Immunol.* **143,** 2120–2126.

Paatero, G. I. L., and Brown, D. L. (1982). Effects of taxol on microtubule organization and on capping of surface immunoglobulin in mouse splenic lymphocytes. *Cell Biol. Int. Rep.* **6,** 1033–1040.

Paulin-Levasseur, M., and Brown, D. L. (1987a). Vimentin dynamics during the mitogenic stimulation of mouse splenic lymphocytes. *Cell Mot. Cytoskel.* **8,** 227–237.

Paulin-Levasseur, M., and Brown, D. L. (1987b). Organizational fate of vimentin during redistribution of surface immunoglobulin in mouse splenic lymphocytes. *Cell Biol. Int. Rep.* **11,** 583–590.

Paulin-Levasseur, M., and Brown, D. L. (1989). Organization of the vimentin system and its spatial relationship to the microtubule complex during the division of mammalian cells growing attached and in suspension. *Eur. J. Cell Biol.* **49,** 189–195.

Piperno, G., and Fuller, M. T. (1985). Monoclonal antibodies specific for an acetylated form of α-tubulin recognize the antigen in cilia and flagella from a variety of organisms. *J. Cell Biol.* **101,** 2085–2094.

Ranney, D. F., and Pincus, J. H. (1976). Supression of stimulating cell activity by microtubule disrupting alkaloids. *J. Supramol. Struct.* **5,** 335–343.

Rieder, C. L., and Borisy, G. G. (1982). The centrosome complex in PtK_2 cells: Asymmetric distribution and structural changes in the pericentriolar material. *Biol. Cell* **44,** 117–132.

Roca, J., and Mezquita, C. (1989). DNA topoisomerase II activity in nonreplicating, transcriptionally inactive, chicken late spermatids. *EMBO J.* **8,** 1855–1860.

Rogers, K. A., Khoshbaf, M. A., and Brown, D. L. (1981). Relationship of microtubule organization in lymphocytes to the capping of immunoglobulin. *Eur. J. Cell Biol.* **24,** 1–8.

Roy, C., Chaly, N., and Brown, D. L. (1988). Taxol-induced reorganization of the microtubule system in murine splenic lymphocytes inhibits response to allogeneic cells but not concanavalin A. *Biochem. Cell Biol.* **66,** 389–395.

Rudd, C. E., Rogers, K. A., Brown, D. L., and Kaplan, J. G. (1979). Microtubules, colchicine, and lymphocyte blastogenesis. *Can. J. Biochem.* **57,** 673–683.

Sandoval, I. V., Bonifacino, G. S., Klausner, R. D., Henkart, M., and Wehland, J. (1984). Role of microtubules in the organization and localization of the Golgi apparatus. *J. Cell Biol.* **99,** 1135–1185.

Schweitzer, I., and Brown, D. L. (1984). Changes in organization and microtubule assembly activity of the centrosome during lymphocyte stimulation. *Biol. Cell* **52,** 147–160.

Setterfield, G., Hall, R., Bladon, T., Little, J., and Kaplan, J. G. (1983). Changes in structure and composition of lymphocyte nuclei during mitogenic stimulation. *J. Ultrastruct. Res.* **82,** 264–282.

Setterfield, G., Bladon, T., Hall, R., Chaly, N., Brasch, K., El Ansary, H., and Brown, D. L. (1985). Extrachromatin nuclear components and structural changes in nuclei. *In* "Nuclear Envelope

Structure and RNA Maturation" (E. A. Smuckler and G. A. Clawson, eds.), pp. 63–86. Liss, New York.

Springer, T. A. (1990). Adhesion receptors of the immune system. *Nature* **346,** 425–434.

Springer, T. A., Dustin, M. L., Kishimoto, T. K., and Marlin, S. D. (1987). The lymphocyte function-associated LFA-1, CD2, and LFA-3 molecules: Cell adhesion receptors of the immune system. *Ann. Rev. Immunol.* **5,** 223–252.

Steinert, P. M., and Liem, R. K. H. (1990). Intermediate filament dynamics. *Cell* **60,** 521–523.

Stenzel, K. H., Schwartz, R., Rubin, A. L., and Novogrodsky, A. (1978). Potentiation of lymphocyte activation by colchicine. *J. Immunol.* **121,** 863–865.

Stewart, M. (1990). Intermediate filaments: Structure, assembly and molecular interactions. *Curr. Opin. Cell Biol.* **2,** 91–100.

Suthanthiran, M., Stenzel, K. H., Rubin, A. L., and Novogradsky, A. (1980). Augmentation of proliferation and generation of specific cytotoxic cells in human mixed lymphocyte culture reactions by colchicine. *Cell Immunol.* **50,** 379–391.

Sweet, S. C., Rogers, C. M., and Welsh, M. J. (1989). Calmodulin is associated with microtubules forming in PtK_1 cells upon release from nocodazole treatment. *Cell Mot. Cytoskel.* **12,** 113–122.

Thyberg, J. (1984). The microtubular cytoskeleton and the initiation of DNA synthesis. *Exp. Cell Res.* **155,** 1–8.

Thyberg, J., Moskalewski, S., and Friberg, U. (1977). Effects of colchicine and vinblastine on the phytohaemagglutinin-induced transformation of lymphocytes. *J. Cell Sci.* **27,** 183–198.

Thyberg, J., and Moskalewski, S. (1985). Microtubules and the organization of the Golgi complex. *Exp. Cell Res.* **159,** 1–16.

Turner, J. R., and Tartakoff, A. M. (1989). The response of the Golgi complex to microtubule alterations: The roles of metabolic energy and membrane traffic in Golgi complex organization. *J. Cell Biol.* **109,** 2081–2088.

Verde, F., Labbe, J.-C., Doree, M., and Karsenti, E. (1990). Regulation of microtubule dynamics by cdc2 protein kinase in cell-free extracts of *Xenopus* eggs. *Nature (London)* **343,** 233–238.

Vorobjev, I. A., and Chentsov, Y. S. (1982). Centrioles in the cell cycle. I. Epithelial cells. *J. Cell Biol.* **93,** 938–949.

Yahara, I. and Edelman, G. M. (1973). Modulation of lymphocyte receptor redistribution by concanavalin A, anti-mitotic agents, and alterations of pH. *Nature (London)* **236,** 152–155.

Yahara, I., and Kakimoto-Sameshima, F. (1978). Microtubule organization in lymphocytes and its modulation by patch and cap formation. *Cell,* **15,** 251–259.

Young, J. D.-E., Lui, C.-C., Persechini, P. M., and Cohn, Z. A. (1988). Perforin-dependent and independent pathways of cytotoxicity mediated by lymphocytes. *Immunol. Rev.* **103,** 161–202.

Waterhouse, P. D., Anderson, P. J., and Brown, D. L. (1983). Increases in microtubule assembly and in tubulin content in mitogenically stimulated mouse splenic T lymphocytes. *Exp. Cell Res.* **144,** 367–376.

11

The Centrosome in Stationary and Migrating Endothelial Cells

V. I. KALNINS* AND K. A. ROGERS†

*Department of Anatomy University of Toronto
Toronto, Ontario M5S 1A8 Canada

†Department of Anatomy
University of Western Ontario
London, Ontario N6A 5C1 Canada

I. INTRODUCTION

Blood vessels are lined by a monolayer of highly flattened spindle-shaped endothelial cells (ECs) with a long axis that is oriented parallel to the direction of blood flow (Fig. 1). The ECs at the interface of blood and tissues regulate the

Copyright © 1992 Academic Press, Inc.
All rights of reproduction in any form reserved.

permeability of blood vessels and injury to the endothelial lining is thought to induce changes in arterial walls that eventually lead to the development of atherosclerotic lesions (Ross, 1988). Because of the relative inaccessibility of the endothelium to experimental manipulation, relatively little is known about the factors that maintain the integrity of the endothelial lining and the role of the centrosome and the associated microtubules (MTs) in the organization and function of ECs.

In most animal cells, including ECs, many MTs radiate from a single focal center, the centrosome, through the cytoplasm. The centrosome generally consists of a pair of centrioles and surrounding material containing MT organizing centers (MTOCs), capable of initiating the assembly of MTs and thereby playing a major role in determining their distribution. The cytoplasmic MTs, the minus ends of which are embedded in the centrosome and the plus ends of which extend toward the cell periphery, in turn determine the organization of many other cell organelles in the cytoplasm, including the Golgi apparatus (Kreis, 1990) and the

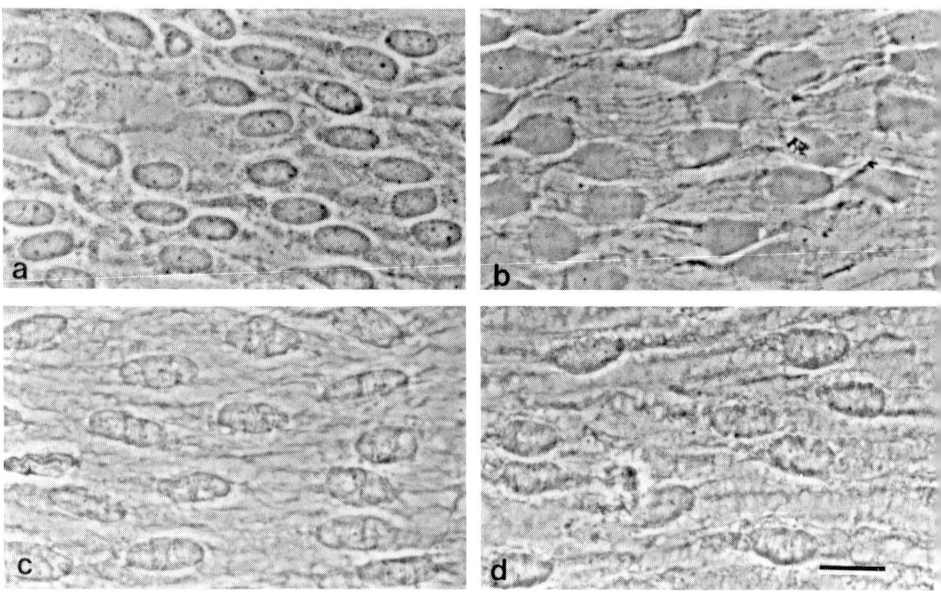

Fig. 1. Phase-contrast photomicrographs of typical *en face* preparations of endothelium from pig thoracic aorta (a) and inferior vena cava (b) and from rabbit thoracic aorta (c) and inferior vena cava (d). Preparations consist of single layers of endothelial cells with their long axis oriented in a direction parallel to the blood flow. The heart was on the left. Blood flow is from left to right for a and c and from right to left for b and d. Bar: 40 μm. (Reprinted from Rogers *et al.*, 1985, with permission.)

endoplasmic reticulum (Terasaki, 1990). By providing pathways along which organelles such as membrane-bound vesicles can travel (Kelly, 1990) and by interacting with other components of the cytoskeleton, such as intermediate filaments and microfilaments (MFs), the MTs radiating from the centrosome could control the organization and activity of the cytoplasm, including the cell cortex.

The MT system is also important in determining cell shape. Because ECs will undergo dramatic changes in shape and in orientation of their longitudinal axis in response to altered blood flow (Reidy and Langille, 1980; Dewey *et al.,* 1981); and because the centrosome is a major site for MT assembly, we have investigated whether a particular relationship exists between the direction of blood flow and the position of centrosomes in ECs.

The position of centrosomes has long been thought to define the polarity of mammalian cells by establishing an axis running from the center of the nucleus through the centrosome (Van Beneden, 1883). In most cells the centrosomes are found in a characteristic position (Wheatley, 1982). Generally this position is on one side of the nucleus, close to the Golgi apparatus, and leads to an asymmetric distribution of MTs. For example, in epithelial cells the centrosomes are located toward the apical end where one of the centrioles often gives rise to a cilium. In cytotoxic T lymphocytes, the centrosome faces the target cells before the mechanism that kills the target cell is activated (Geiger *et al.,* 1982; Kupfer *et al.,* 1985). These observations suggest that the MTs radiating from the centrosome could also control activities such as migration of cells in a specific direction.

The migrating cell has a distinct polarity and its anterior or leading edge can be easily distinguished from the posterior or trailing one because of differences in the organization of the cytoskeleton, which also provides the force and the organization necessary for cell migration. The ability of animal cells to migrate in a particular direction probably requires mechanisms that can control and coordinate the interactions of MFs with other components of the cytoskeleton. Several studies have suggested that the centriole or the centrosome may be responsible for this coordination and that the MTs emanating from the centrosome are required for directional cell migration (see subsequent text). We therefore also examined changes in the position of the centrosome in ECs induced to migrate in a specific direction by wounding the endothelial monolayer.

In this review of our work, we will first examine the position of centrosomes in ECs *in situ* and then compare it with that in ECs in cell and organ cultures before and after experimentally wounding the endothelium to induce cell migration. The possible role of the centrosome and the associated MT system in defining cell polarity and how it relates to the directional migration of cells will be discussed. Our work on ECs and centrosomes will be related to that of investigators working with other cell types and model systems.

II. DISTRIBUTION OF CENTROSOMES IN ENDOTHELIAL CELLS

A. In Aorta and Vena Cava

To determine whether a relationship exists between the position of centrosomes in ECs and the direction of blood flow, the endothelial layers of the larger blood vessels were examined in a number of species. To do so, methods were developed for making *en face* preparations of the endothelium in which various components of the cytoskeleton can be visualized in a large number of cells by immunofluorescence (Rogers and Kalnins, 1983a). For these preparations, 0.5–1.0-cm segments of the blood vessels are fixed in cold methanol and acetone, partly air dried, and their endothelial surfaces are pressed against a coverslip. The endothelium that adheres to the coverslip is then examined by immunofluorescence using nonimmune rabbit sera containing antibodies that react with centrioles (Connolly and Kalnins, 1978; Turksen *et al.,* 1982) or antibodies to tubulin to visualize MTs. Typical *en face* preparations of ECs after immunofluorescence staining with an anticentriole serum showing the distribution of centrosomes is shown in Fig. 2. In such preparations, the position of the centrosome relative to the nucleus, which occupies the central part of the spindle-shaped EC, can readily be determined. After dividing each EC into three regions with respect to the nucleus and the direction of blood flow, the centrosomes are classified as being on the side "toward the heart," in the "center," or on the side "away from heart."

Using this approach to examine the position of centrosomes in ECs *in situ,* we found that, in ECs lining the aorta and vena cava of the pig, the centrosomes are preferentially located on the heart side of the nucleus (Rogers *et al.,* 1985). In the pig aorta and inferior vena cava, 77–79% of the ECs have centrosomes located on the heart side of the nucleus, 12–14% have them on the side away from the heart, and 7–11% have them in the central position along the side of the nucleus (Table I). Since in these two vessels the blood flows in opposite directions, the centrosome is preferentially located on the upstream side of the nucleus in the former and on the downstream side in the latter. The position of the centrosomes in ECs of the larger blood vessels is thus independent of the direction of blood flow. These results show, nevertheless, that many ECs lining the pig aorta and vena cava are polarized in the same anterior–posterior direction along the vessel wall.

If, in ECs lining the branches of the pig aorta, the centrosomes are also preferentially oriented toward the heart, then, at flow dividers, ECs with opposite polarities would have to meet and form a continuous sheet over the flow divider. Indeed, the results obtained indicate that, in ECs on both sides of a typical flow divider between the left subclavian artery and the aortic arch in the pig, the centrosomes are preferentially located on the heart side of the nucleus (Rogers *et al.,* 1985; Table I).

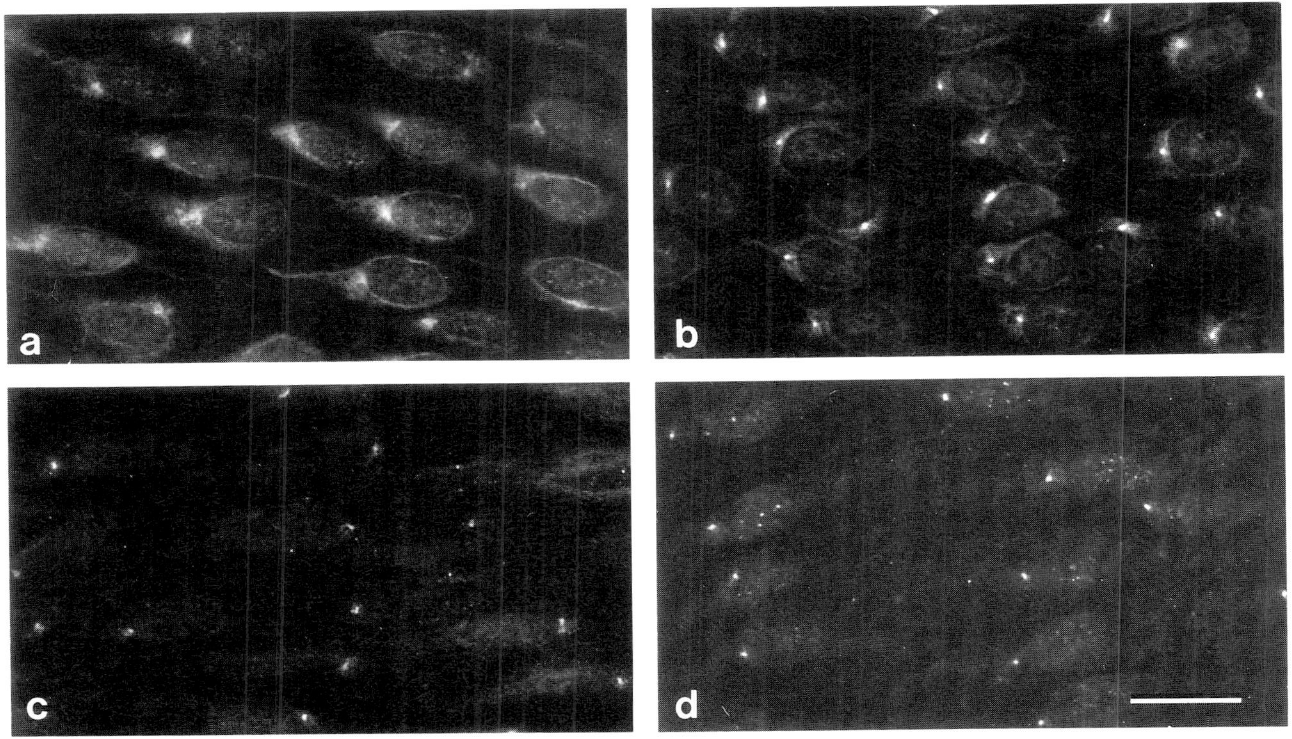

Fig. 2. Immunofluorescence staining of *en face* preparations of vascular endothelium of pig thoracic aorta (a) and inferior vena cava (b) and of rabbit thoracic aorta (c) and inferior vena cava (d) with sera specific for centrioles. In endothelial cells (EC) in a, b, and d, centrosomes are preferentially located toward the heart (on the left). However, in EC in the rabbit thoracic aorta (c), no preferential orientation toward or away from the heart was detected. Blood flow is from left to right for a and c and from right to left for b and d. Bar: 18 μm. (Reprinted from Rogers *et al.*, 1985, with permission)

TABLE I

Distribution of Centrosomes in Vascular Endothelial Cells *in Vivo*[a,b]

Vessel	Endothelial cells with centrosomes (%)		
	Toward heart	Center	Away from heart
Pig			
Thoracic aorta	79 ± 3	7 ± 1	14 ± 6
Inferior vena cava	77 ± 3	11 ± 2	12 ± 2
Branch point			
Aortic arch	69 ± 4	13 ± 2	18 ± 4
Left subclavian artery	69 ± 2	9 ± 1	22 ± 3
Rabbit			
Thoracic aorta	33 ± 5	28 ± 5	39 ± 7
Inferior vena cava	61 ± 5	15 ± 2	24 ± 3

[a]Reprinted from Rogers *et al.* (1985) with permission.
[b]Values represent mean ± SEM calculated from at least three animals.

A similar preferential distribution of centrosomes toward the heart was seen in the inferior vena cava of the rabbit. In the ECs of adult rabbit aorta, however, approximately equal numbers of centrosomes are located on the side of the nucleus toward and away from the heart (Rogers *et al.*, 1985). The position of centrosomes in rabbit aorta is being investigated in more detail to determine if the centrosome is preferentialy oriented toward the heart in younger rabbits but gradually loses this orientation with increasing age.

To determine whether the preferential orientation of the centrosomes toward the heart in ECs of the larger blood vessels is re-established after changing it, surgical techniques were used to reverse a 3-cm segment of the inferior vena cava in the region between the renal veins and the common iliac veins of the rabbit. The position of centrosomes in the reversed segment was determined at various times after the surgery and compared with that in ECs in a region distal to the reversed segment and with that in ECs in segments removed and then grafted back in the correct orientation. The results obtained (Fig. 3) show that by 1 wk after surgery, most of the centrosomes have moved to the central part of the EC along the side of the nucleus and after 3 wk, most of the centrosomes were again located on the heart side of the nucleus. By 12 wk, control values normally found in this region and in regions distal to the graft had been reached (Rogers *et al.*, 1985). The reorientation of centrosomes toward the heart in ECs of the reversed segment strongly suggests that this particular orientation of the centrosome is important and necessary for the normal functioning of endothelium.

In segments of the inferior vena cava removed and then anastomosed back in their original orientation, the centrosomes in ECs remained preferentially oriented toward the heart. Transiently, however, the number of centrosomes in this orientation decreased slightly shortly after surgery when compared with the number in unoperated controls or in control regions distal to the segment, but returned to control levels by 6 wk. This indicates that the position of the centrosomes in ECs can be easily altered and is influenced to some extent by the trauma associated with surgery (Rogers *et al.,* 1985).

We are now trying to determine whether a preferential orientation of the centrosomes is also present in the ECs of smaller blood vessels all the way down to the capillaries. A more detailed knowledge of the position of centrosomes in ECs throughout the circulatory system may provide clues about the significance of the observed polarization of ECs in the anterior–posterior direction along the walls of larger blood vessels.

As previously mentioned, in most cell types the centrosome is closely associated with the Golgi apparatus. We have found that this is also true in ECs lining the aorta and vena cava in the rabbit (Kiosses, B. W., unpublished observations). This linkage between the two organelles and their orientation in the same direction along the vessel wall is probably necessary for the normal functioning and behavior of ECs.

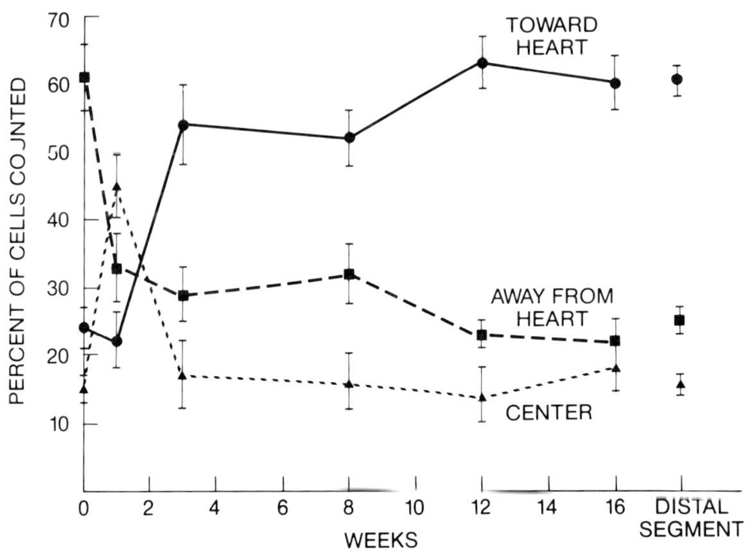

Fig. 3. Orientation of centrosomes in endothelial cells (EC) in the reversed segment of the rabbit inferior vena cava at various times after surgery. Control values for the segment distal to the reversed segment are shown on the right. Within 1 wk, centrosomes in the majority of EC are located in the central part; by 3 wks, in most EC, they are again oriented toward the heart; by 12 wks, values seen in the nonreversed segment are reached. (Reprinted from Rogers *et al.,* 1985, with permission.)

B. In Cell and Organ Cultures

The spindle-shaped ECs become flat and polygonal when they are removed from blood vessels, dissociated, and grown in cell cultures. The number of MTs emanating from the centrosome also increases greatly and, in the resulting monolayer, the centrosomes are randomly oriented (Rogers and Kalnins, 1983b).

In aortic organ cultures (Gotlieb and Boden, 1984), in contrast to cell cultures, the ECs remain on their normal substratum and maintain contact with neighboring cells. We examined what happens to centrosome position when pig aortic ECs with preferentially oriented centrosomes are maintained in organ cultures. During the first 24 hr the centrosomes in the majority of ECs in the cultured segment of the aorta remain preferentially oriented in the same direction as *in situ.* Between 24 and 96 hr, as the cells gradually lose their elongated shape and become polygonal, the position of the centrosomes becomes more random and the endothelium in organ cultures starts to resemble an EC monolayer in cell culture (Rogers *et al.,* 1986).

Apparently the absence of shear force is responsible for the shape change in organ cultures, since similar polygonal ECs are observed not only in cell cultures but also in the walls of vessels in which flow has been interrupted (Reidy and Langille, 1980). Although not observed, it is possible that some ECs are lost during the culture period, creating space that is filled by the surrounding ECs. Spreading and/or migration of the ECs into this space could lead to the increased randomization that is observed in the position of the centrosomes (Rogers *et al.,* 1986).

III. MICROTUBULES AND POLARITY OF MIGRATING CELLS

Treatment of cells with reagents that depolymerize MTs have been used to examine the role of MTs in determining the polarity of migrating cells. Earlier studies in single mouse embryo fibroblasts and BHK-21 cells in culture indicate that MTs are essential for determining the direction of the leading edge and that coordinated movement in one direction is either reduced (Vasiliev *et al.,* 1970) or inhibited (Goldman, 1971) when MTs are depolymerized. Now it appears that the polarization of fibroblasts during the first 48 hr in primary cultures can occur without MTs (Middleton *et al.,* 1989). Likewise, MTs do not seem to be required for the polarization of epithelial cells (Euteneuer and Schliwa, 1984; Middleton *et al.,* 1988), except those from continuous cell lines (Domnina *et al.,* 1985). Although drugs that depolymerize MTs have little effect on random migration of neutrophils, they markedly inhibit their directional migration in response to a chemoattractant (Bandmann *et al.,* 1974; Malech *et al.,* 1977). In peritoneal macrophages, colchicine induced a shift from gliding to ameboid movement

(Bhisey and Freed, 1971). Although colchicine enhances random migration of lymphoblasts, these cells lose the capacity to respond to a chemotactic gradient (Russel *et al.,* 1975). These results, however, are difficult to reconcile with those of other investigators, who found that, in colchicine-treated neutrophils migrating in response to a chemotactic gradient, neither the morphological nor the behavioral polarity is dependent on the presence of MTs (Zigmond *et al.,* 1981; Keller *et al.,* 1984). In neutrophils, the loss of MTs also stimulates motility and random locomotion (Keller *et al.,* 1984). In the absence of MTs, the radius of turning when the direction of the source of chemoattractant is changed does, however, increase (Devreotes and Zigmond, 1988). It seems that in leukocytes the MTs are not essential for migration although they can certainly modulate it (Keller and Zimmermann, 1987; Haston and Wilkinson, 1988), perhaps by limiting the ability of cells to change direction.

The migration of epithelial sheets into an *in vitro* wound is also insensitive to colchicine (Dipasquale, 1975; Dunlap and Donaldson, 1978; Gipson *et al.,* 1982). It could be that, in epithelial sheets, the polarization of cells at the wound edge results from the loss of both cell–cell contacts and contact inhibition rather than from the organization of MTs. In contrast, colchicine does inhibit the migration of ECs in culture, although the ECs are likewise linked to each other by junctions and can be contact inhibited (Selden *et al.,* 1981).

IV. POSITION OF CENTROSOMES IN MIGRATING CELLS

Many studies also suggest that the centrosome may be important in defining and even determining the direction of cell migration. For example, in most 3T3 fibroblasts migrating in cell culture (Albrecht-Buehler and Bushnell, 1979) and in ameboid cells of *Dictyostelium* migrating in response to a gradient of cAMP (Rubino *et al.,* 1984), the centrosome is located in front of the nucleus in the direction of cell migration. Moreover, ultrastructural studies of migrating 3T3 fibroblasts indicate that one centriole of the pair is generally perpendicular and the other parallel to the substratum (Albrecht-Buehler and Bushnell, 1979). This indicates that centrioles can maintain well-defined orientation not only with respect to the direction of cell migration but also with respect to the plane of the substratum. Neutrophils (Malech *et al.,* 1977) and peritoneal macrophages (Nemere *et al.,* 1985) induced to migrate in response to a chemoattractant also orient their centrosomes and the Golgi apparatus in the direction of cell migration. In the neutrophils, an increase in MFs on the side of the cell toward the chemoattractant, the formation of pseudopods, and migration were not generally seen before reorientation of the centrosome had occurred (Malech *et al.,* 1977). In migrating newt eosinophils, the centrosomes are also oriented toward the leading edge. If the centrosome in these cells is damaged by laser microirradiation, directional cell migration is blocked and motility becomes uncoordinated, directionless, and much slower (Koonce *et al.,* 1984). The irradiation appears to damage the centrioles rather than the surrounding MTOCs, since MT assembly is restored soon after irradiation stops. The widespread occurrence of centrosomes at the front end of many kinds of migrating cells suggests that the centrioles or centrosomes play a central role in controlling the direction of cell migration after the stimulus for migration is received.

There are, however, also some interesting exceptions to the general rule that centrosomes are preferentially oriented toward the front end of migrating cells. In migrating lymphocytes, they are located instead at the posterior end of the cell (Gudima *et al.,* 1988). In *Dictyostelium* amebas the position of centrosomes can be varied by experimentally changing the conditions under which the cells migrate (Sameshima, 1988). In amebas migrating in response to a chemotactic gradient of cAMP, the centrosome is located in the front end of the cell, whereas in amebas induced to migrate toward folic acid, aggregating on agar plates, and moving randomly, the centrosomes are preferentially oriented toward the posterior end of the cell. In amebas that are aggregating while submerged in the medium there is no preferential orientation in either direction. These examples indicate that the position of the centrosome in migrating amebas is readily altered by changes in the environment. The results also suggest that in these relatively rapidly migrating cells other factors may play a role in maintaining polarity. It is probably significant that even in these few but important exceptions to the general rule the position of the centrosome is rarely random.

Earlier studies of microplasts, small motile fragments of cytoplasm prepared from fibroblasts, suggest that some type of coordinating or supervising mechanism must exist in intact cells for directional migration (Albrecht-Buehler, 1980). The centrosome could provide such a coordinating mechanism. Cytoplasts prepared from the anterior ends of rapidly migrating leukocytes and fish epidermal keratocytes, however, are able to migrate in a specific direction and to change their direction of migration even in the absence of centrosomes (Malawista and DeBoisfleury, 1982; Euteneuer and Schliwa, 1984). It has been suggested that the cytoplasts from the keratocytes may be permanently polarized (Singer and Kupfer, 1986). This, however, would not explain how the cytoplasts can change their direction of migration in response to new stimuli.

Several studies on multinucleated cells and their formation support the idea that centrosomes are important in cell migration. During myotube formation by fusion of myoblasts, both the centrosomes and the centrioles disappear and the site of initiation of MTs changes from the centrosome to the nuclear envelope (Connolly *et al.,* 1985; Tassin *et al.,* 1985). This suggests that the centrosomes may be necessary for myoblast migration, alignment, and perhaps even fusion, but not for the function of the resulting myotube. Other multinucleated cells that retain centrosomes, in contrast, remain motile (Lewis and Albrecht-Buehler, 1987;

Moudjou *et al.*, 1989). In BHK syncytia, which have numerous centrosomes, the position of the centrosomes determines the direction of cell migration (Lewis and Albrecht-Buehler, 1987).

Interesting speculations also have been made about the role of centrioles in cell migration. These include suggestions that centrioles may act as navigational devices (Bornens, 1979; see Chapter 1) or sensors for locating the direction of signal sources from the environment (Albrecht-Buehler, 1981; see Chapter 3).

If one of the functions of the centrioles and/or the associated MT system is to receive signals from the environment and transmit them to the interior of the cell, it would seem appropriate to have such devices oriented toward the front end. Since the centrosome is generally located close to the Golgi apparatus, the MTs radiating from this region could also transport various components of the cytoplasm toward the front end of the cell, including vesicles for incorporation into cell membrane as suggested by some models of cell migration (Singer and Kupfer, 1986; Hay, 1989). According to this attractive hypothesis, the centrosome–MT–Golgi complex should play a critical role in determining the polarity of migrating cells.

The previous studies dealt with the position of centrosomes in single cells migrating in cell cultures. We wanted to determine whether the centrosome is also found at the front end of cells migrating as a sheet while interacting with neighboring cells in a more complex, less artificial environment. We therefore examined how the position of the centrosome in ECs changes when cell sheets are induced to migrate in a specific direction by wounding the EC monolayer. This was done in cell cultures in which neither normal substratum nor blood flow is present, in organ cultures in which the substratum is presumably normal but blood flow is absent, and *in situ* on normal substratum in the presence of blood flow. The results obtained from these studies will now be described.

V. CHANGES IN CENTROSOME POSITION IN AORTIC ENDOTHELIAL CELLS INDUCED TO MIGRATE

A. In Cell Cultures

Migration of ECs in a well-defined direction in cell cultures was induced by mechanical denudation of part of a confluent monolayer. The position of the centrosome in relation to the direction of migration and the nucleus at different times after wounding was recorded. As mentioned previously, in the initial monolayer the centrosomes of ECs are randomly positioned in relation to the nucleus. In migrating sheets of ECs, however, the centrosome in the great majority of cells along the wound edge is oriented toward the wound, that is, in front of the nucleus in the direction of migration. In addition, numerous MTs radiate from these centrosomes toward the wound edge (Fig. 4). A preferential orientation of the centrosomes toward the wound was first noticeable as early as 20 min after wounding. By 4 hr after wounding, about 80% of the ECs in the first row along the wound edge had centrosomes oriented toward the wound (Fig. 5). A similar reorientation of centrosomes toward the wound developed more slowly and to a lesser degree in the second and third rows of ECs behind the wound edge. These results show that in ECs that migrate as a sheet, as in many kinds of cells that migrate separately, the centrosome is preferentially located toward the anterior end in the direction of cell migration (Gotlieb *et al.,* 1981).

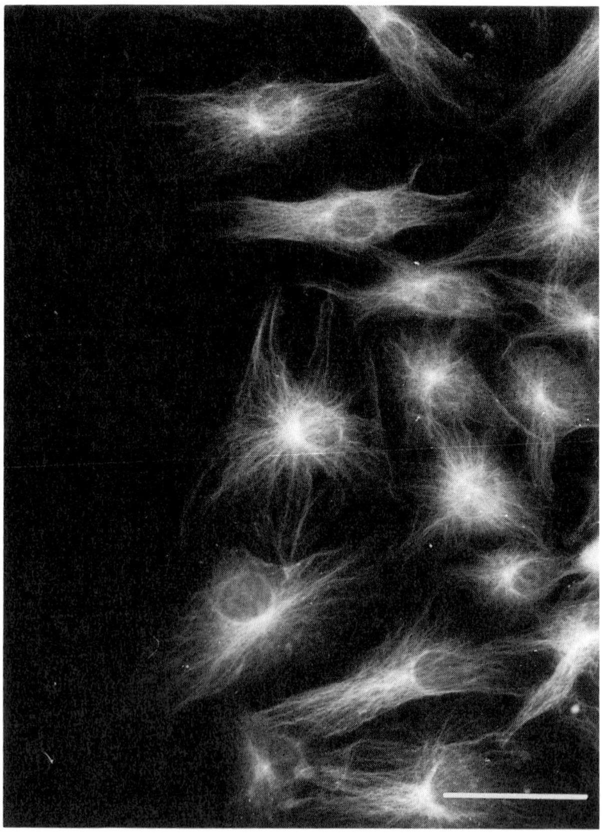

Fig. 4. Immunofluorescence staining of endothelial cells (EC) at the wound edge with antitubulin serum, 44 hr after wounding. Note that the centrosomes, which initially are randomly distributed relative to the wound edge and the position of the nuclei, become oriented so that, in most ECs, they face the wound (left side) toward which the endothelial sheet is migrating. Bar: 50 μm.

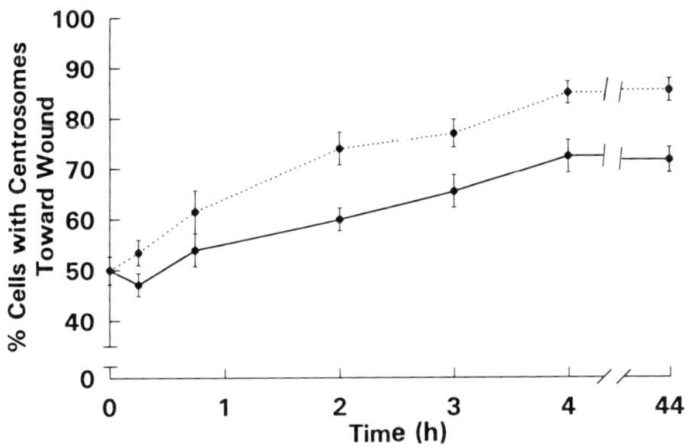

Fig. 5. Graph showing the percentage of migrating endothelial cells (EC) with centrosomes oriented toward the wound edge in the first row (· · ·) of cells along the edge and in the second and third rows (—) of cells at various times after wounding. Note that gradually increasing numbers of EC show a reorientation of the centrosomes toward the wound during the first 4 hr. (Reprinted from Gotlieb *et al.,* 1981, with permission.)

After the EC sheet had extended into the wound and the cells had become flatter, the position of the centrosome could be determined by time-lapse cinemicrophotography. These studies indicated that in migrating ECs the centrosomes could change positions rapidly relative to the nucleus and the wound edge. In the first row of ECs, some of the centrosomes oriented toward the leading edge at one particular time were not the same as those oriented toward this edge at another time. The time-lapse studies also showed that some ECs, especially those in the second and third rows, advanced or were pulled forward by the actively migrating cells even when their centrosomes were located behind or on the side of the nucleus (Gotlieb *et al.,* 1981).

Mascardo and Sherline (1984) have confirmed these reuslts in cultures of bovine ECs and have shown in addition that the reorientation of centrosomes toward the wound edge requires serum, high-energy phosphates, and Ca^{2+}–calmodulin interactions. Specific humoral factors such as multiplication-stimulating activity and insulin could also induce reorientation toward the wound edge, but somatostatin had the opposite effect, causing centrosomes to orient away from the wound edge. A similar reorientation of the centrosomes toward the leading edge, as well as of the Golgi apparatus, has been observed also in sheets of cultured NRK fibroblasts induced to migrate by wounding monolayers (Kupfer *et al.,* 1982).

Interesting results that complement those just described also have been obtained from studies on small wound re-endothelialization in monolayers of ECs. Wong and Gotlieb (1988) showed that spreading of ECs into the wound could occur without the reorientation of the centrosomes toward the wound edge. In their system, centrosome reorientation occurred only when the wounds were made large enough to induce cell migration.

B. In Organ Cultures

To determine if centrosomes would also reorient toward the wound when ECs are induced to migrate on their natural substratum, we studied the response of ECs to wounding in segments of pig aorta in organ culture (Rogers *et al.,* 1986). The ECs were examined along the wound edge at different times after denuding the endothelium in the distal half (with respect to the heart) of a rectangular piece of aorta. We found that by 24 hr the ECs had started to spread into the wound, by 72 hr most of cells in the first row had acquired an elongated shape with the long axis oriented perpendicular to the wound edge, and by 96 hr this change in shape and orientation had extended to several rows of ECs along the wound edge.

As mentioned previously, the centrosomes in most of the ECs were initially oriented toward the former position of the heart in the direction away from the wound. This preferential orientation of centrosomes is lost by about 24 hr after wounding; by 72 hr, most of the ECs along the wound edge had their centrosomes oriented in the opposite direction, that is, toward the wound edge.

This study (Rogers *et al.,* 1986) demonstrates that ECs in organ culture, migrating in response to wounding on a substratum similar to that found *in vivo,* but in the absence of blood flow, show a similar reorientation of their centrosomes toward the wound edge as do ECs in cell cultures. Moreover, in organ cultures the original orientation of the centrosome could be reversed by wounding. In organ cultures, the reorientation of centrosomes, however, occurred much more slowly, reaching maximal values by 24 hrs, whereas in cell cultures, maximal values were already reached by 4 hr (Gotlieb *et al.,* 1981). The reason that the reorientation of centrosomes is much slower after wounding in organ culture may be related to differences in the substratum. This explanation would be consistent with the observations showing that the subendothelial matrix has an effect on the orientation of ECs (Buck, 1979). It would not be surprising, therefore, if the substratum also affected the rate of reorientation of centrosomes toward the wound.

C. In Aorta

The behavior of ECs in response to wounding *in situ* has been extensively studied in the aorta after mechanical removal of the endothelium (Schwartz *et al.,* 1978; Reidy *et al.,* 1982). The wounding is done by inserting a balloon catheter into the thoracic or abdominal aorta via a distributing artery, inflating the balloon, and pulling it along the vessel to remove the endothelium. Within hours, ECs from either end of the denuded segment begin to migrate into the wounded region. The changes in cell shape at the wound edge and the time course of re-endothelialization have been carefully documented (Schwartz *et al.*, 1978). After wounding, the ECs at the wound edge flatten and send out thin sheets of cytoplasm (lamellipodia) over the denuded surface. Later, a similar flattening of cells is seen in a region extending over 100 rows of ECs from the wound edge. After balloon-catheter wounding in this system, in contrast to cell and organ cultures, the ECs are migrating *in situ* on their normal substratum in the presence of blood flow while interacting with blood-borne molecules and cells.

We determined the position of the centrosomes in the ECs with respect to the nucleus and the wound edge, identified by the presence of adhering platelets, at 0, 6, 12, 24, and 48 hr after wounding in *en face* preparations by immunofluorescence with sera specific for centrioles (Fig. 6). Regions 0.4 mm wide located up to 2 mm from the wound edge were examined. The changes in the orientation of centrosomes in ECs at various distances from the wound edge during the time course of the experiment are summarized in Fig. 7 (Rogers *et al.*, 1991, unpublished results).

Initially, about equal numbers of ECs in the rabbit aorta have centrosomes oriented toward and away from the wound edge. By 12 hr after wounding, the percentage of cells with centrosomes oriented toward the wound doubles from 24 to 48% in the field that extends 0.4 mm from the wound edge and includes the wound edge itself. This number remains at about this level up to at least 48 hr after wounding. A similar but slightly lower increase in the number of ECs with centrosomes oriented toward the wound edge is seen in the adjacent field located 0.4–0.8 mm from the wound edge. In the 6 and 12 hr samples, the number of cells with centrosomes in the central part stays fairly constant at about 40%, whereas the percentage of cells with centrosomes oriented away from the wound decreases from 40 to 14% in the field next to the wound and to a slightly lesser degree in the adjacent one. The simplest explanation for these observations is that, during the first 12 hr after wounding, the centrosomes from the central part of the cell migrate toward the wound edge and are replaced by others that move from the side away from the wound to the central part of the cell. Although the number of cells with centrosomes toward the wound edge shows no further increase in the period from 12 to 24 hr, there is a continuing increase in the number of cells with centrosomes in the central part, from 40 to 55%, not only in the two fields next to the wound edge, but in the fields further away as well. During this increase, the number of ECs with centrosomes away from the wound edge continues to decrease as the centrosomes continue to move toward the center of the cell. No further changes in the position of centrosomes were observed between 24 and 48 hr after wounding in any of the fields examined.

These results show that centrosomes in most migrating ECs along the wound edge *in situ,* as well as in cell and organ cultures, reorient toward the wound and the leading edge. Changes in the orientation of the centrosomes *in situ,* as in organ

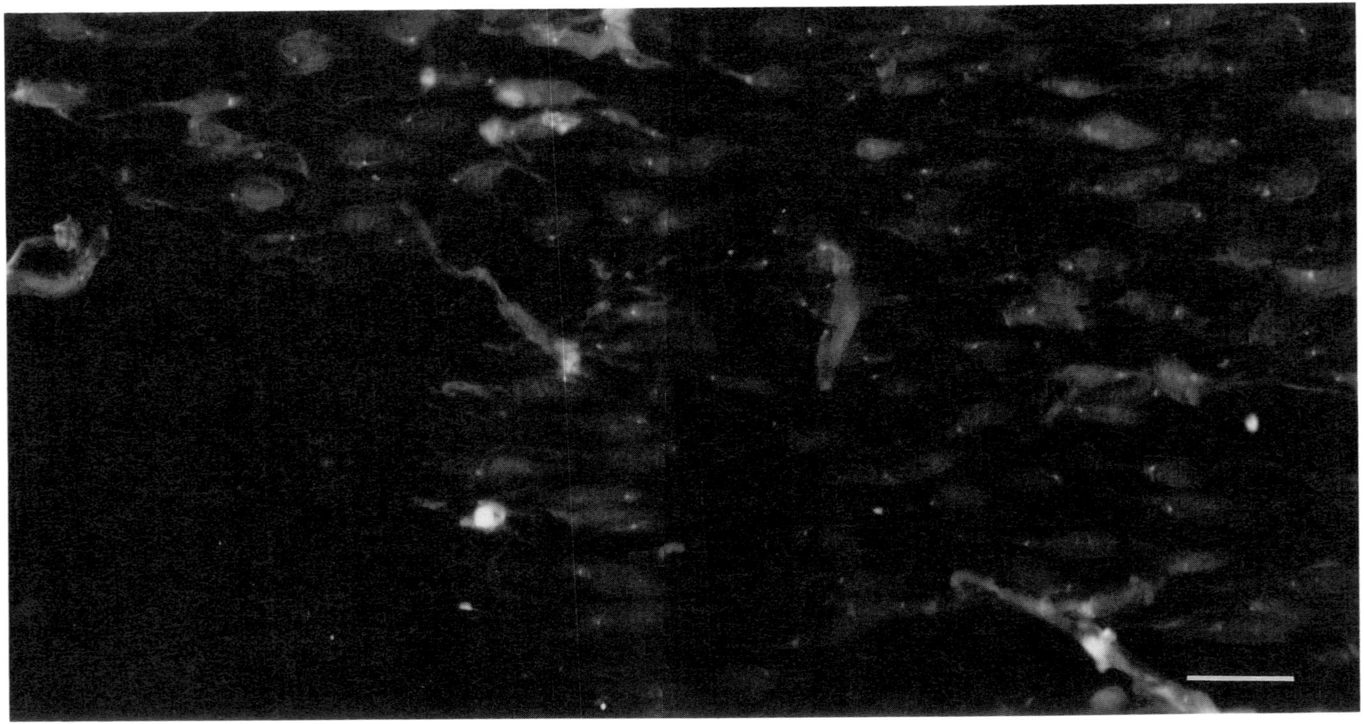

Fig. 6. Immunofluorescence staining of an *en face* preparation of endothelial cells from the distal wound edge of rabbit thoracic aorta harvested 12 hr after wounding to determine the position of the centrosomes, seen as brightly staining dots, in relation to the nucleus. At this time the number of cells near the wound edge that have centrosomes oriented toward the wound has increased greatly. Blood flow is from left to right. Bar: 36 μm.

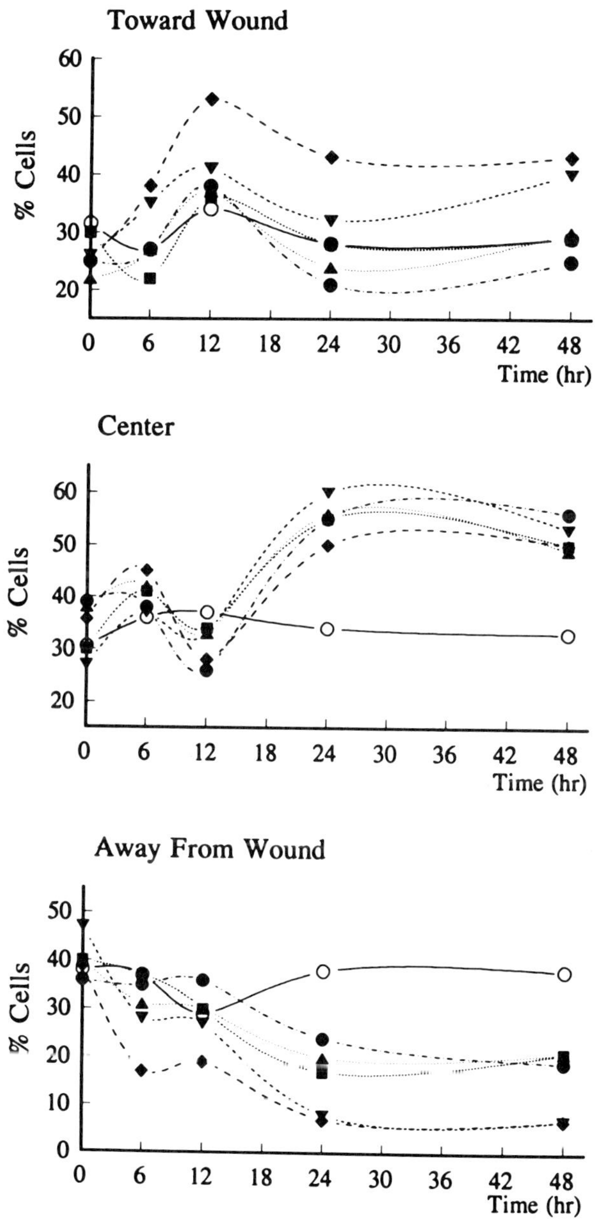

Fig. 7. Percentage of endothelial cells (EC) with centrosomes toward the wound (*top*), in the center (*middle*), and away from the wound (*bottom*) at various times after wounding. Each graph depicts the position of centrosomes in EC in fields 1–5 located 0–400, 400–800, 800–1200, 1200–1600, and 1600–2000 μm away from the wound edge, respectively. Reorientation of centrosomes toward the wound relative to controls can be detected as early as 12 hr after wounding in fields 1 and 2. Note that, in the majority of EC, 24 hr and 48 hr after wounding the centrosomes are located in the central part of EC at distances of up to 2000 μm from the wound edge, while relatively few remain on the side away from the wound. Control (○); field 1 (♦); field 2 (▼); field 3 (■); field 4 (▲); field 5 (●). Controls from an unwounded region of the aorta distal to the wound.

cultures, occur more slowly than in cell cultures, but extend over a much wider region from the wound edge than in cell cultures. *In situ,* a reorientation occurred up to 2 mm from the wound edge 24 hr after wounding whereas in cell cultures it occurred only in the first few rows of cells. It is interesting that, in the rat aorta, increases in the rate of EC replication have been recorded as far as 1 and 2.4 mm from the wound edge 24 and 48 hr after wounding, respectively (Schwartz *et al.*, 1978). These studies showing changes in ECs located relatively far from the wound edge suggest that some form of communication between cells plays a role in the response of endothelium to injury.

In the wounded rabbit aorta, the time course of the reorientation of the centrosomes in ECs reaching maximal values by 12 hr suggests that the observed changes in the distribution of centrosomes in ECs are related to the onset of cell migration in the direction of the wound. A similar centrosome reorientation toward the wound edge reached maximal values much earlier (4 hr) in porcine aortic ECs migrating in cell cultures, but somewhat later (24 hr) during EC migration in organ cultures of porcine aortas. The slower reorientation of the centrosomes *in situ* and in organ cultures in comparison with cell cultures may again reflect the extent of organization of the underlying extracellular matrix. The even slower reorientation seen in organ culture, despite the fact that the substratum *in situ* and in organ cultures is presumably very similar, may be due to the fact that, in the organ culture of the pig aorta but not in the rabbit aorta, the centrosomes initially are preferentially oriented away from the wound edge. Thus, it may take longer for the centrosomes of ECs in organ cultures of pig aorta to reorient in the opposite direction, that is, toward the wound edge. The slower reorientation may also be partly due to species differences or to trauma induced by the removal of the aorta from its normal environment to culture it in an artificial one.

In summary, these results show that, not only in cell and organ cultures but also *in situ,* ECs on their normal substratum in the presence of blood flow reorient their centrosomes in the direction of cell migration. This reorientation *in situ* occurs more slowly but over a much wider region along the wound edge than in cell cultures.

VI. SIGNIFICANCE OF THE PREFERENTIAL ORIENTATION OF CENTROSOMES IN ENDOTHELIAL CELLS OF AORTA AND VENA CAVA

The functional significance of the preferential orientation of centrosomes in the direction toward the heart in the unwounded adult rabbit vena cava, and in pig aorta and vena cava as described previously (Rogers *et al.*, 1985), is unknown. We have suggested, on the basis of our results from migrating ECs in which the centrosome is generally found at the front end in the direction of cell migration, that the

ECs lining the larger blood vessels may be similarly migrating in the direction of their centrosomes, that is, toward the heart (Rogers *et al.*, 1985).

Several *in vivo* studies using Evans blue, a dye that stains regions of the vessel wall permeable to macromolecules, have shown that injury to the endothelium is more common in some regions of the aorta than in others and that the former have a higher turnover of ECs (Friedman and Byers, 1969; Caplan and Schwartz, 1973). In the pig aorta, Evans blue staining is most prominent in the aortic arch (Caplan and Schwartz, 1973). If the turbulence of blood flow in the pig aortic arch causes injury to the endothelium and loss of ECs, migration of ECs from more distal regions of the aorta could re-establish the normal density of ECs. Thus the preferential orientation of centrosomes may reflect a steady migration of ECs to areas of high cell turnover (Rogers *et al.*, 1985). Since flow dividers (Table I) are sites subjected to high shear stress at which injury to the endothelium is more likely to occur as well as sites at which atherosclerotic lesions commonly develop, the migration of ECs toward the flow dividers as suggested by the position of centrosomes in ECs on either side may also serve to replace lost or damaged ECs in this region. It is interesting that in the thoracic aorta of mature rabbits, where no preferential orientation of centrosomes in ECs was found, exposure to Evans blue does not stain the aortic arch (Friedman and Byers, 1963). Thus migration of ECs to the aortic arch from more distal parts of the aorta may not be necessary to maintain an intact endothelium in this species. The distribution of centrosomes in the ECs of the inferior vena cava of the pig and the rabbit suggests that these cells are also migrating toward heart. We are not aware, however, of studies showing sites of damage to the endothelium in the venous system that could account for such migration.

More direct evidence for migration of ECs in the direction of their centrosomes in the unwounded larger blood vessels is now required to test the hypothesis. Although there is no evidence for such migration in the adult, there are indications that such migration may occur during embryonic development. Using quail limb buds grafted into a chick host, it has been shown that quail cells, which can be identified by their prominent nucleoli, reach and become incorporated into the endothelial lining of the chick aorta and then migrate as part of it in the direction of the heart (Christ *et al.*, 1990). However, until evidence for migration of EC populations with preferentially oriented centrioles becomes available in the adult, this hypothesis must remain highly speculative.

VII. EFFECT OF CYTOCHALASIN B AND COLCEMID ON CENTROSOME REORIENTATION

It is not clear from the studies on the response of ECs to wounding whether the reorientation of the centrosome precedes cell migration or whether the migration of the cell causes a reorientation of the centrosome in the direction of cell migration. Studies to determine the orientation of centrosomes in cells in which MT and MF systems are disrupted to inhibit migration therefore have been undertaken in cultures of ECs to determine whether the centrosome reorientation occurs independently or as a result of cell migration. These studies would also provide information on whether the observed reorientation of centrosomes requires intact MT and/or MF systems.

Results show that centrosome reorientation toward the wound edge occurs, although at a slightly reduced level, even when EC migration is prevented by cytochalasin B (Gotlieb *et al.,* 1983), a drug that inhibits MF assembly. Similar results are obtained in neutrophils (Malech *et al.,* 1977) and in bovine ECs (Mascardo and Sherline, 1984). Thus, the reorientation of the centrosome toward the leading edge in ECs occurs independent of cell migration. These results suggest that the migration itself is neither the cause of nor the signal for the reorientation of the centrosomes. It is interesting that the second row of cells does not show a similar reorientation of their centrosomes in the presence of cytochalasin B. However, once the sheet begins to move forward after the cytochalasin B is washed out, the centrosomes in cells of the second row also reorient toward the wound edge. This would be expected if the signal for the reorientation somehow comes from a reduction in cell–cell contacts along the wound side of the ECs. Such a reduction in contacts between the first and second rows of ECs may occur only after the cells in the first row have begun to translocate.

The reorientation of centrosomes in ECs can also be inhibited by Colcemid, which disrupts MTs (Gotlieb *et al.,* 1983). After the Colcemid is removed, reorientation of centrosomes and cell migration occurs only after the reappearance of most cytoplasmic MTs. This suggests that a Colcemid-sensitive mechanism is responsible for the reorientation of the centrosomes and that here, as in other systems (Malech *et al.,* 1977; Mascardo and Sherline, 1984), the reorientation requires MTs and/or structures dependent on MTs for distribution and function.

VIII. MECHANISMS FOR REORIENTATION OF CENTROSOMES

Several mechanisms can be envisaged by which the observed reorientation of centrosomes could take place. For example, the entire cell could rotate to bring its centrosome toward the wound edge or another source of migration signal. This mechanism, however, seems unlikely, since it would require the breaking and reformation of a large number of junctions between ECs forming the monolayer. The nucleus to which the centrosome is attached (Bornens, 1977) could also rotate. No evidence for either of these mechanisms was obtained by observing centrosomal regions in migrating ECs in culture by time-lapse cinemicrophotography (Gotlieb *et al.,* 1981). A more likely possibility is that the MTs radiating from the centrosome in the direction away from the wound edge elongate while those radiating toward the wound edge shorten. Such an elongation of MTs on one side and a shortening on the other would bring the centrosome to a position in front of the nucleus. There is evidence in other systems that the MFs play a major role in the positioning of centrosomes (Euteneuer and Schliwa, 1985; Buendia *et al.*, 1990). This does not seem to be the case in ECs, since the centrosomes can reorient toward the wound edge in the presence of cytochalasin B.

IX. POLARITY AND DIRECTIONAL MIGRATION OF CELLS

After the reorientation of the centrosome, the polarity could be maintained by the selective stabilization of MTs radiating toward the leading edge of the migrating ECs. Such stabilization of MTs in the direction of cell migration has been observed in fibroblasts induced to migrate by wounding a monolayer (Gundersen and Bulinski, 1988). Knowledge of the dynamics of MT assembly and of the relative stability of MTs in different parts of the ECs during centrosome reorientation and cell migration may help us understand the mechanism by which polarity in migrating cells develops and is maintained.

The observation that MTs are associated with focal contacts forming behind the leading edge of migrating fibroblasts suggests that another function of the MTs radiating from the centrosome toward the leading edge may be to stabilize these sites of cell–substratum interaction (Rinnerthaler *et al.*, 1988) and other parts of the cell cortex. In lymphoblasts, an intrinsic oscillatory activity of the cortical actomyosin system is confined by the centrosome and associated MTs to the side of the cell on which the centrosome is located and results in a polarized extension of the cytoplasm on that side (Bornens *et al.*, 1989; see Chapter 1). This occurs without any apparent polarizing signal from the environment. It appears that the centrosome is somehow able to respond to events at the cell cortex as well as modulate them through the MT system by some type of feedback mechanism. Occupying, as it does, the center of the cell, the centrosome would be ideally situated to perform this task.

The results presented in this chapter, and those of others discussed previously, are consistent with the idea that the centrosome and associated MTs, if they do not determine it, at least play an important role in defining the polarity of migrating ECs. They suggest that the following general steps may be required to induce directional migration in ECs and in other cell types. Once a cell receives a signal to migrate in a given direction, intracellular processes are activated that orient the centrosome and structures associated with it, for example, the Golgi apparatus, in that direction by a process that requires MTs. After the centrosome is oriented in the direction of migration, the mechanism responsible for moving the cell forward is activated. This last step is sensitive to cytochalasin B and probably requires

MFs. In the presence of cytochalasin B, the two steps are dissociated, that is, the centrosome can turn toward the wound but the cell is unable to move forward because its system of MFs is disorganized. Since the reorientation of centrosomes precedes cell migration, the position of centrosomes could define the polarity of the migrating cell and the direction of cell migration.

In summary, studies on ECs suggest that directional cell migration is a very complex process in which a sensory system that can respond to external signals must somehow be linked to the force-generating machinery, the activation of which results in the actual translocation of the cell. The centrosome, through the associated MTs, could insure that the two remain coupled as the cell migrates and that the material required for continued cell migration is efficiently transported toward the front end of the cell.

It seems that the intrinsic polarity of cells is determined in large part by the position of the centrosome. Its direction, however, can be altered readily by signals from the environment. The highly complex centrosome and the associated MT system could have evolved to give the motile cell a more precise control over the force-generating step provided by the cortical MFs. More information on the evolution of the centrosome and its behavior in cells displaying different types of motility would be helpful in understanding the function of this organelle in cell translocation. It may help to explain why some rapidly moving cytoplasts can migrate in the absence of centrosomes and MTs and why, in slower moving cells, the centrosomes are found in specific positions in relation to the direction of cell migration to define cell polarity. Finally, several mechanisms may exist for controlling polarity in migrating cells, so if one is inactivated another could replace it. If this is true, further studies of cell migration *in situ* would help us decide which of these mechanisms is actually used by the migrating cells in the intact animal.

ACKNOWLEDGMENTS

This work was supported by grants from the Medical Research Council of Canada.

REFERENCES

Albrecht-Buehler, G. (1980). Autonomous movements of cytoplasmic fragments. *Proc. Natl. Acad. Sci. U.S.A.* **77,** 6639–6643.

Albrecht-Buehler, G. (1981). Does the geometric design of centrioles imply their function? *Cell Motil.* **1,** 237–245.

Albrecht-Buehler, G., and Bushnell, A. (1979). The orientation of centrioles in migrating 3T3 cells. *Exp. Cell Res.* **120,** 111–118.

Bandmann, V., Rydgren, L., and Norberg, B. (1974). The difference between random movement and chemotaxis. Effects of anti-tubulins on neutrophil granulocyte locomotion. *Exp. Cell Res.* **88,** 63–73.

Bhisey, A., and Freed, J. J. (1971). Ameboid movement induced in cultured macrophages by colchicine or vinblastine. *Exp. Cell Res.* **64,** 419–429.

Bornens, M. (1977). Is the centriole bound to the nuclear membrane? *Nature (London),* **270,** 80–89.

Bornens, M. (1979). The centriole as a gyroscopic oscillator: Implications for cell organization and some other consequences. *Biol. Cell* **35,** 115–132.

Bornens, M., Paintrand, M., and Celati, C. (1989). The cortical microfilament system of lymphoblasts displays a periodic oscillatory activity in the absence of microtubules: Implications for cell polarity, *J. Cell Biol.* **109,** 1071–1083.

Buck, R. C. (1979). The longitudinal orientation of structures in the subendothelial space of the rat aorta. *Am. J. Anat.* **156,** 1–13.

Buendia, B., Bré, M.-H., Griffiths, G., and Karsenti, E. (1990). Cytoskeletal control of centrioles movement during the establishment of polarity in Madin-Darby canine kidney cells. *J. Cell Biol.* **110,** 1123–1135.

Caplan, B. A., and Schwartz, C. J. (1973). Increased endothelial cell turnover in areas of *in vivo* Evans blue uptake in the pig aorta. *Atherosclerosis* **17,** 401–417.

Christ, B., Poelmann, R. E., Mentnik, M. M. T., and Gittenberger de Groot, A. C. (1990). Vascular endothelial cells migrate centripetally within embryonic arteries. *Anat. Embryol.* **181,** 333–339.

Connolly, J. A., and Kalnins, V. I. (1978). Visualization of centrioles and basal bodies by fluorescent staining with nonimmune rabbit sera. *J. Cell Biol.* **79,** 526–532.

Connolly, J. A., Kiosses, B. W., and Kalnins, V. I. (1985). Centrioles are lost as embryonic myoblasts fuse into myotubes *in vitro. Eur. J. Cell Biol.* **39,** 341–345.

Devreotes, P. N., and Zigmond, S. H. (1988). Chemotaxis in eucaryotic cells: A focus on leucocytes and *Dictyostelium. Ann. Rev. Cell Biol.* **4,** 649–686.

Dewey, C. F., Jr., Bussolari, S. R., Gimbrone, M. A., and Davies, P. F. (1981). The dynamic response of vascular endothelial cells to fluid shear stress. *J. Biomech.* **103,** 177–185.

Dipasquale, A. (1975). Locomotion of epithelial cells. Factors involved in extension of the leading edge. *Exp. Cell Res.* **95,** 425–439.

Domnina, L. V., Rovensky, J. A., Vasiliev, J. M., and Gelfand, I. M. (1985). Effect of microtubule-destroying drugs on the spreading and shape of cultured epithelial cells. *J. Cell Sci.* **74,** 267–282.

Dunlap, M. K., and Donaldson, D. J. (1978). Inability of colchicine to inhibit newt epidermal cell migration or prevent concanavalin A-mediated inhibition of migration. *Exp. Cell Res.* **116,** 15–19.

Euteneuer, U., and Schliwa, M. (1984). Persistent, directional motility of cells and cytoplasmic fragments in the absence of microtubules. *Nature (London)* **310,** 58–61.

Euteneuer, U., and Schliwa, M. (1985). Evidence for involvement of actin in the positioning and motility of centrosomes. *J. Cell Biol.* **101,** 96–103.

Friedman, M., and Byers, S. O. (1963). Endothelial permeability in atherosclerosis. *Arch. Path.* **76,** 99–105.

Geiger, B., Rosen, D., and Berke, G., (1982). Spatial relationships of microtubule-organizing centers and the contact area of cytotoxic T lymphocytes and target cells. *J. Cell Biol.* **95,** 137–143.

Gipson, I. K., Westcott, M. J., and Brooksby, N. G. (1982). Effect of cytochalasins B and D and colchicine on the migration of the corneal epithelium. *Invest. Ophthalmol. Visual Sci.* **22,** 633–642.

Goldman, R. D. (1971). The role of three cytoplasmic fibers in BHK-21 cell motility. I. Microtubules and the effects of colchicine. *J. Cell Biol.* **51,** 752–762.

Gotlieb, A. I., McBurnie-May, L., Subrahmanyan, L., and Kalnins, V. I. (1981). Distribution of microtubule organizing centers in migrating sheets of endothelial cells. *J. Cell Biol.* **91,** 589–594.

Gotlieb, A. I., Subrahmanyan, L., and Kalnins, V. I. (1983). Microtubule-organizing centers and cell migration: Effect of inhibition of migration and microtubule disruption in endothelial cells. *J. Cell Biol.* **96,** 1266–1272.

Gotlieb, A. I., and Boden, P. (1984). Porcine aortic organ culture: A model to study the cellular response to vascular injury. *In Vitro* **20,** 535–542.

Gudima, G. O., Vorobjev, I. A., and Chentsov, Y. S. (1988). Centriolar location during blood cell spreading and motion *in vitro:* An ultrastructural analysis. *J. Cell Sci.* **89,** 225–241.

Gundersen, G. G., and Bulinski, J. C. (1988). Selective stabilization of microtubules oriented toward the direction of cell migration *Proc. Natl. Acad. Sci. U.S.A.* **85,** 5946–5950.

Haston, P. S., and Wilkinson, P. C. (1988). Locomotion and chemotaxis of leucocytes: Gradient perception and locomotor capacity. *Curr. Opinion Immunol.* **1,** 5–9.

Hay, E. D. (1989). Theory of epithelial–mesenchymal transformation based on the "fixed cortex" cell motility model. *Cell Motil. Cytosk.* **14,** 455–457.

Keller, H. U., Naef, A., and Zimmerman, A. (1984). Effects of colchicine, vinblastine, and nocodazole on polarity, motility, chemotaxis and cAMP levels of human polymorphonuclear leukocytes. *Exp. Cell Res.* **153,** 173–185.

Keller, H. U., and Zimmermann, A. (1987). Shape, movement and function of neutrophil granulocytes. *Biomed. Pharmacother.* **41,** 285–289.

Kelly, R. B. (1990). Microtubules, membrane traffic and cell organization. *Cell* **61,** 5–7.

Koonce, M. P., Cloney, R. A., and Berns, M. W. (1984). Laser radiation of centrosomes in newt eosinophils: Evidence of centriole role in motility. *J. Cell Biol.* **98,** 1999–2010.

Kreis, T. E. (1990). Role of microtubules in the organization of the Golgi apparatus. *Cell Motil. Cytosk.* **15,** 67–70.

Kupfer, A., Louvard, D., and Singer, S. J. (1982). Polarization of the Golgi apparatus and the microtubule organizing center in cultured fibroblasts at the edge of an experimental wound. *Proc. Natl. Acad. Sci. U.S.A.* **79,** 2603–2607.

Kupfer, A., Dennert, G., and Singer, S. J. (1985). The reorientation of the Golgi apparatus and the microtubule-organizing center in the cytotoxic effector cell is a prerequisite in the lysis of bound target cells. *J. Mol. Cell Immunol.* **2,** 37–49.

Lewis, L., and Albrecht-Buehler, G. (1987). Distribution of multiple centrospheres determines migration in BHK syncitia. *Cell Motil. Cytosk.* **7,** 282–290.

Malawista, S. E., and DeBoisfleury, A. (1982). The cytokineplast: Purified, stable and functional motile machinery from human polymorphonuclear leukocytes. Possible formative role of heat induced centrosomal dysfuncion. *J. Cell Biol.* **95,** 960–973.

Malech, H. R., Root, R. K., and Gallin, J. I. (1977). Structural analysis of human neutrophil migration. Centriole, microtubule, and microfilament orientation and function during chemotaxis. *J. Cell Biol.* **75,** 666–693.

Mascardo, R. N., and Sherline, P. (1984). Insulin and multiplication-stimulating activity induce a very rapid centrosomal orientation response to wounding in endothelial cell monolayers. *Diabetes* **33,** 1099–1105.

Middleton, C. A., Brown, A. F., Brown, R. M., and Roberts, D. J. (1988). The shape of cultured epithelial cells does not depend on the integrity of their microtubules. *J. Cell Sci.* **91,** 337–3.

Middleton, C. A., Brown, A. F., Brown, R. M., Karavanova, I. D., Roberts, D. J. H., and Vasiliev, J. M. (1989). The polarization of fibroblasts in early primary cultures is independent of microtubule integrity. *J. Cell Sci.* **94,** 25–32.

Moudjou, M., Lanotte, M., and Bornens, M. (1989). The fate of the centrosome-microtubule network in monocyte-derived giant cells. *J. Cell Sci.* **94,** 237–244.

Nemere, I., Kupfer, A., and Singer, S. J. (1985). Reorientation of the Golgi apparatus and the microtubule-organizing center inside macrophages subjected to a chemotactic gradient. *Cell Motil.* **5,** 17–29.

Reidy, M. A., and Langille, B. L. (1980). The effect of local blood flow patterns on endothelial cell morphology. *Exp. Mol. Pathol.* **32,** 276–289.

Reidy, M. A., Standaert, D., and Schwartz, S. M. (1982). Inhibition of endothelial cell regrowth. Cessation of aortic endothelial cell replication after balloon catheter denudation. *Arteriosclerosis* **32,** 3216–3220.

Rinnerthaler, G., Geiger, B., and Small, J. V. (1988). Contact formation during fibroblast locomotion: Involvement of membrane ruffles and microtubules. *J. Cell Biol.* **106,** 747–760.

Rogers, K. A., and Kalnins, V. I. (1983a). A method for examining the endothelial cytoskeleton *in situ* using immunofluorescence. *J. Histochem. Cytochem.* **31,** 1317–1320.

Rogers, K. A., and Kalnins, V. I. (1983b). Comparison of the cytoskeleton in aortic endothelial cells *in situ* and *in vitro. Lab Invest.* **49,** 650–654.

Rogers, K. A., McKee, N. H., and Kalnins, V. I. (1985). Preferential orientation of centrioles towards the heart in endothelial cells of major blood vessels is reestablished after reversal of a segment. *Proc. Natl. Acad. Sci. U.S.A.* **82,** 3272–3276.

Rogers, K. A., Boden, P., Kalnins, V. I., and Gotlieb, A. I. (1986). The distribution of centrosomes in endothelial cells of non-wounded and wounded aortic organ cultures. *Cell Tissue Res.* **243,** 223–227.

Rogers, K. A., Sandig, M., McKee, N. H., and Kalnins, V. I. (1991). The distribution of centrosomes in migrating endothelial cells during wound healing *in situ.* In preparation.

Ross, R. (1988). Endothelial injury and atherogenesis. *In* "Endothelial Cell Biology" (N. Simionescu and M. Simionescu, eds.) pp. 371–384. Plenum Press, New York.

Rubino, S., Fighetti, M., Unger, E., and Cappuccinelli, P. (1984). Location of actin, myosin, and microtubular structures during directed locomotion of *Dictyostelium* amebae. *J. Cell Biol.* **98,** 382–390.

Russel, R. J., Wilkinson, P. C., Sless, F., and Parrot, D. M. V. (1975). Chemotaxis of lymphoblasts. *Nature (London)* **256,** 646–648.

Sameshima, M., Imai, Y., and Hashimoto, Y. (1988). The position of the microtubule-organizing center relative to the nucleus is independent of the direction of cell migration in *Dictyostelium discoideum. Cell Motil. Cytosk.* **9,** 111–116.

Schwartz, S. M., Haudenschild, C. C., and Eddy, E. M. (1978). Endothelial regeneration. I. Quantitative analysis of initial stages of endothelial regeneration in rat aortic intima. *Lab. Invest.* **38,** 568–580.

Selden, S. C., III, Rabinovitch, P. S., and Schwartz, S. M. (1981). Effects of cytoskeletal disrupting agents on replication of bovine endothelium. *J. Cell Physiol.* **108,** 195–211.

Singer, S. J., and Kupfer, A. (1986). The directed migration of eurkaryotic cells. *Ann. Rev. Cell Bio.* **2,** 337–365.

Tassin, A.-M., Maro, B., and Bornens, M. (1985). Fate of microtubule-organizing centers during myogenesis *in vitro. J. Cell Biol.* **100,** 35–46.

Terasaki, M. (1990). Recent progress on structural interactions of the endoplasmic reticulum. *Cell Motil. Cytosk.* **15,** 71–75.

Turksen, K., Aubin, J. E., and Kalnins, V. I. (1982). Identification of a centriole-associated protein by antibodies present in normal rabbit sera. *Nature (London)* **298,** 763–765.

Van Beneden, E. (1883). Recherches sur la maturation de l'oeuf, la fecondation et la division cellule. *Arch. Biol.* **4,** 265–638.

Vasiliev, J. M., Gelfand, I. M., Domnina, L. V., Ivanova, O. Y., Komm, S. G., and Olshevskaya, T. A. (1970). Effect of colcemid on the locomotory activity of fibroblasts. *J. Embryol. Exp. Morphol.* **24,** 625–640.

Wheatley, D. N. (1982). The Centriole: A Central Enigma of Cell Biology. Elsevier/North-Holland Biomedical Press, Amsterdam.

Wong, M. K. K., and Gotlieb, A. I. (1988). The reorganization of microfilaments, centrosomes, and microtubules during *in vitro* small wound reendothelialization. *J. Cell Biol.* **107,** 1777–1783.

Zigmond, S. H., Levitsky, H. I., and Kreel, B. J. (1981). Cell polarity: An examination of its behavioral expression and its consequences for polymorphonuclear leukocyte chemotaxis. *J. Cell Biol.* **89,** 585–592.

12

The Centrosome: Contributions to Cell Form

MARK A. MCNIVEN* AND KEITH R. PORTER†

*Department of Biochemistry and Molecular Biology
Center for Basic Research in Digestive Diseases
Mayo Clinic Foundation
Rochester, Minnesota 55905

†Department of Biology
University of Pennsylvania
Philadelphia, Pennsylvania 19104

I. INTRODUCTION

It is well established that microtubules play an important role in cellular morphogenesis. The contributions made by specific subpopulations of cytoplasmic microtubules to the formation of distinctive cell shapes have been well documented. These contributions include the elongation of epithelial cells of the developing lens rudiment (Byers and Porter, 1964), the elongation of the spermatid and its caudal sheath into mature sperm (Fawcett *et al.*, 1971), and the extension of cytoplasmic processes in neuronally derived cells such as hippocampal neurons (Bartlett and Banker, 1984), pigment cells (Bikle *et al.*, 1966; Schliwa and Bereiter-Hahn, 1973), and pheochromocytoma or PC-12 cells (Black and Greene, 1982). It is generally accepted that the formation of mi-

Copyright © 1992 Academic Press, Inc.
All rights of reproduction in any form reserved.

crotubule patterns responsible for this morphological diversity is initiated at the central apparatus or centrosome. For over 20 years, this structure has been described to include a pair of centrioles and a complex of dense amorphous material from which the assembly of microtubules is initiated (Fig. 1). Through a combination of biochemical, cell biological, and genetic applications described elsewhere in this volume, we now have a limited list of the components that constitute this complex. However, it remains unclear how structural information is conveyed from the nucleus to the centrosome in a way that generates and conserves the diverse microtubule patterns seen in differentiated cells, including those just described. Furthermore, the mechanism by which the centrosome influences microtubule assembly and disposition in distant cytoplasmic locations such as neurites is also a mystery. The goal of this chapter is not to review the wealth of information gathered on centrosomes over the past several decades. This has already been skillfully done by Peterson and Berns (1980), Tucker (1979, 1984), McIntosh (1983); Mazia (1984), Brinkley (1985), Karsenti and Maro (1986), and others. Instead, our aim is to reflect on some novel observations that have been made which lend insight into how cells control the disposition and organization of their microtubules. These observations suggest that (1) the shape and orientation of the centrosomal complex contribute to the formation and organization of specific microtubule patterns; (2) noncentrosomal cytoplasmic components contribute to microtubule organization in peripheral cytoplasm; and (3) both the centrosome and peripheral components of the cytoplasmic matrix are interconnected, thereby forming a unit structure, or cytoplast, that acts to polymerize, orient, and organize cytoplasmic microtubules throughout the cytoplasm.

II. THE CENTROSOME AS AN ORGANIZATIONAL TEMPLATE INFLUENCING CELL FORM

Since the centrosome was first observed with the electron microscope (Bernhard and deHarven, 1958) it has presented investigators with a paradox. Primarily, how can a seemingly amorphous structure generate well-organized and reproducible microtubule patterns? With this question in mind, many investigators made extensive morphological studies of vertebrate cell centrosomes in an attempt to find specialized organizing templates persisting in the centrosomal haze. None were clearly resolved. At the same time, other investigators were examining the configurations and distribution of dense bodies in cells that were long known to exhibit tremendous morphological diversity, namely, the protozoans. From studies of microtubule organization and cell structure in *Ochromonas* (Bouck and Brown, 1973), *Polytomella* (Stearns and Brown, 1979), *Nassula* (Tucker, 1970a,b), and others, it was learned that in many protozoa the orientation and topography of a nucleating element, or dense body, can define the initial orientation of the mi-

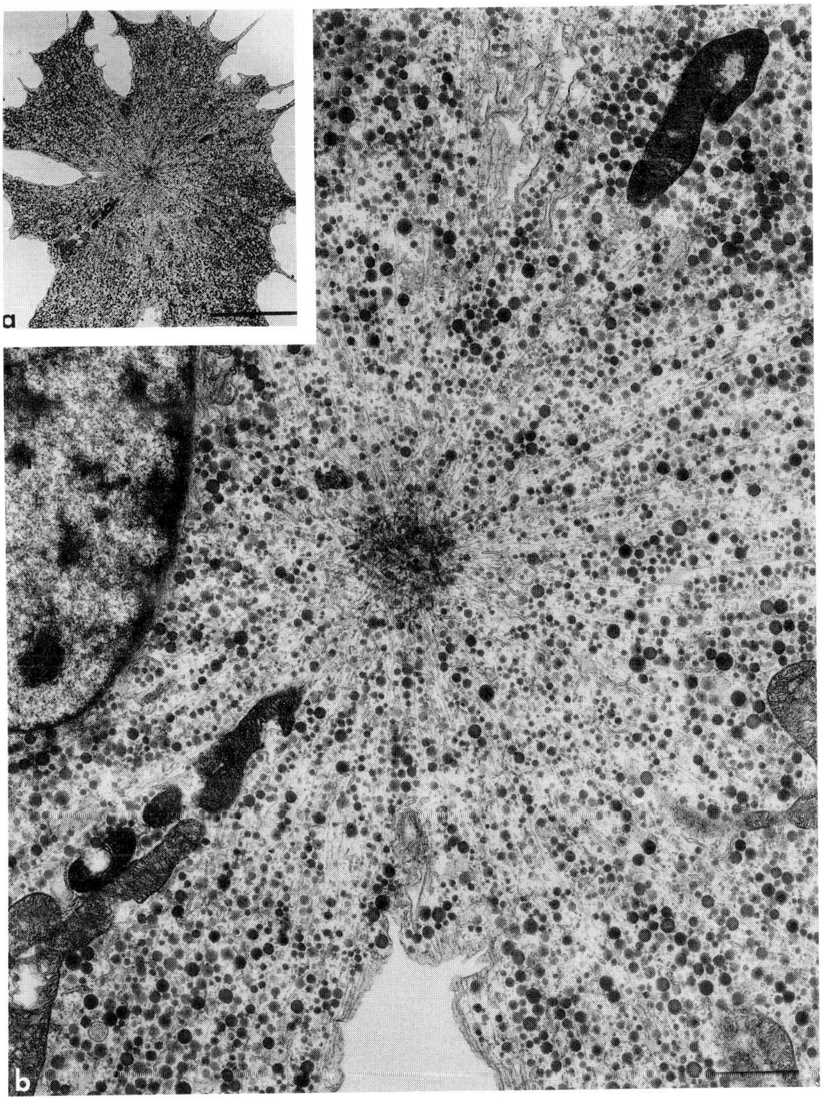

Fig. 1. The centrosome is continuous with the cytoplasmic matrix. A low (a) and high (b) magnification electron micrograph of the central organizing complex from a teleost pigment cell (erythrophore). The cytoplasm is organized around an extensive radial array of microtubules which emanate outward from the central apparatus. In the erythrophore, this complex is composed of large clusters of dense bodies and a centriole pair. Under higher magnification, the dense bodies are seen as clusters of fuzzy dense material that have no clear physical boundaries but, instead, appear to be continuous with the surrounding cytoplasmic ground substance. Each dense body ensheathes one or several microtubules along the proximal or minus ends. It does not appear that the microtubules end directly in a dense body; they may instead extend through it for several microns before they eventually terminate. Bars: (a) 5 μm; (b) 2 μm.

crotubules it nucleates (Tucker, 1979). From these studies, it became evident that in many instances cells possess a program for the distribution of dense centrosomal material that nucleates the specific microtubule scaffolding and ultimately dictates cellular form. Some of the first experimental information indicating that such mechanisms exist in vertebrate cells was provided indirectly through a series of provocative studies conducted by Solomon (1979) and co-workers using mouse neuroblastoma cells. Through the growth and extension of microtubule-filled neurites, these cells exhibit remarkable morphological diversity that is lost as the cells enter mitosis, disassemble microtubules, retract their neurites, and subsequently round up. Following mitosis, the cells can be stimulated exogenously to regrow their neurites, allowing one to observe and compare the morphology of sister cells. Solomon found, remarkably, that over 60% of the sister mitotic pairs displayed similar morphologies, either as mirror images or identical twins (Fig. 2a–f). Further, if these cells were treated with a microtubule-disrupting agent, nocodazole, to induce the complete retraction of their neurites and allowed to recover in drug-free media, over 58% of the cells regained their original neurite morphology in detail (Solomon, 1980) (Fig. 2g–i). These observations not only suggest that mammalian cells can inherit specific determinants of shape that are independent of external cues but that these determinants survive an interruption in their expression during mitosis. Although these studies did not address where, or how, such morphological determinants are stored in the cell, it seems likely that such information persists in the centrosome itself. Even when cytoplasmic microtubules are disassembled, the structure of the centrosome is maintained and recreates the microtubule organization existing in the cell prior to drug treatment. Concomitant studies by Spiegelman and co-workers, using tubulin antibodies and immunofluorescence techniques, demonstrated that these same mouse neuroblastoma cells, and several other mammalian cell types, contain multiple sites for the initiation of microtubule assembly (Marchisio *et al.*, 1979; Spiegelman *et al.*, 1979b). Most importantly, they also observed that these multiple initiating sites in the neuroblastoma cells, 24 hr prior to neurite outgrowth, aggregate to the cytoplasmic area from which a neurite will eventually extend (Spiegelman *et al.*, 1979a). This result supports the view that the centrosome provides an initial orientation for directed microtubule growth that may later be stabilized or modified through the participation of peripheral microtubule-associated components.

These studies contribute some of the first evidence suggesting that, in higher eukaryotic cells as well as in the protozoa, the overall shape and positioning of centrosomal material provides a structural memory that has the capacity to control microtubule growth spatially, and, subsequently, to control cell form. Thus, in this model, one might expect to see a direct correlation between the shape of a centrosome and that of the cell from which it came. With this in mind, we have completed a detailed study on the three-dimensional organization of the cen-

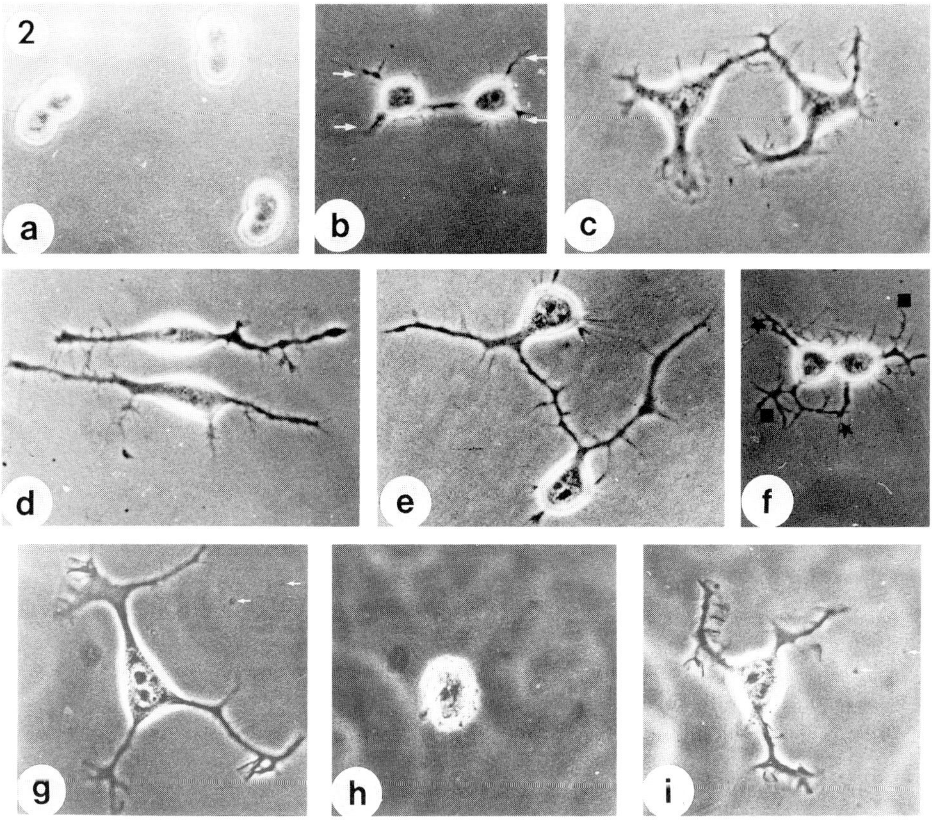

Fig. 2. Storage of morphological determinants: neurite morphologies of sister neuroblastoma cells are related. (a) Mitotic pairs of mouse neuroblastoma cells in culture photographed 30 min after shift to low serum. (b) Over a 2 hr period in low serum, the cells initiate neurite growth and become asymmetric in shape. (c–f) Examples of four different mitotic pairs photographed after 4–12 hr in low serum. Note the similarity in form between sister cells. (g–i) Recapitulation of detailed neurite morphologies after reversible microtubule assembly. During a 30-min incubation period in nocodozole, neuroblastoma cells lose their characteristic shape (g) by disassembling their microtubules, retracting their neurites, and rounding up (h). Three hours after drug removal, the neurites re-extend and the original shape of the cell is restored (i). (Reproduced from Solomon, 1979, with permission.)

trosomal complex from a highly specialized, neuronally derived vertebrate cell, the teleost chromatophore. These cells have been utilized as models to study the mechanisms of agonist-stimulated organelle transport (McNiven and Porter, 1984; Haimo and Rozdzial, 1989) and make excellent subjects for studying the origins of cell shape. Chromatophores have a characteristic conserved morphology both *in situ* and in culture. The red erythrophores of *Holocentrus* are small (30–50 μm in diameter) and disk-shaped (Fig. 3a) whereas the larger melanophores (50–200 μm) are stellate in form, possessing long cytoplasmic arms that give them a more neuron-like appearance (Fig. 3b). Finally, both the active organelle transport and the distinctive shapes of these cells are dependent on large populations of microtubules that extend radially from an elaborate centrosomal complex consisting of one pair of centrioles and an intricate array of surrounding dense bodies (Fig. 1). Through serial thick sectioning, high-voltage electron microscopy, and standard or computer-aided morphometric techniques, we have examined closely the three-dimensional organization of both erythrophore and melanophore centers to look for similarities between centrosomal shape and cell form. Figure 3c,d summarizes the observations gathered from 12 serially sectioned cells. Fig. 3c depicts the configuration of dense bodies in the erythrophore as a circular mass surrounding the centriole pair which, when viewed in three dimensions (not shown), has a depth of 1–2 μm, making it disk-shaped. In contrast, the central complex in the melanophore (Fig. 3d) is stellate and possesses dense bodies that extend outward from the centrioles, thereby mimicking the distribution of cytoplasmic extensions of the parent cell.

These observations provided us with one of the first direct correlations between cell form and the shape and distribution of centrosomal material. Whether such correlations exist in other differentiated vertebrate cells remains to be seen. We find it attractive to suggest that a conserved structural organization of centrosomal components could provide the stored "morphological determinants" observed in the neuroblastoma cells recovering from nocodazole (Solomon, 1980). Whether the central complex of the chromatophore can maintain its organizational memory during similar drug treatments needs to be explored further.

We have conducted further studies to examine whether the shape of the centrosome in a particular cell type is conserved and expressed in cell shape. One might predict that the morphological determinants residing in a centrosome could be introduced into a foreign cell to affect dramatically the microtubule display and shape of the recipient cell. To test this concept, we again utilized the erythrophore with its unique centrosomal complex and radial microtubule array. Our plan was to fuse chemically, using polyethylene glycol (PEG), the erythrophore to an immortal cell line that differed markedly in shape from the erythrophore, and to observe the effects on microtubule organization through phase and immunofluorescence microscopy. We used normal rat kidney cell (NRK) cells, which are rectangular in shape and possess the confused disorganized microtubule population characteristic of many transformed cells (Brinkley *et al.*, 1975, 1978). When NRK cells and erythrophores are fused with PEG and viewed immediately with tubulin immunofluorescence, the resulting hybrid cells initially show a mixed microtubule organization that is characteristic of both cell types. However, after 4–6 hr this mixed disposition is completely reorganized until the hybrid cells exhibit an overall shape and microtubule pattern strickingly similar to that of the

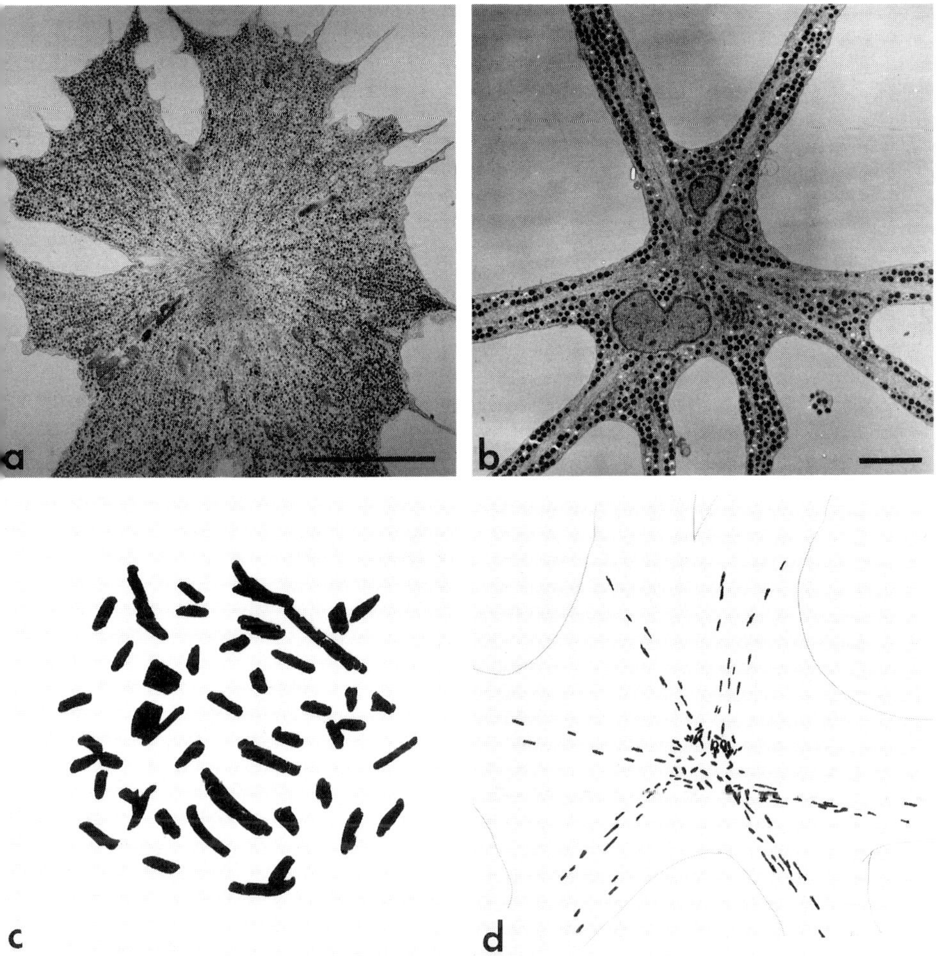

Fig. 3. Coincidence between centrosomal shape and cell form. (a,b) Low magnification electron micrographs of two types of teleost chromatophores from Holocentrus. The erythrophore (a), 25–50 μm in diameter, is smaller and discoid in shape whereas the melanophore (b), 50–200 μm in diameter, is larger and stellate with long cytoplasmic processes. The centrosomes of these cells (and 10 others) were serial sectioned, photographed with the high-voltage electron microscope, and reconstructed using either conventional or computer-aided morphometric techniques. The resulting images (c,d) show a remarkable parallel between the shape of a centrosome and the form of the cells from which they came. Bars: (a) 5 μm; (b) 10 μm.

erythrophore alone (Fig. 4). Thus, structural information stored in the erythrophore is somehow conserved in the hybrid and dominates over that of the NRK cells. This information cannot come from external cues, nor is it likely to reside in the cell membrane, since the reorganization process occurs when only one erythrophore is fused with more than six NRK cells. Instead, it is likely that the centrosome and its associated proteins provide this structural information.

III. MICROTUBULE ORGANIZATION IN PERIPHERAL CYTOPLASM

The protozoa again provide graphic examples of microtubule organization via peripheral noncentrosomal components that have the capacity to cross-link microtubules into unique cell structures such as axopod (Tilney and Porter, 1965; Tilney, 1971), axoneme (Mooseker and Tilney, 1973), or cytopharyngeal basket (Tucker, 1970). The contributions these distally cross-linking components make to protozoan morphogenesis have been reviewed well by Tucker (1979, 1984). Not surprisingly, it is now evident that many specialized vertebrate cells also have evolved alternative methods of assembling and organizing subcortical arrays of microtubules independent of a centrosome. Some of the best studied examples include the marginal microtubule band in chicken erythrocytes (Miller and Solomon, 1984; Swan and Solomon, 1984), the cortical arrays of microtubules in many root tip cells (Gunning *et al.,* 1978; Hardham and Gunning, 1978; Gunning, 1979), the preprophase band in meristem cells (Busby and Gunning, 1980; Brown and Lemmon, 1984), and proteins that nucleate, bundle, and capture microtubules in the epidermal wing cells of *Drosophila* (Mogensen *et al.,* 1989). The morphological determinants that act to position the cortical nucleating sites in all of these cells are presently undefined.

Although the cell types just listed may present somewhat unusual cases, since they lack true centrosomes, it has become increasingly apparent that the distribution and spatial organization of microtubules in most, if not all, cells is regulated through a synergistic relationship between the centrosome and ancillary proteins

---

Fig. 4. Chromatophore centrosomes maintain characteristic microtubule displays when introduced into foreign cells. Tubulin immunofluorescence micrographs depicting the organization of microtubules in erythrophores, normal rat kidney (NRK) cells, and in erythrophore–NRK cell hybrids. The microtubules in unfused cultured erythrophores (a) form a striking radial aster which differs dramatically from the disorganized array found in the NRK cells (b). The phase micrograph (c) shows the distribution of pigment within an erythrophore–NRK cell hybrid. Note that, even in the absence of aggregating agents (epinephrine), the pigment is restricted to a central region of the cell. The arrowheads mark the cell boundaries. When the same cell is stained with antibodies to tubulin, long, well-ordered microtubules can be seen reaching outward from the cell center to the cell margins. (e,f) Two additional fluorescence micrographs of microtubules in hybrid cells. It is clear that the organization of microtubules in the hybrids mimics that found in erythrophores and not NRK cells.

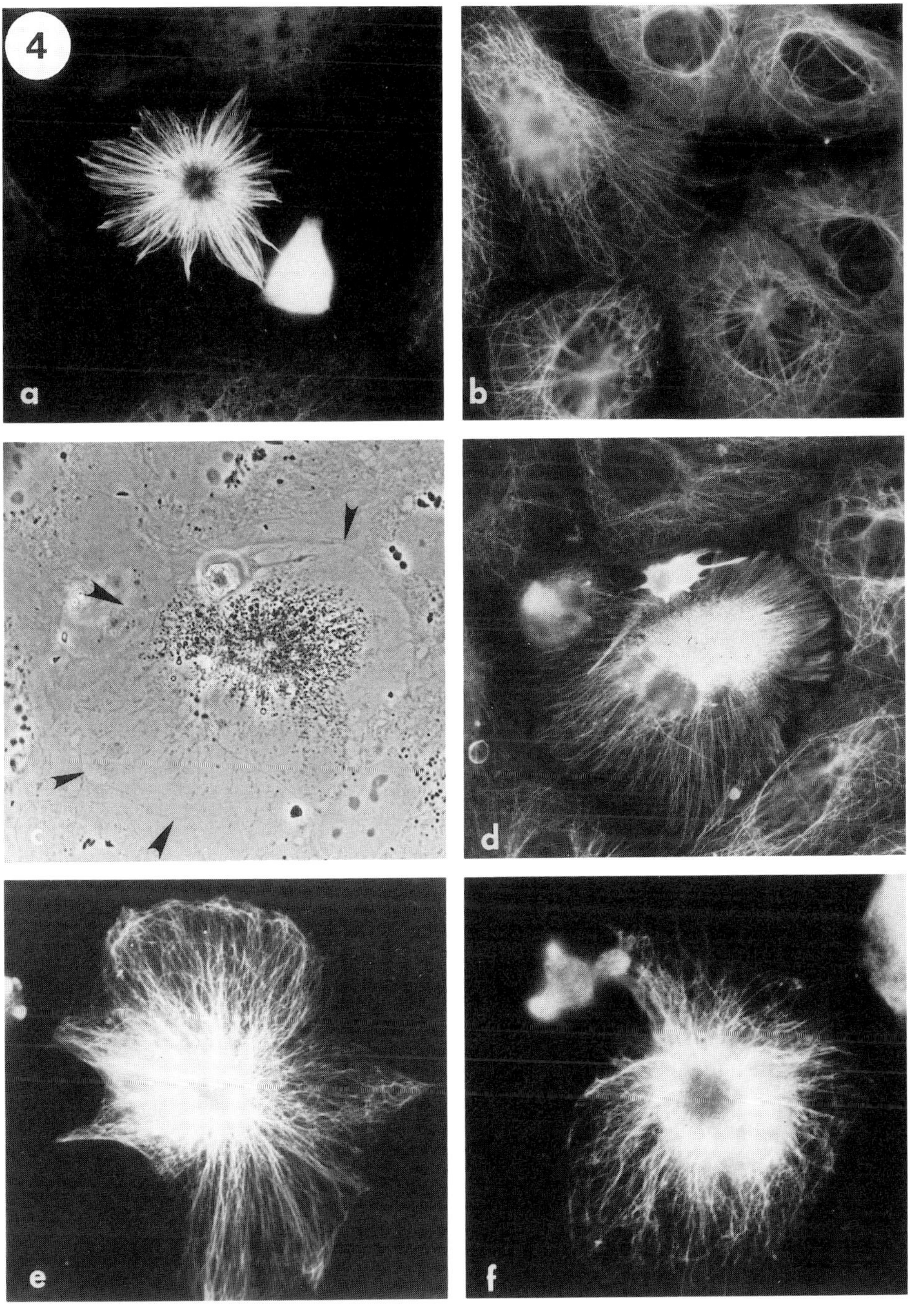

located in the peripheral cytoplasm. This is not surprising if one views the centrosome and the material it comprises as a continuum, unlimited by membranes, that may extend throughout the entire cytoplasm (Fig. 1). Our inability to recognize nondescript microtubule nucleating material in outlying cytoplasm does not necessarily mean that it is absent. Over a decade ago, investigators postulated that the centrosome and its microtubules are cross-linked to both intermediate and actin filament systems through an extensive collection of associated proteins. The result of these extensive interactions is a gel or structural continuum termed the "cytoplast" (Porter and McNiven, 1982). Thus, while many of the proteins that participate in the assembly and spatial organization of microtubules are conjugated at the cell center, it is very likely that these proteins extend throughout the entire unit cytoplast as well. This concept is now supported by a substantial collection of morphological and experimental observations. From our close electron microscopic examinations of melanophore cytoplasmic extensions, we have observed microtubule-associated dense bodies extending up to 15–20 μm from the cell center until they eventually disappear into the surrounding cytoplasmic ground substance. This phenomenon may be similar to that found in the neurons of the nematode in which microtubules, extending discontinuously over the entire neurite length, are each associated with a tuft of dense centrosome-like material (Chalfie and Thomson, 1979). Are these microtubules simply collecting and binding soluble associated proteins in the peripheral cytoplasm, or could these proteins be prepositioned to affect microtubule assembly and/or organization? Provocative evidence from Bernhardt and Matus (1982) supports the latter. Through immunohistochemical staining of frozen slices of neonatal rat cerebellum they have observed a substantial accumulation of high molecular weight microtubule-associated proteins (HMW-MAPs) in the tips of growing Purkinje neurites prior to the presence of assembled microtubules. They suggest that "these HMW-MAP proteins could act as a framework upon which the microtubules are subsequently built." Further support for a peripheral HMW-MAP framework that could maintain its integrity independent of microtubules comes from very recent studies of the dynamic properties of fluorescently tagged microtubules in living PtK_2 cells. Using low-light-detecting video cameras and fluorescently tagged microinjected tubulin subunits, Schultz and Kirschner (1988) were able to observe individual microtubules, each with a distinctive path and form, growing toward a cell edge, then receding through disassembly. Remarkably, they found that, within seconds, nearby microtubules grew out to follow the same path as the first. From this result they also concluded that peripheral components, possibly MAPs, may remain localized in the cell cortex subsequent to microtubule disassembly to continually redirect microtubule assembly and insure a conserved microtubule organization. The biolgoical ramifications of such an arrangement are displayed perhaps most graphically by the scleroblasts in teleost scales. When stained with antibodies to tubulin and viewed *in situ,* the microtubules are seen to extend out radially from the organizing centers of the cells and then curve abruptly to become aligned with the collagenous fibers of the underlying scale. It is impressive that microtubules in adjacent cells are similar in number, orient in the same manner, and share a nearly identical pattern (Byers *et al.,* 1980; Dane and Tucker, 1986; McBeath and Fujiwara, 1989). It seems unlikely that the centrosome alone could induce changes in the direction of microtubule growth without the involvement of other organizing components in the peripheral cytoplasm. Additional graphic experimental evidence for peripheral microtubule organization is presented in the following text.

IV. POSITIONING THE CENTROSOME

The influence of peripheral microtubule-associated elements on cell function is likely to be significant and, at the very least, affect two important cellular processes. First, as initially predicted by Tucker (1979), peripheral elements could influence cell shape through the cross-linking of nascent microtubules. This concept has been expanded by Kirschner and Mitchison (1986) to predict how peripheral elements, positioned at specific cortical domains, could cap or stabilize a population of dynamically unstable microtubules, thereby recruiting them to either maintain or expand a cytoplasmic domain. Second, peripheral microtubule-associated elements could mediate the assembly, orientation, and positioning of microtubules in distal cytoplasmic domains that extend great distances from the cell body, for example, axons, which have well-ordered uniformly polarized microtubule populations. Several studies have examined the length, number, and continuity of axonal microtubules and concluded that these microtubules are not continuous structures but show numerous starts and stops over the length of the axon (Nadelhaft, 1974; Bray and Bunge, 1981; Sasaki *et al.,* 1983). In fact, the peripheral axon may possess more microtubules than the cell body itself (Zenker and Hohnberg, 1973). These observations suggest that an axon is capable of initiating microtubule assembly at points along its length, quite independent of a single centrosome. Recently, we set out to test whether peripheral cytoplasm in neuronally derived cells (melanophores) can assemble uniformly polarized and well-ordered microtubule bundles in the absence of a centrosome. With a glass microneedle, we surgically isolated the cytoplasmic arms containing microtubules from the melanophore cell body and found that, over time, the microtubules in severed arms change their original polarity and orientation. By the end of a short incubation period, microtubules reorganize with their minus ends gathered into the centers of the arms and their plus ends oriented toward the plasma membrane, as if to form new microcells (McNiven *et al.,* 1984). Following this major alteration in microtubule organization, there comes a reversal in the direction of pigment transport in response to caffeine or epinephrine. Rather than transporting pigment to and from the cut end, which is the case immediately following severing, an arm aggregates pigment to and disperses it from an area near its center. Thus, over time, these cytoplasmic fragments establish what we have termed "a new functional center" that performs the same role as a centrosome in the intact melanophore. From these observations one is led to consider what factors in an isolated arm are involved in determining where a reorganization of microtubules occurs. How is the final position of the new center determined? To approach these questions we subjected different severed arms that had established new centers to computer-aided image analysis that marked the position of each arm centroid. This image was compared with the position of its functional center, visualized as an aggregated pigment mass. Remarkably, the cross-hairs in each analyzed arm marking the centrosome are situated over the center of each pigment mass with near precision (Fig. 5). Even in situations in which the pigment mass has favored one side of the arm (Fig. 5b) or a large branch over a small branch (Fig. 5d), the cross-hairs and the centers of the aggregated pigment mass occupy the same location. From this striking correlation it would appear that factors in a severed arm favor this positioning, suggesting that this centroid location is the region in which tension forces exerted on the new center from all regions of the arm are equal. Indeed, one may interpret the migratory behavior of the new center as an attempt to find a stable position similar to the one occupied by a centrosome in a whole cell. It has been well documented by Bray (1979) that cells subjected to pulling forces will subsequently alter their morphology when such forces are thrown into imbalance. Bray calculated the amount of tension that a neurite exerts on its parent cell center and consistently found that a neurite pulling on one side of a cell body was always counterbalanced by a neurite of appropriate size on the other side. When several neurites were amputated from a single cell, the cell responded by repositioning its remaining neurites or forming new ones, resulting in an eventual return to equilibrium. In contrast to neurons, melanophore arms may respond to the sudden force imbalance generated by severing by reforming a new center at the arm centroid instead of undergoing extensive cytoplasmic regrowth. The result is the same, however, that is, an eventual equilibrium of initially unbalanced pulling forces.

Although there are several plausible explanations for this phenomenon, we favor a model implicating the action of a unit cytoplast. The cytoplast extending through the body and arms of a melanophore could generate and transduce pulling

Fig. 5. Cytoplasmic fragments, devoid of centrosomes, organize microtubules outward from their centroids. (a–d) Light micrographs of four different melanophore arms that have been severed from their parent cells and that subsequently reorganized microtubules from and aggregated pigment into their centers. The arrow depicts the cut end of each arm. (a′–d′) The circumference of each arm was outlined, digitized, and used to calculate the cytoplasmic center of mass or "centroid" (+). Note that the calculated center in each arm corresponds remarkably well with the point at which pigment aggregates.

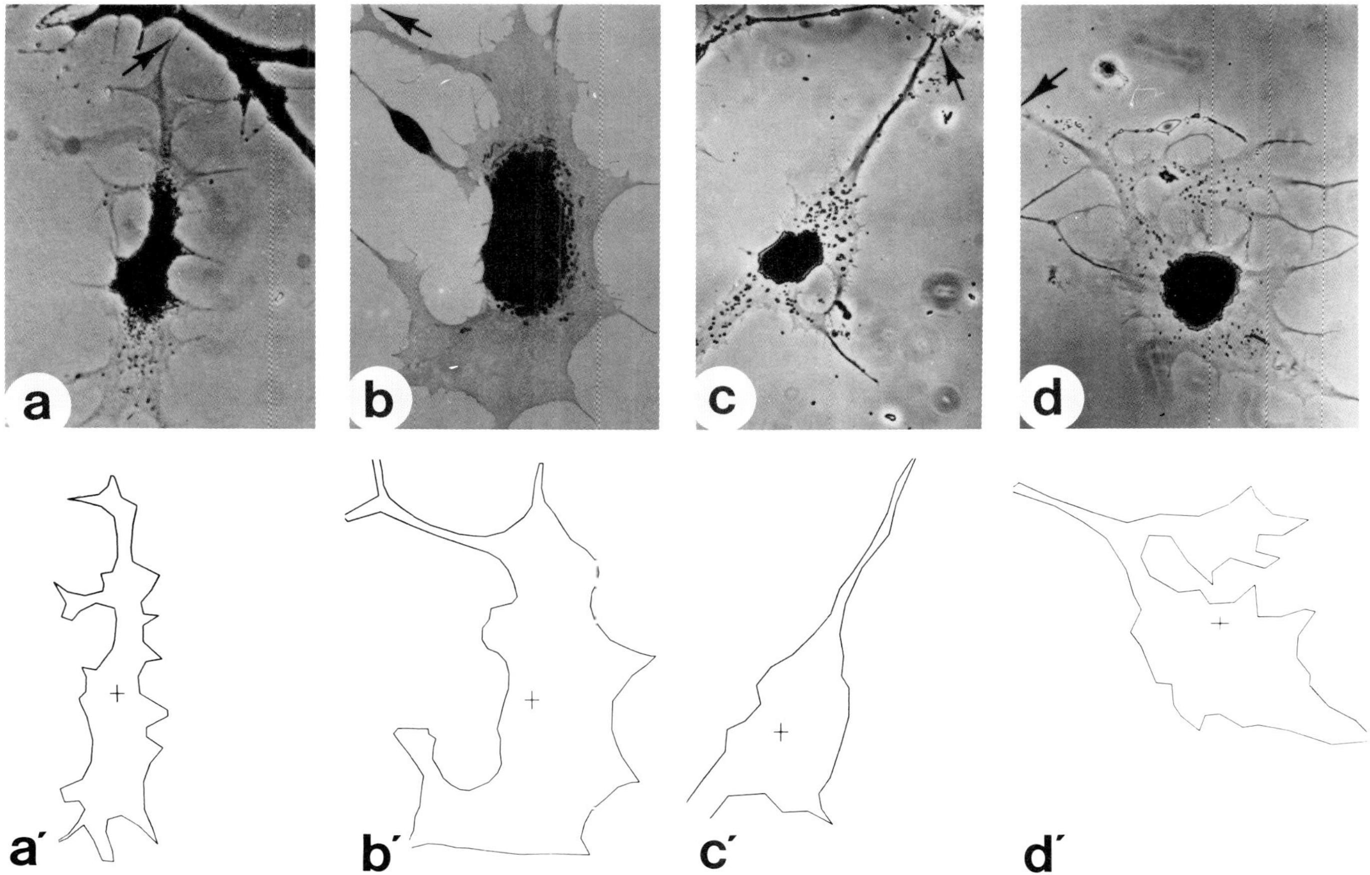

forces on the cell center. When the continuity of the cytoplast in an arm is broken by severing, it is placed in a dynamically unfavorable state and responds by contracting isotropically into the arm middle. This contraction would be represented by the formation and subsequent migration of a new functional center into the arm centroid.

Thus, these studies suggest that, in intact cells, the cytoplast itself may play a role in the positioning of the centrosomal complex by following the simple biophysical properties of a gel. One may conclude that certain highly polarized cells such as lens ectodermal cells or the intestinal epithelium, which have apically rather than centrally positioned centrosomes, are an exception to this model, although recent studies with a popular polarized tissue culture cell line appear to support, rather than repudiate, this model. Madin–Darby canine kidney cells (MDCK), when initially plated on a substrate, are apolar and possess centrally situated centrosomes. However, once the cells have reached confluence, thereby promoting the establishment of apical cell contacts, the cells become polarized with a concomitant migration of the centrosomes to the cell apices. If the apical cell junctions are disrupted by the removal of calcium from the medium, the centrosomes reposition themselves at the cell center (Buendia *et al.,* 1990). Thus, apical asymmetric centrosomal distributions may be initiated and maintained through extracellular pulling by adjacent cells on the unit cytoplast. One could predict that any disruption of the structural gel constituting the cytoplast would dissipate the extracellular pulling forces, thereby sending the centrosome back to its central position. This is what occurs when MDCK cells and other cells are treated with actin antagonists (Euteneuer and Schliwa, 1985). Such findings are consistent in theory with observations made on the severed melanophore arms and manipulated neurites.

V. PROSPECTIVE

It has become apparent that the diverse and conserved microtubule patterns characteristic of most differentiated cells are not generated soley from amorphous randomly arranged nucleating material located at the cell center. From the studies reviewed in this chapter, it appears that the shape and disposition of centrosomal nucleating elements in higher eukaryotic cells are more important than first predicted, and that additional constituents, possibly located throughout the cytoplasm or at the plasma membrane, act synergistically with the centrosome to regulate microtubule growth and, subsequently, cell form.

Many questions of seminal importance remain unanswered. These include how the basic structure and form of the centrosomal complex in specific cell types is conserved over countless generations. What ancillary proteins influence microtubule nucleation and growth in distal cytoplasm? How are these proteins positioned in the cytoplasm and how might they assist in maintaining microtubule organization in neurites that extend great distances from their cell bodies? How do these proteins insure such remarkably homogeneous microtubule polarity in the distal neurites? Answers to these questions will provide important insight into our understanding of cellular differentiation, morphogenesis, and associated pathologies.

ACKNOWLEDGMENTS

The authors are pleased to acknowledge the assistance of Karen Anderson in the preparation of the illustrations.

REFERENCES

Bartlett, W. P., and Banker, G. (1984). Electron microscopic studies of the development of axons and dendrites by hippocampal neurons in culture. I. Cells which develop without intercellular contacts. *J. Neurosci.* **4,** 1944–1953.

Bernhard, W., and deHarven, E. (1960). L'ultrastructure du centriole et les autres eléments d'appareil achromatique. *Proc. 4th Int. Conf. Electron Microsc. Berlin* **2,** 217–227.

Bernhardt, R., and Matus, A. (1982). Initial phase of dendrite growth: Evidence for the involvement of high molecular weight microtubule-associated proteins (HMWP) before the appearance of tubulin. *J. Cell Biol.* **92,** 589–593.

Bikle, D., Tilney, G. G., and Porter, K. R. (1966). Microtubules and pigment migration in the melanophores of *Fundulus heteroclitus L. Protoplasma* **61,** 322.

Black, M. M., and Greene, L. A. (19820. Changes in the colchicine susceptibility of microtubules associated with neurite outgrowth: Studies with nerve growth factor responsive PC12 pheochromocytoma cells. *J. Cell Biol.* **95,** 379–386.

Bouck, G. B., and Brown, D. L. (1973). Microtubule biogenesis and cell shape in Ochromonas. I. The distribution fo cytoplasmic and mitotic microtubules. *J. Cell Biol.* **56,** 340–359.

Bray, D. (1979). Mechanical tension produced by nerve cells in tissue culture. *J. Cell Sci.* **37,** 391–410.

Bray, D., and Bunge, M. B. (1981). Serial analysis of microtubules in cultured rat sensory axons. *J. Neurocytol.* **10,** 589–605.

Brinkley, B. R. (1985). Microtubule organizing centers. *Ann. Rev. Cell Biol.* **1,** 145–172.

Brinkley, B. R., Fuller, G. M., and Highfield, D. P. (1975). Cytoplasmic microtubules in normal and transformed cells in culture. Analysis of tubulin immunofluorescence. *Proc. Natl. Acad. Sci. U.S.A.* **73,** 4981–4985.

Brinkley, B. R., Miller, C. L., Fuseler, J. W., Pepper, D. A., and Wible, L. J. (1978). Cytoskeletal changes in cell transformation to malignancy. *In* "Cell Differentiation and Neoplasia" (G. F. Saunders, ed.), pp. 419–450. Raven, New York.

Brown, R. C., and Lemmon, B. E. (1984). Plastid apportionment and preprophase microtubule bands in monoplastidic root meristem cells of *Isoetes* and *Selaginella* protoplasma. (in press).

Buendia, B., Bre, M.-H., Griffiths, G., and Karsenti, E. (1990). Cytoskeletal control of centrioles movement during the establishment of polarity in Madin-Darby canine kidney cells. *J. Cell Biol.* **110,** 1123–1135.

Busby, C. H., and Gunning, B. E. S. (1980). Observations on preprophase bands of microtubules in uniseriate hairs, stomatal complexes of sugar cane, and *Cyperus* root meristems. *Eur. J. Cell Biol.* **21,** 214–223.

Byers, H. R., Fujiwara, K., and Porter, K. R. (1980). Visualization of microtubules of cells *in situ* by indirect immunofluorescence. *Proc. Natl. Acad. Sci. U.S.A.* **77,** 6657–6661.

Byers, B., and Porter, K. R. (1964). Oriented microtubules in elongating cells of the developing lens rudiment after induction. *Proc. Natl. Acad. Sci. U.S.A.* **52,** 1091–1099.

Chalfie, M., and Thomson, J. N. (1979). Organization of neuronal microtubules in the nematode *Caenorhabditis elegans*. *J. Cell Biol.* **82,** 278–289.

Dane, P. J., and Tucker, J. B. (1986). Supracellular microtubule alignments in cell layers associated with the secretion of certain fish scales. *J. Cell Sci. Suppl.* **5,** 273–291.

Euteneuer, U., and Schliwa, M. (1985). Evidence of an involvement of actin in the positioning and motility of centrosomes. *J. Cell Biol.* **101,** 96–103.

Fawcett, D. W., Anderson, W. A., and Phillips, D. M. (1971). Morphogenetic factors influencing the shape of the sperm head. *Dev. Biol.* **26,** 220–251.

Gunning, B. E. S., Hardham, A. R., and Hughes, J. E. (1978). Evidence for initiation of microtubules in discrete regions of the cell cortex in *Azolla* root-tip cells, and an hypothesis on the development of cortical arrays of microtubules. *Planta (Berlin)* **143,** 161–179.

Haimo, L. T., and Rozdzial, M. M. (1989). Lysed chromatophores: A model system for the study of bidirectional organelle transport. *In* "Methods in Cell Biology," Vol. 31, pp. 3–24. Academic Press, San Diego.

Hardham, A. R., and Gunning, B. E. S. (1978). Structure of cortical microtubule arrays in plant cells. *J. Cell Biol.* **77,** 14–34.

Karsenti, E., and Maro, B. (1986). Centrosomes and the spatial distribution of microtubules in animal cells. *Trends Biochem. Sci.* **11,** 460–463.

Kirschner, M., and Mitchison, T. (1986). Beyond self-assembly: From microtubules to morphogenesis. *Cell* **45,** 329–342.

McBeath, E., and Fujiwara, K. (1989). Coalignment of microtubules, cytokeratin intermediate filaments, and collagen fibrils in a collagen-secreting cell system. *Eur. J. Cell Biol.* **50,** 510–521.

McIntosh, J. R. (1983). The centrosome as an organizer of the cytoskeleton. *In* "Modern Cell Biology" (J. R. McIntosh, Ed.), Vol. 2, pp. 115–142. Liss, New York.

McNiven, M. A., and Porter, K. R. (1984). Chromatophores—Models for studying cytomatrix translocations. *J. Cell Biol.* **99,** 152s–158s.

McNiven, M. A., Wang, M., and Porter, K. R. (1984). Microtubule polarity and the direction of pigment transport reverse simultaneously in surgically severed melanophore arms. *Cell* **37,** 753–765.

McNiven, M. A., and Porter, K. R. (1988). Organization of microtubules in centrosome-free cytoplasm. *J. Cell Biol.* **106,** 1593–1605.

Marchisio, P. C. (1979). Identification of multiple microtubule initiating sites in mouse neuroblastoma cells. *Eur. J. Cell Biol.* **20,** 45–50.

Mazia, D. (1984). Centrosomes and mitotic poles. *Exp. Cell. Res.* **153,** 1–15.

Miller, M., and Solomon, F. (1984). Kinetics and intermediates of marginal band reformation: Evidence for peripheral determinants of microtubule organization. *J. Cell Biol.* **99,** 70s–75s.

Mogensen, M. M., Tucker, J. B., and Stebbings, H. (1989). Microtubule polarities indicate that nucleation and capture of microtubules occurs at cell surfaces in Drosophila. *J. Cell Biol.* **108,** 1445–1452.

Mooseker, M. S., and Tilney, L. G. (1973). Isolation and reactivation of the axostyle; evidence of a dynein like ATPase in the axostyle. *J. Cell Biol.* **56,** 13–26.

Nadelhaft, I. (1974). Microtubule densities and total numbers in selected axons of the crayfish abdominal nerve cord. *J. Neurocytol.* **3,** 73–86.

Peterson, S. P., and Berns, M. W. (1980). The centriolar complex. *In* "International Review of Cytology," Vol. 64, pp. 81–106. Academic Press, New York.

Porter, K. (1966). Cytoplasmic microtubules and their functions. *CIBA Foundation Symposium on Principles of Biomolecular Organization.*

Porter, K. R., and McNiven, M. A. (1982). The cytoplast; a unit structure in chromatophores. *Cell* **29,** 23–32.

Sasaki, S., Stevens, J. K., and Bodick, N. (1983). Serial reconstruction of microtubular arrays within dendrites of the cat retinal ganglion cell: The cytoskeleton of a vertebrate dendrite. *Brain Res.* **259,** 193–206.

Schliwa, M., and Bereiter-Hahn, J. (1973). Pigment movements in fish melanophores: Morphological and physiological studies. II. Cell shape and microtubules. *Z. Zellforsch.* **147,** 107–125.

Schultz, E., and Kirschner, M. (1988). New features of microtubule behaviour observed *in vivo. Nature (London)* **334,** 356–359.

Solomon, F. (1979). Detailed neurite morphologies of sister neuroblastoma cells are related. *Cell* **16,** 165–169.

Solomon, F. (1980). Neuroblastoma cells recapitulate their detailed neurite morphologies after reversible microtubule disassembly. *Cell* **21,** 333–338.

Spiegelman, B. M., Lopata, M. A., and Kirschner, M. W. (1979a). Aggregation of microtubule initiation sites preceding neurite outgrowth in mouse neuroblastoma cells. *Cell* **16,** 253–263.

Spiegelman, B. M., Lopata, M. A., and Kirschner, M. W. (1979b). Multiple sites for the initiation of microtubule assembly in mammalian cells. *Cell* **16,** 239–252.

Stearns, M. E., and Brown, D. L. (1979). Purification of cytoplasmic tubulin and microtubule organizing center proteins functioning in microtubule initiation from the alga Polytomella. *Proc. Natl. Acad. Sci. USA,* **76,** 5745–5749.

Swan, J. A., and Solomon, F. (1984). Reformation of the marginal band of avian erythrocytes *in vitro* using calf-brain tubulin: Peripheral determinants of microtubule form. *J. Cell Biol.* **99,** 2108–2113.

Tilney, L. G. (1971). Microtubule patterns are generated. Relative importance of nucleation and bridging of microtubules in the formation of the axoneme of *Raphidiophrys. J. Cell Biol.* **51,** 837–854.

Tilney, L. G., and Porter, K. R. (1965). Studies on microtubules in heliozoa. I. Fine structure of *Actinosphaerium* with particular reference to axial rod structure. *Protoplasma* **60,** 317.

Tiwari, S. C., Wick, S. M., Williamson, R. E., and Gunning, B. E. S. (1984). Cytoskeleton and integration of cellular function in cells of higher plants. *J. Cell Biol.* **99,** 63s–69s.

Tucker, J. B. (1970a). Initiation and differentiation of microtubule patterns in the ciliate *Nassula. J. Cell Sci.* **7,** 793–821.

Tucker, J. B. (1970b). Morphogenesis of a large microtubular organelle and its association with basal bodies in the ciliate *Nassula. J. Cell Sci.* **6,** 385–429.

Tucker, J. B. (1979). Spatial organization of microtubules. *In* "Microtubules" (K. Roberts and J. S. Hyams, eds.), pp. 316–357. Academic Press, London.

Tucker, J. B. (1984). Spatial organization of microtubule-organizing centers and microtubules. *J. Cell Biol.* **99,** 55s–62s.

Zenker, W., and Hohnberg, E. (1973). A-α-nerve-fibre. Number of neurotubules in the stem fibre and in the terminal branches. *J. Neurocytol.* **2,** 143–148.

13

Cell Polarity and Centrosomes

MANFRED SCHLIWA

Department of Molecular and Cell Biology
University of California
Berkeley, California 94720

I. INTRODUCTION

The centrosome has always occupied a special place in our concept of cellular architecture and function. Since its first description over 100 years ago (Flemming, 1875; van Beneden, 1876), generations of cell biologists have become absorbed with the question of its role in the life of eukaryotic cells. Early cytologists noted that centrosomes not only are key components of dividing cells, but also persist in interphase cells in which they are part of a "central apparatus" or "microcentrum" (Boveri, 1888) believed to be associated with cellular activities that have no immediate connection with cell division (Wilson, 1925). In particular, their juxtaposition to the nucleus and their spatial relationship with other cell organelles gave rise to the notion that centrosomes are the "dynamic centers" of cells and the cytoplasm's true organizers (Boveri, 1901).

Copyright © 1992 Academic Press, Inc.
All rights of reproduction in any form reserved.

Over a century ago, van Beneden (1883) noted that a line through the centrosome and the center of the nucleus establishes an axis that gives a cell a polar organization. In subsequent years, the idea of the centrosome as a determinant of cell polarity has received little attention. Interest in the question of the role of centrosomes in cell polarity was rekindled by the finding that microtubule depolymerization abolishes polarity in fibroblasts (Vasiliev *et al.*, 1970). Since then, a correlation between polar cell asymmetry and the position of the centrosome–microtubule complex was discovered in many cell types. This chapter examines whether this apparent correlation implies that centrosomes, in fact, determine cell polarity. Three exemplary model systems that each have a distinctly polar organization will be discussed: epithelial cells, neurons, and migrating cells. However, it may be useful first to consider some properties of the centrosome–microtubule complex and how these properties relate to cytoplasmic organization.

II. FEATURES OF THE CENTROSOME

A. Most Interphase Cells Possess One Centrosome

This feature is characteristic of most tissue cells, but, of course, there are exceptions to this rule. In plant cells, certain microtubule arrays are not generated by a known centrosomal entity: cortical microtubule arrays, the preprophase band, and the phragmoplast (for reviews, see Lloyd, 1982). In certain protozoa, microtubules apparently self-organize into ordered arrays (e.g., see Koonce *et al.*, 1986) or become ordered by membrane association (Hunt and Ellar, 1974, Murray, 1984). Membrane-associated sites may also operate in the establishment of the marginal band in nucleated erythrocytes (Miller and Solomon, 1984; Swan and Solomon, 1984). Typically, however, microtubule organization is astral, and the center of the aster is centrosomal.

B. Centrosomes as Predominant Origin of Microtubules in Cells

This feature is particularly well illustrated in many cultured cells. If microtubules are depolymerized and allowed to regrow, they first appear as a small aster around a structure that, by electron microscopy, is identified as a pair of centrioles surrounded by pericentriolar material (Porter, 1966; Osborn and Weber, 1976). Although this may be the standard for multicellular organisms, it is by no means the rule. Centrosomes come in a variety of sizes, shapes, and morphologies.

Particularly in protozoa, they may appear as electron-dense bars, spheres, plates, or ribbons. Centrioles may or may not be associated with these structures, so they are not essential for the function of these centers as microtubule-organizing entities (Tucker, 1979).

C. Organization of Microtubules into Radial Arrays

In determining the site of growth of microtubules, centrosomes exert some initial control over microtubule distribution throughout the cytoplasm. "Typical" centrosomes organize microtubule asters. Changes in that pattern may occur at some distance from the centrosome and are brought about by association with microtubule-associated proteins (MAPs), other filament systems, and the cell membrane (Griffith and Pollard, 1978; Weihing, 1979; Schliwa and van Blerkom, 1981). Such interactions might also be of retroactive influence on centrosome position. By placing the organizing center in a strategic position in the cytoplasm, cells might determine a number of features related to their overall organization (see Raff, 1979, for a review). Whether the decision to execute these changes in centrosome placement is autonomous will be considered in this chapter. Centrosomes exert control not only over microtubule patterns, but also over microtubule number, polarity, and structural fidelity. For example, centrosomes isolated from Chinese hamster ovary cells (Mitchison and Kirschner, 1984) or lymphoblasts (Bré and Karsenti, 1990) initiate microtubule asters with a more-or-less constant number of microtubules; in most cases in which microtubule polarity has been determined, it is unipolar (plus ends distal) with respect to the centrosome (e.g., Burton and Paige, 1981; Euteneuer and McIntosh, 1981); and the number of protofilaments in microtubules initiated by centrosomes is uniform (Evans *et al.*, 1985).

D. Microtubules as the Only Filament System with a Single Focal Center

Again, this feature is a generalization that applies to "typical" animal cells. Nonfocal microtubule arrays are largely the result of the activity of noncentrosomal factors (linkers, bridges). Examples include the microtubule coils, ribbons, sheets, baskets, and axostyles found in a wide variety of protozoan cells (Tucker, 1984). Focal patterns of microtubule organization may vanish in cells that lose centrosomes as part of their cytomorphogenetic program (e.g., myoblasts fusing to form myotubes; Tassin *et al.*, 1985) or as a result of experimental manipulation (e.g., virus-induced syncytia; Wang *et al.*, 1979). Multiple centrosomes may be found in certain developing embryos (Maro *et al.,* 1985; Raff and Glover, 1989) or tumorigenic cells (Spiegelman *et al.,* 1979a).

In contrast, intermediate filaments form networks of meandering interwoven bundles (Osborn *et al.,* 1977) and more-or-less radial arrays of single fibers (Franke *et al.,* 1978). It is striking that, in the cell periphery, vimentin filaments are coextensive with microtubules (Geiger and Singer, 1980; Geuens *et al.,* 1983). However, vimentin forms neither an autonomous network nor does it rely on the presence of an organizing center (Eckert *et al.,* 1984). Rather, the extension of the vimentin network depends on an intact microtubule system (Gordon *et al.,* 1978).

Focal organizing centers also are not seen as a prominent feature of actin microfilament arrays. Actin superstructures are built either from bundles or networks. The formation of both is determined by specific cross-bridging proteins and their restructuring is governed by severing, capping, and nucleation-controlling factors (Stossel *et al.,* 1985). They are expressed to varying degrees in different cell types and in different regions of the same cell, such as microvilli and filopodia, stress fibers, circumferential girdles in certain epithelia, and cortical structures such as lamellae, ruffles, veils, and pleats (Byers *et al.,* 1984; Stossel, 1984). The only place at which radial "foci" of actin filaments have been observed is the ventral surface of leukocytes and macrophages (Trotter, 1981; Pryzwansky *et al.,* 1983), but they are small, numerous, and dispersed, and it is unclear whether they are the result of focal nucleation or local contraction. There is no master organizer of actin filament structures in the cell. On the contrary, reorganization of, in particular, cortical actin is controlled and executed locally (Stossel, 1989), often in response to external stimuli.

Collectively, these features make the centrosome unique among the cellular components concerned with the organization of cytoplasmic space. After all, the centrosome is the sole origin of the entire microtubule apparatus. However, exactly how the centrosome fits into the texture of interacting and interdependent cytoskeletal components is largely unknown. This chapter critically evaluates the question whether the centrosome can be considered an autonomous mastermind that dictates cell architecture, polarity, and, therefore, cell behavior, in short, whether it is the overall organizer of the cell. Surprisingly, the three model systems discussed in the following sections do not support this view. There is no conclusive evidence for a determinative role of the centrosome in epithelial and neuronal polarity and in directional cell migration. The establishment and maintenance of these features is the result of a concerted effort of many effectors and involves transmembrane stimuli, cortical activity, and cytoskeletal interactions. The centrosome, via the microtubules it initiates, participates in the effort in a cell-type- and situation-dependent manner and, again via the microtubules it initiates, is subject to external forces and constraints.

III. DEVELOPMENT AND MAINTENANCE OF CELL POLARITY: THREE CASE STUDIES

A. Epithelial Cells

Pronounced structural and functional polarity is the hallmark of epithelial cells. Particularly with respect to cell-surface-associated features, epithelial cells are subdivided into an apical layer facing the epithelial exterior (e.g., the external environment or the lumen of an internal cavity or duct) and a basolateral domain in contact with neighboring epithelial cells or the basal lamina. The composition and properties of the apical membrane are unique to each epithelium, whereas many of the properties of the basolateral domain are shared with other non-epithelial cells. The border between the two is marked by tight junctions—linear domains of close apposition of the plasma membrane that exclude intercellular space. Tight junctions appear to function as a "fence" separating apical and basolateral membrane proteins and lipids (Simons and Fuller, 1985; Gumbiner, 1987). The asymmetric distribution of membrane constituents is matched by an apical–basal polarity of intracellular components (Hull and Staehelin, 1979). The nucleus and mitochondria are typically located near the basal pole of the cell, whereas the Golgi apparatus is found apical to the nucleus. A membrane-associated spectrin–ankyrin skeleton is often apposed to basolateral membranes only (Nelson and Veschnok, 1986), whereas a dense network of actin filaments, the terminal web, underlies microvilli at the apical pole of the cell (Mooseker and Tileny, 1975; Hull and Staehelin, 1979; DeRosier and Tilney, 1989). A contractile belt composed of a bundle of actin filaments encircles the cell at the level of the zonulae adherentes near the apical pole (Keller and Mooseker, 1982); intermediate filament bundles interconnect hemidesmosomes at the basal and desmosomes at the lateral membrane, forming a system of "rivets" that supports the columnar shape of the cell. How are the extraordinary features of this polarity established and maintained? To what extent, if at all, are microtubules and centrosomes involved in the remarkable process that leads to cell polarization?

The differences in the composition of basal and apical plasma membrane domains are maintained by a combination of two mechanisms: intracellular sorting and anchoring to specific submembranous cytoskeletal assemblies (for reviews, see Simons and Fuller, 1985; Rodriguez-Boulan and Salas, 1989). The role of microtubules in this process is, at best, controversial. Microtubule disruption does not impair the correct targeting of vesicular stomatitis virus (VSV) G protein to the basolateral membrane (Salas *et al.,* 1986; Rindler *et al.,* 1987). However, the same groups report conflicting observations for the targeting of influenza virus hemagglutinin to the apical membrane. Antimicrotubule agents interfere with the distribution and sorting of some integral apical membrane proteins (Achler *et al.,*

1989; Eiler *et al.,* 1989) but not others (Ojakian and Schwimmer, 1988). The correct distribution of the tight junction protein ZO-1 in mouse embryos (Fleming *et al.,* 1989) and the membrane glycoprotein PAS-O in mammary epithelial cells (Parry *et al.,* 1990) seems to lack involvement of microtubules. However, observations based on the use of microtubule inhibitors are difficult to interpret because they do not distinguish between an effect on the sorting machinery (for example, the Golgi apparatus disperses upon microtubule depolymerization; Rogalski and Singer, 1984), the microtubular tracks for vesicle transport, or the insertion process. Therefore, none of these experiments speak very clearly of the role of the centrosome–microtubule complex in the maintenance of cell asymmetry. However, cell shape and intracellular organization are not grossly distorted on microtubule disassembly, suggesting that factors other than microtubules are more important.

Clues about the nature of these factors may come from a closer look at the development of a polar phenotype. In most cases, particularly during development, polar epithelial cells arise from nonpolarized precursor cells. The expression of the program that leads to the establishment of polarity clearly depends on specific external cues. The series of events is best illustrated in cultured epithelial cells which, when maintained in suspension, are essentially nonpolar (Pisam and Ripoche, 1976; Rodriguez-Boulan *et al.,* 1983). When these cells settle on substratum, the membrane in contact with the substratum becomes the basal plasma membrane (Rodriguez-Boulan *et al.,* 1983). Under these conditions, when cells do not establish lateral contacts, they polarize with respect to some, but not all, cell-surface markers. For example, an apical cell-surface marker is excluded from the attached surface, whereas several basolateral markers are not polarized (Balcarova-Stander *et al.,* 1984: Herzlinger and Ojakian, 1984; Vega-Salas *et al.,* 1987). Exclusion of apical membrane markers may be triggered by basal accumulation of extracellular matrix receptors during the process of attachment (Salas *et al.,* 1987) which, in turn, organizes a basal membrane cytoskeleton (Nelson and Veshnok, 1986). The extracellular matrix, in particular the adhesion molecule laminin that is involved in cell–substratum contact, appears to play an inductive role during the conversion of mesenchymal cells to epithelial cells in the developing kidney, since antibodies against laminin inhibit the polarization of mesenchymal precursor cells (Klein *et al.,* 1988). The morphogenesis of the polarized state in mammary epithelial cells is also regulated by the formation of a basement membrane (Li *et al.,* 1987; Parry *et al.,* 1987). Collectively, these and similar studies suggest that the establishment of epithelial cell polarity is initiated by cell–substratum contact, although cell–cell contact alone may be sufficient to trigger the segregation of certain marker proteins in the absence of a basal lamina (Wang *et al.,* 1990a).

Whereas cell–substratum interactions play an inductive role in the genesis of polarity in single cells, expression of full polarity correlates with the formation of cell–cell contacts and the appearance of tight junctions (see Fleming and Johnson, 1988, for review). Cell interactions clearly are required for the polarization of certain membrane markers (Vega-Salas *et al.,* 1987, 1988). In particular, cell adhesion molecules such as uvomorulin play an important role in the process of polarity establishment. Disruption of uvomorulin function not only disturbs cell–cell contacts but also redistributes integral membrane proteins (Nelson and Veshnok, 1987). The establishment of a full epithelial axis, including polarization of the entire cytoplasm, requires both cell–substratum and cell–cell contacts (Wang *et al.,* 1990a). Cell contact formation is accompanied by a reorganization of microtubules from a radial arrangement to a crisscross pattern without a clear focal organizing center (Bacallao *et al.,* 1989). By the time a full columnar epithelium has formed, centrosomes are located apically at about the same level as the Golgi apparatus and the tight-junction-specific protein ZO-1. Most microtubules run parallel to the apical–basal axis; their plus ends point toward the basal end of the cell, but it is unclear whether they originate from a focal apical centrosome or whether the centrosome is scattered across the apical cell pole. Immunofluorescence microscopy with antibodies against a centrosomal antigen suggests that the latter is the case, and further shows that the antigen spreads out after cell–cell contact formation (Bacallao *et al.,* 1989). During the transition from an unpolarized to a columnar epithelial cell, centrosomes split and, with the Golgi apparatus, move around the nucleus to reach their apical position (Buendia *et al.,* 1990). Clearly, more work will be required to determine the precise sequence of events leading to the polarization of the microtubule system, but the available evidence suggests an inductive role for the cell cortex.

The sequence of events shaping the development of polar organization may also be revealed by experiments in which cell polarity is reversed experimentally. Unfortunately, not many experiments of this sort have been done. If a collagen gel is overlaid on the apical surface of epithelium in culture, or added to apical-out epithelial cysts formed in suspension, the epithelium is remodeled so the surface in contact with the collagen becomes the basal cell membrane (Hall *et al.,* 1982; Wang *et al.,* 1990b). In the process, plasma membrane components are disassembled and reassembled in the correct orientation according to a seemingly preprogrammed pattern. Even more remarkable is the reversal of thyroid epithelial cell polarity in culture. This process occurs when hollow epithelial cysts grown in low serum are exposed to high serum. The cells undergo complete reversal of polarity, from apical-in to apical-out, in a few days (Nitsch and Wollman, 1980). Ultrastructural analysis clearly shows that the earliest changes recognized are cell-surface changes in which tight junctions and microvilli appear at the lateral margins. Changes in the distribution of intracellular organelles, such as the location of the Golgi apparatus, occur later. The deployment of microtubules and centrosomes has not been studied in these cells. However, in otic epithelial cells undergoing a similar reversal of polarity during development of the auditory system, the centrosome also shifts from an apical to a basal position (Carney and Couve, 1989).

Much remains to be learned about the role of the centrosome–microtubule complex in the establishment of epithelial cell polarity. The available evidence suggests that epithelial polarity is shaped by the environment of cell. Attachment determines the basal surface. Cell–cell contacts further reinforce the reorganization of plasma membrane components and lead to the formation of the tight junction fence that effectively separates apical from basolateral domains. The microtubule organizing center (MTOC) and the microtubules reorient in response to cues originating at the cell periphery. Thus, events at the cell cortex trigger cell polarization. Localized changes in the properties of the cell cortex that take place on cell–substrate or cell–cell contact may be propagated beyond the contact point, perhaps through changes in the cortical cytoskeleton, and may lead ultimately to a reorganization of the entire cytoplasmic space. This view is supported also by the finding that reversal of the polarity of cell cortex components precedes reversal of the polarity of intracellular components. The nature of the signals and the effectors must still be elucidated.

B. Neurons

Form determines function in neurons. The distinctive feature of the nerve cell is the neurite, an elongated often tortuous extension sometimes many thousand times the length of the cell body. Neurites are backed by an extensive cytoskeletal array of parallel microtubules, neurofilaments, and actin, all of which are highly interconnected. Neurites fall into two distinct classes—axons and dendrites. A typical nerve cell has one of the former and many of the latter, all emerging from the cell body. Neuronal polarity is specified by the axon. However, neurons begin as simple spherical cells that are the products of cell division of neuronal precursor cells. What sequence of events transforms apolar neuroblasts to the most highly polarized of cells? Which intrinsic and extrinsic factors determine the site of emergence of a growth cone from the cell body? Are centrosomes part of the polarity-determining process in neurons?

The dynamics of neurite development has been studied primarily in tissue culture, although under certain circumstances it is possible to observe the growth of axons *in situ.* Two such studies make particular reference to the spatial relationship of neurites and centrosomes. One is a study of the development of retinal ganglion cells. Unipolar mouse retinal ganglion cells at stage 13 of embryonic development evolve from an essentially spherical postmitotic precursor cell. When this cell begins to extend an embryonic axon, there is no clear relationship between the site of axon emergence and the position of the centriolar complex (Hinds and Hinds, 1974). In fact, the centrioles often are located opposite the base of the axon. A second study examines the arrangement of cytoskeletal elements during axonogenesis in the pair of Ti1 pioneer neurons of the grasshopper. These cells arise at the distal tip of the limb bud epithelium and extend an axon toward the thorax to establish contacts with the central nervous system (CNS) (Bate, 1976). On completion of cytokinesis of the pioneer mother cell, the pioneer axon of the proximal Ti1 neuron stereotypically extends from the proximal face of the cell where, in addition to other cell constituents, the centrosome apparently resides (Lefcort and Bentley, 1989). This spatial relationship suggests a causal relationship between growth cone emergence and the organization of cytoskeletal elements, particularly centrosomes. However, the data do not support such a causal link. The position of the centrosome at the proximal pole of the proximal Ti1 cell is the consequence of the positioning of the two pioneer daughter cells, which is the consequence of spindle orientation of the pioneer mother cell, which is always perpendicular to the surface of the limb bud epithelium. Thus, the correlation of centrosome location and growth cone emergence might be coincidental, since centrosomes are automatically placed in that position. A closer look at the distal Ti1 cell shows that it is not simply the position of the centrosome that determines polarity development. In over 99% of the distal pioneer neurons, the axon emerges not from the distal end, as one might expect, but from the proximal face in contact with the proximal cell. Conceivably, environmental cues, such as cell contact with the proximal cell, influence positional decisions of the distal cell.

Information about spatial and causal relationships during axonogenesis is more accessible experimentally in cells in tissue culture. Surprisingly few studies have addressed this issue, however, despite the importance of the cellular basis of neuronal polarity. Neuroblastoma cells are one of the cell types that have served as important model systems for the study of early events in neurite formation. In contrast to other neuronal cells, neuroblastoma cells contain multiple centrioles and centrosomes (Marchisio *et al.,* 1979; Spiegelman *et al.,* 1979a; Sharp *et al.,* 1981). It has been suggested that multiple centrosomes might be related to the ability of neuroblastoma cells to extend multiple processes (Spiegelman *et al.,* 1979b). Subsequent more detailed studies, however, showed no consistent relationship between centrosome number and position and cell process formation (Sharp *et al.,* 1982). Thus multiple centrosomes are not obligatory for multiple process formation. Indeed, several neuronal cell types, including PC-12 cells, dorsal root ganglion cells, and hippocampal neurons, which all possess only one pair of centrioles, are capable of multiple process formation. Serial section reconstruction of PC-12 cells shows that the pair of centrioles is located near the nucleus, never near the base of a neurite; it may or may not be found on the same side as an extending neurite (Stevens *et al.,* 1988). Hippocampal neurons in culture initially establish several apparently identical short processes (Dotti *et al.,* 1988) with no apparent spatial relationship to centrosome position (C. Dotti and G. Banker, unpublished results). These processes are not specified as axons or dendrites when they arise. Therefore, it is conceivable that the centrosome is associated with, and perhaps "earmarks," the process that is destined to become the axon; all other processes will become the dendrites of the cell. This possibility is not confirmed by recent studies (Dotti and Banker, 1991). A preferential positioning of the centrosome (and the Golgi apparatus associated with it) in relation to the site from which the axon emerged is not found, and changes in cytoplasmic organization preceding or accompanying the outgrowth of the axon are not obseved. Moreover, the fate of a process can be changed experimentally; a process that would have become a dendrite instead becomes the axon (Dotti and Banker, 1987). This has led to the hypothesis that, during normal development, the establishment of neurite specificity and, therefore, neuronal polarity may involve a dynamic interaction among essentially pluripotent nerve cell processes; if one outcompetes the others because of extrinsic cues, it is specified as the axon (Goslin and Banker, 1989).

In summary, there is insufficient evidence to support the idea of a correlation or a causal relationship between centrosome position and neurite extension or neuronal polarity. What, then, are the determinants of neuronal shape? The answer to this question does not appear to be simple and straightforward. The determinants may be both extrinsic and intrinsic. The existence of certain intrinsic determinants is suggested by experiments that show the ability of cultured neuroblastoma cells to re-establish the pattern of neurite extensions after an experimentally induced loss of asymmetry (Solomon, 1980). Although both the collapse of cytoskeletal asymmetry and the recapitulation of neurite morphologies are incomplete in these experiments, endogeneous factors that specify cell morphology cannot be dismissed. The site(s) and nature of these determinants still remain unknown.

C. Migrating Cells

Cells move, and moving cells have a distinct polarity. In contrast to polarized stationary epithelial or neuronal cells, however, the polarity of a locomoting cell is dynamic and subject to change on a time scale of minutes. The polarity of moving cells arises from the fact that protrusive activity is confined to a small portion of the cell perimeter, the lamellipodium, while retraction occurs at the opposite end. This external polarization is supplemented in many cell types by an internal asymmetry, namely, the deployment of the MTOC and Golgi apparatus between the advancing lamellipod and the cell nucleus. How is that polarity established and maintained? What is the role of microtubules and centrosomes in this process? We will briefly consider these questions here, but they are dealt with in more detail elsewhere in this volume (see Chapter 11).

A number of studies show that microtubules are required for the expression of a locomotory phenotype. Vasiliev and Gelfand (1970) reported that microtubule depolymerization causes locomoting cells to lose their polarization, a finding subsequently confirmed in numerous studies (e.g., Goldman, 1971; Vasiliev and Gelfand, 1976; reviewed in Trinkaus, 1985). An intact system of microtubules was also found to be required for the migration of certain epithelial cells (e.g., Gotlieb *et al.,* 1981) and leukocytes (Malech *et al.,* 1977). However, intact microtubules are not a universal requirement for cell movement; in fact, some of the fastest locomoting cell types do very well without them. For example, in leukocytes moving randomly on glass surfaces, microtubule disruption stimulates cell migration and even may elicit polarity (Rich and Hoffstein, 1981; Keller *et al.,* 1984). Colcemid and related agents do not inhibit ruffling, lamellar extension, and movement within a gradient of chemoattractant, although cells may make wider turns after a change in the direction of the gradient (Keller *et al.,* 1984; Devreotes and Zigmond, 1988). These and other findings (reviewed in Gallin, 1980; Haston and Wilkinson, 1988) have led to the belief that microtubules are not an essential component of motor control in leukocytes, although they may modulate locomotory activity (Keller and Zimmermann, 1987). A second example includes certain cells of epithelial origin in which polarization is independent of microtubules (DiPasquale, 1975; Dunlap and Donaldson, 1978; Middleton *et al.,* 1988). Some of these cell types, notably epidermal keratocytes in culture, can move persistently and directionally in the absence of microtubules at rates far higher than those observed in any other cell type (Euteneuer and Schliwa, 1984; Cooper and Schliwa, 1985). These cells are not simply "permanently polarized" (Singer and Kupfer, 1986), since they can develop a polar phenotype from an apolar rounded state in the complete absence of microtubules. Finally, many free-living ameboid cells are capable of persistent directional locomotion in the absence of a microtubule system.

This first look at the involvement of microtubules in directional cell movement reminds us of the dangers of generalization. Nevertheless, exceptions confirm the rule and, as a rule, an intact microtubule system is involved in directional migration of many tissue cell types.

However, the mere presence of microtubules may not be sufficient to also guarantee cell polarization. In several cell types, an intriguing correlation exists between the position of the MTOC and the direction of cell movement: the centrosome is positioned between the leading edge and the cell nucleus (for reviews, see Bornens and Karsenti, 1984; Singer and Kupfer, 1986). This fixed spatial relationship not only lends support to van Beneden's idea that a line through the centrosome and the center of the nucleus defines an axis of cellular polarity (van Beneden, 1883), but is interpreted to mean that the centrosome actually determines the direction of movement. A logical consequence of this proposal is that any change in the direction of cell movement, or the initiation of movement from a quiescent state, is preceded by a reorientation of the centrosome in the intended direction of migration.

Several findings during the past decade are consistent with this interpretation. First, electron microscopy of migrating 3T3 fibroblasts located the centrosomes ahead or on the side of the nucleus in the majority of cells (Albrecht-Buehler and Bushnell, 1979). Second, cells at the margins of linear wounds inflicted on monolayer cultures reposition the centrosome to the side facing the wound (Gotlieb *et al.,* 1981, 1983), a process accompanied by relocation of the Golgi apparatus (Kupfer *et al.,* 1982). Third, fused multinucleated BHK cells only translocate if the majority of centrosomes is positioned in the anterior of the giant cell (Lewis and Albrecht-Buehler, 1987). Finally, in newt eosinophils the centrosome is almost always found near the lamellipod, ahead of the nucleus (Koonce *et al.,* 1984).

The conclusion from these findings is that the centrosome serves as a steering aid for directional cell locomotion. In addition, the relocation of the Golgi apparatus along with the centrosome was interpreted to reflect a requirement for polarized membrane insertion at the leading edge. The delivery of new membrane mass would be accomplished by the polarized transport of membrane vesicles along microtubules from the pericentrosomal Golgi apparatus to the leading edge (Bergmann *et al.,* 1983; Singer and Kupfer, 1986).

However, other findings are at odds with this conceptually very attractive model. In addition to the fact that leukocytes and keratocytes are capable of persistent locomotion in the absence of microtubules (see previous text), centrosome-free fragments derived from the anterior pseudopods of these cells move directionally in a manner similar to intact cells (Keller and Bessis, 1975; Malawista and de Boisfleury Chevance, 1982; Euteneuer and Schliwa, 1984; Malawista and van Blaricom, 1987). These fragments are not only capable of persistent directional movement for many cell diameters, but respond to external stimuli, such as changes in the direction of a chemotactic gradient or collisions with other cells, with a change in the direction of movement. They are, therefore, not simply preprogrammed to move in one direction, but endowed with a behavioral repertoire that allows them to make decisions in a manner indistinguishable from that of intact centrosome-containing cells.

The position of the centrosome ahead of the nucleus as a hallmark of actively moving cells can no longer be considered universally applicable. For example, in several types of motile blood cells, centrosome position is vastly different: anterior to the nucleus in newt eosinophils (Koonce *et al.,* 1984), on the side or anterior in macrophages (Gudima *et al.,* 1988), posterior in lymphocytes (Bessis and De Boisfleury, 1976; Gudima *et al.,* 1988), and variable in leukocytes (Anderson *et al.,* 1982; Schliwa *et al.,* 1982; Gudima *et al.,* 1988). In *Dictyostelium,* the position of the MTOC was found to vary considerably under five different conditions of directional cell movement (Sameshima, 1988). In cells aggregating on agar plates, the centrosome is positioned behind the nucleus in the vast majority of the cells. In cells aggregating submerged in medium, anterior and posterior positions occur with approximately equal frequency. Cells moving chemotactically toward folic acid favor a posterior position 2:1 over an anterior position. Cells derived from dissociated pseudoplasmodia migrating chemotactically toward cAMP show a predominantly anterior position. Finally, randomly moving cells favor a posterior location of the centrosome. The conclusion from these findings is that a general requirement for an anterior location of the centrosome simply does not exist. Other factors seem to contribute to the establishment of a locomotory phenotype.

Two factors to consider are environmental conditions and "developmental history" of the cells. Middleton *et al.* (1989) found that fibroblasts polarize independent of microtubules (and, hence, of centrosomal position) in primary cultures but develop microtubule dependency during the first 48 hr in culture. The same cell type shows no correlation between centrosome position and the direction of cell movement in a three-dimensional collagen network, its natural environment, whereas homologous assay conditions on glass yield a predominantly anterior orientation of the centrosome (Schuetze *et al.*, 1989). Again, these findings show that generalizations may be misleading.

Another argument for a requirement of centrosomal reorientation during cell movement is that it places the centrosome-associated Golgi apparatus (Kreis *et al.*, 1990) in a strategically favorable position for the directed insertion of new membrane mass into the leading edge via Golgi-derived vesicles (Bergmann *et al.*, 1983). A number of observations (Bretscher, 1983; Nemere *et al.*, 1985) are consistent with this proposal and prompt the suggestion that membrane insertion is actually required for directed migration (Singer and Kupfer, 1986). However, recent studies involving photobleaching of membrane lipids and tracking of small particles attached to the cell surface provide evidence for forward transport of both lipids and membrane glycoproteins on the lamellipods of actively moving cells (Kucik *et al.*, 1989; Lee *et al.*, 1990). This forward transport may be the means by which the cell directs membrane mass to the leading edge. Moreover, vesicles are not observed to penetrate into extending lamellipodia (Euteneuer and Schliwa, 1986; DeBiasio *et al.*, 1987; Kucik *et al.*, 1989). In fact, particles as small as 24-nm Ficoll, ribosomes, and even mRNA are excluded from the ruffling edges of lamellipodia (Luby-Phelps and Taylor, 1988; Sundell *et al.*, 1989). Thus it now seems uncertain whether membrane insertion via transport vesicles and, hence, reorientation of the Golgi apparatus, is a prerequisite for lamellar extension during cell locomotion. In fact, isolated keratocyte lamellae have been observed to move directionally in the absence of a Golgi apparatus as visualized by (NBD) ceramide staining (M.S. Cooper, personal communication).

How do microtubules contribute to the process of directional cell locomotion? Cell movement is a complex phenomenon indeed, involving cell membrane dynamics, actin reorganization, force generation, polarity determination, and cell adhesion. Compelling evidence suggests that instructions for the formation of cell protrusions for locomotion originate at the plasma membrane. Agonists that bind to specific cell-surface receptors or loss of cell–cell contact elicit massive changes in the cortical actin network, leading to profound shape changes that culminate in protrusion (Trinkaus, 1985; Nachmias and Yoshida, 1988; Stossel, 1989). These processes no doubt will influence microtubule deployment; conversely, microtubules may modulate contractile events. Indeed, numerous studies demonstrate the interdependency of the actin-based contractile system and microtubules *in vivo* (Domnina *et al.,* 1985; Euteneuer and Schliwa, 1985; Lyass *et al.,* 1988; Bornens *et al.,* 1989; Danowski, 1989; Goslin *et al.,* 1990). In this scenario, microtubules stabilize or attentuate a locomotory morphology. Gundersen and Bulinski (1988) recently showed that, in 3T3 cells, microtubules facing the edge of an *in vitro* wound are rich in detyrosinated tubulin and more resistant to depolymerization, suggesting selective microtubule stabilization as a possible means for consolidation of cellular asymmetry. Unfortunately, however, this feature is not displayed in wounded monolayers of other cell types (G. G. Gundersen, personal communication; U. Euteneuer and M. Schliwa, unpublished observations) and therefore does not appear to be universal. It may be too early to establish a unifying concept, and generalizations may be invalid.

IV. EPILOGUE

Centrosomes occupy a unique position in our concept of the cell. Not unlike a spider in its (microtubule) web, each resides in the cell center, seeming to survey the activities of the cell. That centrosomes and the fibers they organize (the microtubules) interact with and, therefore, are bound into the intracellular and extracellular environment should not come as a surprise to those who view the cell as an interacting structure–function continuum. Through these interactions, centrosomes participate in cell-shape determination, motility, organelle distribution, cortical activities, and cell-polarity determination. Designating a determinative role in these processes would appear to overemphasize the importance of the centrosomes. They are important players, but function only in the context of the entire cast.

ACKNOWLEDGMENTS

I would like to thank Ursula Euteneuer for critical reading of the manuscript and Gary Banker, Mark Cooper, and Gregg Gunderson for allowing me to refer to their unpublished observations. The author's work reported in this chapter was supported by grants from the NIH and NSF.

REFERENCES

Achler, A., Filmer, D., Merte, C., and Drenckhahn, D. (1989). Role of microtubules in polarized delivery of apical membrane proteins to the brush border of the intestinal epithelium. *J. Cell Biol.* **109,** 179–189.

Anderson, D. C., Wible, L. J., Hughes, B. J., Smith, C. W., and Brinkley, B. R. (1982). Cytoplasmic microtubules in polymorphonuclear leukocytes: Effect of chemotactic stimulation and colchicine. *Cell* **31,** 719–729.

Albrecht-Buehler, G., and Bushnell, A. (1979). The orientation of centrioles in migrating 3T3 cells. *Exp. Cell Res.* **120,** 111–118.

Antanitus, D. S., Choi, B., and Lapham, L. W. (1975). Immunofluorescence staining of astrocytes *in vitro* using antiserum to glial fibrillary acidic protein. *Brain Res.* **89,** 363–367.

Bacallao, R., Antony, C., Dotti, C., Karsenti, E., Stelzer, E. H. K., and Simons, K. (1989). The subcellular organization of Madin-Darby canine kidney cells during the formation of a polarized epithelium. *J. Cell Biol.* **109,** 2817–2832.

Balcarova-Stander, J., Pfeiffer, S. E., Fuller, S., and Simons, K. (1984). Development of cell surface polarity in the epithelial Madin Darby canine kidney (MDCK) cell line. *EMBO J.* **3,** 2687–2694.

Bate, C. (1976). Pioneer neurons in an insect embryo. *Nature* **260,** 54–56.

Bessis, M., and De Boisfleury, A. (1976). Catalog of white blood cell movements (normal and pathologic). *Blood Cells* **2,** 365–410.

Bentley, D., and Keshishian, H. (1983). Pathfinding by peripheral pioneer neurons in grasshoppers. *Science* **218,** 1082–1088.

Bergmann, J. E., Kupfer, A., and Singer, S. J. (1983). Membrane insertion at the leading edge of motile fibroblasts. *Proc. Natl. Acad. Sci. U.S.A.* **80,** 1367–1371.

Bornens, M., and Karsenti, E. (1984). The centrosome. *In* "Membrane Structure and Function" (E. E. Bittar, ed.), pp. 99–171. John Wiley, New York.

Bornens, M., Paintrand, M., and Celati, C. (1989). The cortical microfilament system of lymphoblasts displays a periodic oscillatory activity in the absence of microtubules: Implications for cell polarity. *J. Cell Biol.* **109,** 1071–1083.

Boveri, T. (1888). Zellenstudien II. Die Befruchtung und Teilung des Eies von *Ascaris megalocephala. Jena. Zeitschr. Naturw.* **22,** 685–882.

Boveri, T. (1901). Zellenstudien IV. Über die Natur der Centrosomen. *Jena. Zeitschr. Naturw.* **35,** 1–220.

Bré, M. H., and Karsenti, E. (1990). Effects of brain microtubule-associated proteins on microtubule dynamics and the nucleating activity of centrosomes. *Cell Motil. Cytosk.* **15,** 88–98.

Bretscher, M. S. (1983). Distribution of receptors for transferrin and low density lipoprotein on the surface of giant HeLa cells. *Proc. Natl. Acad. Sci. U.S.A.* **80,** 454–458.

Buendia, B., Bré, M.-H., Griffiths, G., and Karsenti, E. (1990). Cytoskeletal control of centrioles movement during the establishment of polarity in Madin-Darby canine kidney cells. *J. Cell Biol.* **110,** 1123–1135.

Burton, P. R., and Paige, J. L. (1981). Polarity of axoplasmic microtubules in the olfactory nerve of the bullfrog. *Proc. Natl. Acad. Sci. U.S.A.* **78,** 3269–3272.

Byers, H. R., White, G. E., and Fujiwara, K. (1984). Organization and function of stress fibers in cells *in vitro* and *in situ:* A review. *Cell Muscle Motil.* **5,** 83–137.

Carney, P. R., and Couve, E. (1989). Cell polarity changes and migration during early development of the avian peripheral auditory system. *Anat. Rec.* **225,** 156–164.

Cooper, M. S., and Schliwa, M. (1985). Motility of cultured fish epidermal cells in the presence and absence of direct current electric fields. *J. Cell Biol.* **102,** 1384–1399.

Danowski, B. A. (1989). Fibroblast contractility and actin organization are stimualted by microtubule inhibitors. *J. Cell Sci.* **93,** 255–266.

DeBiasio, R., Bright, G. R., Ernst, L. A., Waggoner, A. S., and Taylor, D. L. (1987). Five-parameter fluorescence imaging: Wound healing of living Swiss 3T3 cells. *J. Cell Biol.* **105,** 1613–1622.

Devreotes, P. N., and Zigmond, S. H. (1988). Chemotaxis in eukaryotic cells. *Ann. Rev. Cell Biol.* **4,** 649–686.

DeRosier, D. J., and Tilney, L. G. (1989). The structure of the cuticular plate, an *in vivo* actin gel. J. Cell Biol. **109,** 2853–2867.

DiPasquale, A. (1975). Locomotion of epithelial cells. Factors involved in the extension of the leading edge. *Exp. Cell Res.* **95,** 425–439.

Domnina, L. V., Rovensky, J. A., Vasiliev, J. M., and Gelfand, I. M. (1985). Effect of microtubule-destroying drugs on the spreading and shape of cultured epithelial cells. *J. Cell Sci.* **74,** 267–282.

Dotti, C. G., and Banker, G. A. (1987). Experimentally induced alteration in the polarity of developing neurons. *Nature (London)* **330,** 254–256.

Dotti, C. G., Sullivan, C. A., and Banker, G. A. (1988). The establishment of polarity by hippocampal neurons in culture. *J. Neurosci.* **8,** 1454–1468.

Dunlap, M. K., and Donaldson, D. J. (1978). Inability of colchicine to inhibit newt epidermal cell migration or prevent concanavalin A-mediated inhibition of migration. *Exp. Cell Res.* **116,** 15–19.

Eckert, B. S., Caputi, S. E., and Warren, R. H. (1984). Dynamics of keratin filaments and the intermediate filament distribution center during shape change in PtK_1 cell. *Cell Motil.* **4,** 169–181.

Eilers, U., Klumpermann, J., and Hauri, H. P. (1989). Nocodazole, a microtubule-active drug, inter feres with apical protein delivery in cultured intestinal epithelial cells (Caco-2). *J. Cell Biol.* **108,** 13–2.

Euteneuer, U., and McIntosh, J. R. (1981). Polarity of some motility-related microtubules. *Proc. Natl. Acad. Sci. U.S.A.* **78,** 372–376.

Euteneuer, U., and Schliwa, M. (1984). Persistent, directional motility of cells and cytoplasmic fragments in the absence of microtubules. *Nature (London)* **310,** 58–61.

Euteneuer, U., and Schliwa, M. (1985). Evidence for an involvement of actin in the positioning and motility of centrosomes. *J. Cell Biol.* **101,** 96–103.

Euteneuer, U., and Schliwa, M. (1986). The function of microtubules in directional cell movement. *Ann. N.Y. Acad. Sci.* **466,** 867–886.

Evans, L., Mitchison, T., and Kirschner, M. (1985). Influence of the centrosome on the structure of nucleated microtubules. *J. Cell Biol.* **100,** 1185–1191.

Fleming, T. P., and Johnson, M. H. (1988). From egg to epithelium. *Ann. Rev. Cell Biol.* **4,** 459–485.

Fleming, T. P., McConnell, J., Johnson, M. H., and Stevenson, B. R. (1989). Development of tight junctions *de novo* in the mouse early embryo: Control of assembly of the tight junction-specific protein, ZO-1. *J. Cell Biol.* **108,** 1407–1418.

Flemming, W. (1875). Studien über die Entwicklungsgeschichte der Najaden. *Sitzungsber. Akad. Wissensch. Wien* **71,** 81–147.

Franke, W. W., Osborn, M., and Weber, K. (1978). Different intermediate-sized filaments distinguished by immunofluorescence microscopy. *Proc. Natl. Acad. Sci. U.S.A.* **75,** 5034–5038.

Gallin, J. I. (1980). The cell biology of leukocyte chemotaxis. *In* "The Cell Biology of Inflammation" (G. Weissmann, ed.), pp. 299–335. Elsevier, Amsterdam.

Geiger, B., and Singer, S. J. (1980). Association of microtubules and intermediate filaments in chicken gizzard cells as detected by double immunofluorescence. *Proc. Natl. Acad. Sci. U.S.A.* **77,** 4769–4773.

Geuens, G., De Brabander, M., Nuydens, R., and De Mey, J. (1983). The interaction between microtubules and intermediate filaments in cultured cells treated with taxol and nocodazole. *Cell Biol. Int. Rep.* **7,** 35–47.

Goldman, R. D. (1971). The role of three cytoplasmic fibers in BHK-21 cell motility. I. Microtubules and the effects of colchicine. *J. Cell Biol.* **51,** 752–762.

Gordon, W. E., Bushnell, A., and Burridge, K. (1978). Characterization of the intermediate (10 nm) filament of cultured cells using an autoimmune rabbi antiserum. *Cell* **13,** 249–261.

Goslin, K., and Banker, G. A. (1989). Experimental observations on the development of polarity by hippocampal neurons in culture. *J. Cell Biol.* **108,** 1507–1516.

Goslin, K., Birgbauer, E., Banker, G., and Solomon, F. (1990). The role of cytoskeleton in organizing growth cones: A microfilament-associated growth cone component depends on microtubules for its localization. *J. Cell Biol.* **109,** 1621–1631.

Gotlieb, A. I., May, L. M., Subrahmanyan, L., and Kalnins, V. I. (1981). Distribution of microtubule organizing centers in migrating sheets of endothelial cells. *J. Cell Biol.* **91,** 589–594.

Gotlieb, A. I., Subrahmanyan, L., and Kalnins, V. I. (1983). Microtubule-organizing centers and cell migration: Effect of inhibition of migration and microtubule disruption in endothelial cells. *J. Cell Biol.* **96,** 1266–1272.

Griffith, L. M., and Pollard, T. D. (1978). Evidence for actin filament–microtubule interaction mediated by microtubule-associated proteins. *J. Cell Biol.* **78,** 958–965.

Gudima, G. O., Vorobjev, I. A., and Chentsov, Y. S. (1988). Centriolar location during blood cell spreading and motion *in vitro:* An ultrastructural analysis. *J. Cell Sci.* **89,** 225–241.

Gumbiner, B. (1987). Structure, biochemistry and assembly of epithelial tight junctions. *Am. J. Physiol.* **253,** C749–C758.

Gundersen, G. G., and Bulinski, J. C. (1988). Selective stabilization of microtubules oriented toward the direction of cell migration. *Proc. Natl. Acad. Sci. U.S.A.* **85,** 5946–5950.

Hall, H. G., Farson, D. A., and Bissell, M. J. (1982). Lumen formation by epithelial cell lines in response to collagen overlay: A morphogenetic model in culture. *Proc. Natl. Acad. Sci. U.S.A.* **79,** 4672–4676.

Haston, P. S., and Wilkinson, P. C. (1988). Locomotion and chemotaxis of leukocytes: Gradient perception and locomotor capacity. *Curr. Opinion Immunol.* **1,** 5–9.

Herzlinger, D. A., and Ojakian, G. K. (1984). Studies on the development and maintenance of epithelial cell surface polarity with monoclonal antibodies. *J. Cell Biol.* **98,** 1777–1787.

Hinds, J. W., and Hinds, P. L. (1974). Early ganglion cell differentiation in the mouse retina: An electron microscopic analysis utilizing serial sections. *Dev. Biol.* **37,** 381–386.

Hull, B. E., and Staehelin, L. A. (1979). The terminal web. *J. Cell Biol.* **81,** 67–82.

Hunt, R. C., and Ellar, D. J. (1974). Isolation of the plasma membrane of a trypanosomatid flagellate: General characterization and lipid composition. *Biochim. Biophys. Acta* **339,** 173–189.

Keller, H. U., and Bessis, M. (1975). Migration and chemotaxis of anucleate cytoplasmic leukocyte fragments. *Nature (London)* **258,** 73–74.

Keller, H. U., Naef, A., and Zimmermann, A. (1984). Effects of colchicine, vinblastine and nocodazole on polarity, motility, chemotaxis and cAMP levels of human polymorphonuclear leukocytes. *Exp. Cell Res.* **153,** 173–184.

Keller, H. U., and Zimmermann, A. (1987). Shape, movement, and function of neutrophil granulocytes. *Biomed. Pharmacother.* **41,** 285–289.

Keller, T. C., and Mooseker, M. S. (1982). Ca-calmodulin dependent phosphorylation of myosin, and its role in brush border contraction *in vitro. J. Cell. Biol.* **95,** 943–959.

Klein, G., Langegger, M., Timpl. R., and Ekblom, P. (1988). Role of laminin A chain in the development of epithelial cell polarity. *Cell* **55,** 331–341.

Koonce, M. P., Cloney, R. A., and Berns, M. W. (1984). Laser irradiation of centrosomes in newt eosinophils: Evidence of centriole role in motility. *J. Cell Biol.* **98,** 1999–2010.

Koonce, M. P., Euteneuer, U., McDonald, K. L., Menzel, D., and Schliwa, M. (1986). Cytoskeletal architecture and motility in a giant freshwater amoeba. *Reticulomyxa. Cell Motil. Cytosk.* **6,** 521–533.

Kreis, T. E. (1990). Role of microtubules in the organization of the Golgi appratus. *Cell. Motil. Cytosk.* **15,** 67–70.

Kucik, D. F., Elson, E., and Sheetz, M. P. (1989). Forward transport of glycoproteins on leading lamellipodia in locomoting cells. *Nature (London)* **340,** 315–317.

Kupfer, A., Louvard, D., and Singer, S. J. (1982). Polarization of the Golgi apparatus and the microtubule-organizing center in cultured fibroblasts at the edge of an experimental wound. *Proc. Natl. Acad. Sci. U.S.A.* **79,** 2603–2607.

Lee, J., Gustafsson, M., Magnusson, K. E., and Jacobson, K. (1990) The direction of membrane lipid flow in locomoting polymorphonuclear leukocytes. *Science* **247,** 1229–1233.

Lefcort, F., and Bentley, D. (1989). Organization of cytoskeletal elements and organelles preceding growth cone emergence from an identified neuron *in situ. J. Cell Biol.* **108,** 1737–1749.

Lewis, L., and Albrecht-Buehler, G. (1987). Distribution of multiple centrospheres determines migration in BHK syncitia. *Cell Motil. Cytosk.* **7,** 282–290.

Li, M. L., Aggeler, J., Farson, D. A., Hatier, C., Hassell, J., and Bissell, M. J. (1987). Influence of a reconstituted basement membrane and its components on casein gene expression and secretion in mouse mammary epithelial cells. *Proc. Natl. Acad. Sci. U.S.A.* **84,** 136–140.

Lloyd, C. W. (1982). The Cytoskeleton in Plant Growth and Development. Academic Press, London, 457 pp.

Luby-Phelps, K., and Taylor, D. L. (1988). Subcellular compartmentalization by local differentiation of cytoplasmic structure. *Cell Motil. Cytosk.* **10,** 28–37.

Lyass, L. A., Bershadsky, A. D., Vasiliev, J. M., and Gelfand, I. M. (1988). Microtubule-dependent effect of phorbol ester on the contractility of cytoskeleton of cultured fibroblasts. *Proc. Natl. Acad. Sci. U.S.A.* **85,** 9538–9541.

Malawista, S. E., and de Boisfleury Chevance, A. (1982). The cytokineplast: Purified, stable, and functional motile machinery from human blood polymorphonuclear leukocytes. *J. Cell Biol.* **95,** 960–973.

Malawista, S. E., and van Blaricom, G. (1987). Cytoplasts made from human blood polymorphonuclear leukocytes with or without heat: Preservation of both motile function and respiratory burst oxidase activity. *Proc. Natl. Acad. Sci. U.S.A.* **84,** 454–458.

Malech, H. L., Root, R. K., and Gallin, J. I. (1977). Structural analysis of human neutrophil migration: Centriole, microtubule, and microfilament orientation and function during chemotaxis. *J. Cell Biol.* **75,** 666–693.

Marchisio, P. C., Weber, K., and Osborn, M. (1979). Identification of multiple microtubule initiating sites in mouse neuroblastoma cells. *Eur. J. Cell Biol.* **20,** 45–50.

Maro, B., Howlett, S. K., and Webb, M. (1985). Non-spindle microtubule-organizing centers in metaphase II-arrested mouse oocytes. *J. Cell Biol.* **101,** 1665–1672.

Middleton, C. A., Brown, A. F., Brown, R. M., and Roberts, D. J. (1988). The shape of cultured epithelial cells does not depend on the integrity of their microtubules. *J. Cell Sci.* **91,** 337–345.

Middleton, C. A., Brown, A. F., Brown, R. M., Karavanova, I. D., Roberts, D. J. H., and Vasiliev, J. M. (1989). The polarization of fibroblasts in early primary cultures is independent of microtubule integrity. *J. Cell Sci.* **94,** 25–32.

Miller, M., and Solomon, F. (1984). Kinetics and intermediates of marginal band reformation: Evidence for peripheral determinants of microtubule organization. *J. Cell Biol.* **99,** 70s–75s.

Mitchison, T., and Kirscher, M. (1984). Microtubule assembly nucleated by isolated centrosomes. *Nature (London)* **312,** 232–237.

Mooseker, M. S., and Tilney, L. G. (1975). Organization of actin filament—membrane complex. *J. Cell Biol.* **67,** 725–745.

Murray, J. M. (1984). Disassembly and reconstitution of a membrane-microtubule complex. *J. Cell Biol.* **98,** 1481–1487.

Nachmias, V. T., and Yoshida, K.-I. (1988). The cytoskeleton of the blood platelet: A dynamic structure. *Adv. Cell Biol.* **2,** 181–211.

Nelson, W. J., and Veschnok, P. J. (1986). Dynamics of membrane skeleton (fodrin) organization during development of polarity in Madin-Darby canine kidney cells. *J. Cell Biol.* **103,** 1751–1766.

Nelson, W. J., and Veschnok, P. J. (1987). Modulation of fodrin (membrane skeleton) stability by cell–cell contact in Madin-Darby canine kidney epithelial cells. *J. Cell Biol.* **104,** 1527–1537.

Nemere, I., Kupfer, A., and Singer, S. J. (1985). Reorientation of the Golgi apparatus and the microtubule-organizing center inside macrophages subjected to a chemotactic gradient. *Cell Motil.* **5,** 17–29.

Nitsch, L., and Wollman, S. H. (1980). Ultrastructure of intermediate stages in polarity reversal of thyroid epithelium in follicles in suspension culture. *J. Cell Biol.* **86,** 875–880.

Ojakian, G. K., and Schwimmer, R. (1988). The polarized distribution of an apical cell surface glycoprotein is maintained by interactions with the cytoskeleton of Madin-Darby canine kidney cells. *J. Cell Biol.* **107,** 2377–2387.

Osborn, M., and Weber, K. (1976). Cytoplasmic microtubules in tissue culture cells appear to grow from an organizing structure towards the plasma membrane. *Proc. Natl. Acad. Sci. U.S.A.* **73,** 867–871.

Osborn, M., Franke, W. W., and Weber, K. (1977). Visualization of a system of filaments 7–10 nm thick in cultured cells of an epithelioid line (PtK_2) by immunofluorescence microscopy. *Proc. Natl. Acad. Sci. U.S.A.* **74,** 2490–2494.

Parry, G., Cullen, B., Kaetzel, C. S., Kramer, R., and Moss, L. (1987). Regulation of differentiation and polarized secretion in mammary epithelial cells maintained in culture: Extracellular matrix and membrane polarity influences. *J. Cell Biol.* **105,** 2043–2051.

Parry, G., Beck, J. C., Mos, L., Bartley, J., and Ojakian, G. K. (1990). Determination of apical membrane polarity in mammary epithelial cell cultures: The role of cell–cell, cell–substratum, and membrane–cytoskeleton interactions. *Exp. Cell Res.* **188,** 302–311.

Pisam, M., and Ripoche, P. (1976). Redistribution of surface macromolecules in dissociated epithelial cells. *J. Cell Biol.* **71,** 907–920.

Porter, K. R. (1966). Cytoplasmic microtubules and their functions. *In* "Principles of Biomolecular Organization" (G. E. W. Woltenholme and M. O'Connor, eds.), pp. 308–345. Churchill, London.

Pryzwansky, K. B., Schliwa, M., and Porter, K. R. (1983). Comparison of the three-dimensional organization of unextracted and Triton-extracted human neutrophilic polymorphonuclear leukocytes. *Eur. J. Cell Biol.* **30,** 112–125.

Raff, E. C. (1979). The control of microtubule assembly *in vivo. Int. Rev. Cytol.* **59,** 1–96.

Raff, J. W., and Glover, D. M. (1989). Centrosomes, and not nuclei, initiate pole cell formation in *Drosophila* embryos. *Cell* **57,** 611–619.

Rich, A. M., and Hoffstein, S. T. (1981). Inverse correlation between neutrophil microtubule numbers and enhanced random migration. *J. Cell Sci.* **48,** 181–191.

Rindler, M. J., Ivanov, I. E., and Sabatini, D. D. (1987). Microtubule-acting drugs lead to the non-polarized delivery of the influenza hemagglutinin to the cell surface of polarized Madin–Darby canine kidney cells. *J. Cell Biol.* **104,** 231–252.

Rodrigues-Boulan, E., Paskiet, P., and Sabatini, D. D. (1983). Assembly of enveloped visuses in MDCK cells: Polarized budding from single attached cells and from clusters of cells in suspension. *J. Cell Biol.* **96,** 866–874.,

Rodriguez-Boulan, E., and Salas, P. J. (1989). External and internal signals for epithelial cell surface polarization. *Ann. Rev. Physiol.* **51,** 741–754.

Rogalski, A. A., and Singer, S. J. (1984). Association of elements of the Golgi apparatus with microtubules. *J. Cell Biol.* **99,** 1092–1102.

Salas, P. J., Misek, D. E., Vega-Salas, D. E., Gundersen, D., Cereijido, M., and Rodriguez-Boulan, E. (1986). Microtubules and actin filaments are not critically involved in the biogenesis of epithelial cell surface polarity. *J. Cell Biol.* **102,** 1853–1867.

Salas, P. J., Vega-Salas, D. E., and Rodriguez-Boulan, E. (1987). Collagen receptors mediate early events in the attachment of epithelial (MDCK) cells. *J. Membr. Biol.* **98,** 223–236.

Sameshima, M., Imai, Y., and Hashimoto, Y. (1988). The position of the microtubule-organizing center relative to the nucleus is independent of the direction of cell migration in *Dictyostelium discoideum. Cell Motil. Cytosk.* **9,** 111–116.

Schliwa, M., Pryzwansky, K. B., and Euteneuer, U. (1982). Centrosome splitting in neutrophils: An unusual phenomenon related to cell activation and motility. *Cell* **31,** 705–717.

Schliwa, M., and van Blerkom, J. (1981). Structural interaction of cytoskeletal components. *J. Cell Biol.* **90,** 222–235.

Schuetze, K., Maniotis, A., Euteneuer, U., and Schliwa, M. (1989). Cell locomotion and microtubule organization in chicken embryo fibroblasts grown in three-dimensional collagen networks. *J. Cell Biol.* **109,** 73a.

Sharp, G. A., Osborn, M., and Weber, K. (1981). Ultrastructure of multiple microtubule initiation sites in mouse neuroblastoma cells. *J. Cell Sci.* **47,** 1–24.

Sharp, G. A., Weber, K., and Osborn, M. (1982). Centriole number and process formation in established neuroblastoma cells and primary dorsal root ganglion neurones. *Eur. J. Cell Biol.* **29,** 97–103.

Simons, K., and Fuller, D. (1985). Cell surface polarity in epithelia. *Ann. Rev. Cell Biol.* **1,** 243–288.

Singer, S. J., and Kupfer, A. (1986). The directed migration of eukaryotic cells. *An. Rev. Cell Biol.* **2,** 337–365.

Solomon, F. (1980). Neuroblastoma cells recapitulate their detailed neurite morphologies after reversible microtubule disassembly. *Cell* **21,** 333–338.

Spiegelman, B. M., Lopata, M. A., and Kirschner, M. W. (1979a). Multiple sites for the initiation of microtubule assembly in mammalian cells. *Cell* **16,** 239–252.

Spiegelman, B. M., Lopata, M. A., and Kirschner, M. W. (1979b). Aggregation of microtubule initiation sites preceding neurite outgrowth in mouse neuroblastoma cells. *Cell* **16,** 253–263.

Stevens, J. K., Trogadis, J., and Jacobs, J. R. (1988). Development and control of axial neurite form: A serial electron microscopic analysis. *In* "Intrinsic Determinants of Neuronal Form and Function" (R. J. Lasek and M. M. Black, eds.), pp. 115–145. Liss, New York.

Stossel, T. P. (1984). Contribution of actin to the structure of the cytoplasmic matrix. *J. Cell Biol.* **99,** 15s–21s.

Stossel, T. P. (1989). From signal to pseudopod. How cells control cytoplasmic actin assembly. *J. Biol. Chem.* **264,** 18261–18264.

Stossel, T. P., Chaponnier, C., Ezzell, R. M., Hartwig, J. H., Janmey, P. A., Kwiatkowski, D. J., Lind, S. E., Smith, D. B., Southwick, F. S., Yin, H. L., and Zaner, K. S. (1985). Nonmuscle actin-binding proteins. *Ann. Rev. Cell Biol.* **1,** 353–402.

Sundell, C. L., Lawrence, J. B., and Singer, R. H. (1989). Actin mRNA association with the cytoskeleton and role of protein synthesis in its subcellular localization. *J. Cell Biol.* **109.** 269a.

Swan, J. A., and Solomon, F. (1984). Reformation of the marginal band of avian erythrocytes *in vitro* using calf brain tubulin: Peripheral determinants of microtubule form. *J. Cell Biol.* **99,** 2108–2113.

Tassin, A. M., Maro, B., and Bornens, M. (1985). Fate of microtubule-organizing centers during *in vitro* myogenesis. *J. Cell Biol.* **100,** 35–46.

Trinkaus, J. P. (1985). Protrusive activity of the cell surface and the initiation of cell movement during morphogenesis. *Exp. Biol. Med.* **10,** 130–173.

Trotter, J. A. (1981). The organization of actin in spreading macrophages. *Exp. Cell Res.* **132,** 235–248.

Tucker, J. B. (1979). Spatial organization of microtubules. *In* "Microtubules" (K. Roberts and J. S. Hyams, eds.), pp. 315–358. Academic Press, London.

Tucker, J. B. (1984). Spatial organization of microtubule-organizing centers and microtubules. *J. Cell Biol.* **99,** 55s–62s.

Van Beneden, E. (1876). Contribution a l'histoire de la vesiclue germinative et du premier embryonnaire. *Bull. Acad. Roy. Belgique* **42,** 35–97.

Van Beneden, E. (1883). Recherches sur la maturation de l'oeuf, la decondation et la division cellule. *Arch. Biol.* **4,** 265–683.

Vasiliev, J. M., Gelfand, I. M., Domnina, L. V., Ivanova, O. Y., Komm, S. G., and Olshevkaya, L. V.

(1970). Effect of colcemid on the locomotory behavior of fibroblasts. *J. Embryol. Exp. Morphol.* **24,** 625–640.

Vasiliev, J. M., and Gelfand, I. M. (1976). Effect of colcemid on morphogenetic processes and locomotion of fibroblasts. *Cold Spring Harbor Conf. Cell Prolif.* **3,** 279–304.

Vega-Salas, D. E., Salas, P. J., and Rodriguez-Boulan, E. (1987). Formation of the apical pole of epithelial (MDCK) cells: Polarity of an apical protein is of tight junctions while segregation of a baso-lateral marker requires cell–cell interactions. *J. Cell Biol.* **104,** 905–916.

Vega-Salas, D. E., Salas, P. J., and Rodriguez-Boulan, E. (1988). Exocytosis of vacuolar apical compartment (VAC): A cell–cell contact controlled mechanism for the establishment of the apical plasma membrane domain in epithelial cells. *J. Cell Biol.* **107,** 1717–1728.

Wang, A. Z., Ojakian, G. K., and Nelson, W. J. (1990a). Steps in the morphogenesis of a polarized epithelium. I. Uncoupling the roles of cell–cell and cell–substratum contact in establishing plasma membrane polarity in multicellular epithelial (MDCK) cysts. *J. Cell Sci.* **95,** 137–151.

Wang, A. Z., Ojakian, G. K., and Nelson, W. J. (1990b). Steps in the morphogenesis of a polarized epithelium. II. Disassembly and assembly of plasma membrane domains during reversal of epithelial cell polarity in multicellular epithelial (MDCK) cysts. *J. Cell Sci.* **95,** 153–165.

Wang, E., Connolly, J. A., Kalnins, V. I., and Choppin, P. W. (1979). Relationship between movement and aggregation of centrioles in syncytia and formation of microtubule bundles. *Proc. Natl. Acad. Sci. U.S.A.* **76,** 5719–5723.

Weihing, R. R. (1979). The cytoskeleton and plasma membrane. *Meth. Achiev. Exp. Pathol.* **8,** 42–109.

Wilson, E. B. (1925). The Cell in Development and Heredity, 3d. Ed. MacMillan, New York.

Index

F

T